A CONCISE CSE MATHEMATICS

A Greer

Formerly Senior Lecturer
Gloucestershire College of Arts and Technology

Stanley Thornes (Publishers) Ltd

First published in 1983 by

Stanley Thornes (Publishers) Ltd
Old Station Drive
Leckhampton
CHELTENHAM GL53 0DN

Reprinted 1983
Reprinted 1984
Reprinted 1986

British Library Cataloguing in Publication Data:

Greer, A.
A concise CSE mathematics.
1. Mathematics
I. Title
510 QA37.2

ISBN 0-85950-108-6

Typeset by Tech-Set, Gateshead, Tyne & Wear.
Printed and bound in Great Britain at The Bath Press, Avon.

CONTENTS

PREFACE

This book contains all of the topics required for the 'common core' of the Certificate of Secondary Education in Mathematics. Each of the topics has been dealt with in a simple yet concise fashion.

Since this is a revision course the sections on Arithmetic, Algebra, Geometry and New Mathematics have been dealt with separately and a teacher and the class may work through the book chapter by chapter if this is thought to be desirable.

Throughout the text, a large number of carefully graded exercises have been included and at the end of each chapter there is an exercise consisting of problems of the type found in CSE examination papers. Multi-choice exercises have also been included at the end of most of the chapters.

Gloucester 1983 A. GREER

1 OPERATIONS IN ARITHMETIC

- The result obtained by adding numbers is called the SUM. Thus the sum of 3, 5 and 9 is $3+5+9=17$.
- The DIFFERENCE of two numbers is the larger number minus the smaller number. Thus the difference of 16 and 21 is $21-16=5$.
- The result obtained by multiplying numbers is called the PRODUCT. Thus the product of 5 and 7 is $5\times 7=35$.
- When dividing, the DIVIDEND is the number to be divided.
  The DIVISOR is the number by which the dividend is divided.
  The QUOTIENT is the result of the division. Thus

$$\frac{\text{dividend}}{\text{divisor}} = \text{quotient}$$

Exercise 1.1

1. Find the sum of each of the following:
   (a) 8, 12 and 22 (b) 173 and 295 (c) 8357 and 9874
   (d) Four hundred and thirty seven, and two thousand four hundred and twenty five
   (e) One hundred and six thousand and four, nine thousand two hundred and thirty nine, two hundred and fifty three, and ninety eight.
2. Find the difference of:
   (a) 19 and 34 (b) 385 and 279
   (c) Four hundred and eight, and three hundred and five
   (d) Nine thousand nine hundred and nine, and four hundred and fifty eight.
3. Find the product of each of the following:
   (a) 9 and 7 (b) 15, 3 and 4 (c) 283 and 17
   (d) One hundred and five, and seventeen
   (e) Five thousand and ninety, and one hundred and five.

4. Find the quotient for each of the following:
   (a) $176 \div 8$ (b) $46\,348 \div 4$ (c) $3072 \div 32$
   (d) Sixty one thousand three hundred and eighty six divided by thirteen.

- When numbers are combined in a series of arithmetic operations the following SEQUENCE OF OPERATIONS must be observed.
  (i) Work out brackets
  (ii) Multiply and/or divide
  (iii) Add and/or subtract.

$$\begin{aligned} 2\times(7+3)-16\div4+11 &= 2\times10-16\div4+11 \\ &= 20-4+11 \\ &= 31-4 \\ &= 27 \end{aligned}$$

Exercise 1.2

Find the value of each of the following:

1. $8+5\times2$
2. $3\times7-2$
3. $6\times5-2\times3+4$
4. $7\times5-16\div4+7$
5. $3+4\times(3+2)$
6. $8-3\times(5-4)$
7. $11-12\div4+2\times(7-2)$
8. $15\div(3+2)-8\times3+7\times(5+3)$.

- IF 0 IS ADDED TO ANY NUMBER, the sum is the number. Thus

$$36+0 = 36 \quad \text{and} \quad 0+378 = 378$$

- IF ANY NUMBER IS MULTIPLIED BY ZERO, the product is zero. Thus

$$7\times0 = 0 \quad \text{and} \quad 0\times53 = 0$$

- IF ANY NUMBER IS MULTIPLIED BY 1, the product is the number. Thus

$$18\times1 = 18 \quad \text{and} \quad 1\times93 = 93$$

- IT IS IMPOSSIBLE TO DIVIDE BY ZERO. Thus $\frac{5}{0}$ and $\frac{716}{0}$ are meaningless operations. However if 0 is divided by any number the result is 0. Thus $\frac{0}{3}=0$ and $\frac{0}{215}=0$.

Exercise 1.3

Find the value of each of the following:

1. $7000+0$
2. 3050×1
3. 87×0
4. $5\times 7\times 0\times 3$
5. $8\times 7\times 1$
6. $12\times 0+9$
7. $19\times 1-3$
8. $15-8\times 0-6$
9. $14-8\div 4+0\times(5-3)$
10. $0\div 4\times(7-5)\div 2+3$.

- A set of numbers connected by some definite law is called a SERIES or SEQUENCE of numbers. Each of the numbers in the series is called a TERM of the series. Here are some examples.

  (i) 2, 4, 6, 8, ... (each term in the series is formed by adding 2 to the previous term).

  (ii) 3, 12, 48, ... (each term is formed by multiplying the previous term by 4).

  (iii) 1, 2, 3, 5, 8, ... (each term is formed by adding the two previous terms).

- EVEN NUMBERS are numbers which can be divided exactly by 2. They always end in 0, 2, 4, 6 or 8. Thus 34, 630, 58, 6 and 8272 are all even numbers. The sequence of even numbers is 2, 4, 6, 8, 10, 12, 14, ...

- ODD NUMBERS are numbers which do *not* divide exactly by 2. They always end in 1, 3, 5, 7 or 9. Thus 3, 15, 21, 239 and 4217 are all odd numbers. The sequence of odd numbers is 1, 3, 5, 7, 9, 11, 13, 15, ...

- A SQUARE NUMBER can be shown as a pattern of dots in the shape of a square.

  $9 = 3\times 3$ $\qquad$ $25 = 5\times 5$ $\qquad$ $36 = 6\times 6$

  The sequence of square numbers is 1, 4, 9, 16, 25, 36, ...

- A RECTANGULAR NUMBER can be shown as a pattern of dots in the shape of a rectangle.

  $6 = 2\times 3$ $\qquad$ $24 = 6\times 4$ $\qquad$ $24 = 8\times 3$

Note that 1 is not regarded as being a rectangular number. The sequence of rectangular numbers is 4, 6, 8, 9, 10, ...

- A TRIANGULAR NUMBER can be shown as a pattern of dots in the shape of a triangle.

3 6 10

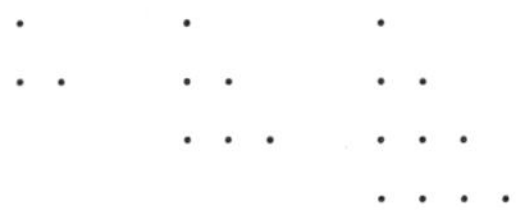

The sequence of triangular numbers is 1, 3, 6, 10, 15, 21, … .

Note that

$$3 = 2+1; \quad 6 = 3+2+1; \quad 10 = 4+3+2+1;$$
$$15 = 5+4+3+2+1$$

Exercise 1.4

Write down the next two terms of each of the following sequences:

1. 2, 10, 50, …

2. 1, 5, 9, 13, …

3. 3, 9, 15, 21, …

4. 176, 88, 44, …

5. 23, 19, 15, …

6. 1, 3, 6, 10, …

7. 1, 1, 2, 3, 5, 8, 13, …

8. 5, 4, 6, 5, 7, 6, 8, … .

Write down the terms denoted by the question marks in the following sequences:

9. 0, 2, 4, ?, ?, 10, 12, 14, …

10. 3, 5, 8, 13, ?, 34, ?, ?

11. 3, 7, 15, ?, 63, 127, ?

12. 2, 4, 10, 28, ?, ?, 730, … .

From the set of numbers 15, 34, 55, 63, 67 and 81, write down one that is:

13. Even

14. A prime number

15. A number exactly divisible by 7

16. A square number

17. A rectangular number which is not a square number

18. A triangular number.

19. Write down the next two terms in the sequence of triangular numbers starting from 21, 28, … .

20. The following is a sequence of rectangular numbers: 12, 14, 15, … . What are the next three terms in the sequence?

- The FACTORS of a number are those numbers which divide exactly into the number. The factors of 24 are 1, 2, 3, 4, 6, 8, 12 and 24.

- If an INTEGER (a whole number) divides exactly into a second integer, the second integer is a MULTIPLE of the first integer. Some multiples of 3 are 3, 6, 9, 12 and 15. Some multiples of 7 are 7, 21, 42 and 84.

- A PRIME NUMBER has only two factors, itself and 1. The first few prime numbers are 2, 3, 5, 7, 11, 17 and 19. Note that a prime number can never be a rectangular number and that 1 is not a prime number.

- In many cases it is often convenient to write a number as a POWER. Thus $81 = 3 \times 3 \times 3 \times 3 = 3^4$, and 81 is said to be the fourth power of 3. The figure 4, which gives the number of 3's to be multiplied together, is called the INDEX (plural: indices).

$$5^3 = 5 \times 5 \times 5 = 125$$

- A PRIME FACTOR is a factor which is a prime number. Every whole number greater than 1 can be written as a product of its prime factors.

Example

Find the prime factors of 420.

2	420
2	210
3	105
5	35
7	7
1	1

$$420 = 2 \times 2 \times 3 \times 5 \times 7 = 2^2 \times 3 \times 5 \times 7$$

- The LOWEST COMMON MULTIPLE (LCM) of a set of numbers is the smallest number into which each of the numbers in the set will divide exactly. Thus the LCM of 4, 5 and 10 is 20, because 20 is the smallest number into which 4, 5 and 10 will divide exactly. The LCM of a set of numbers may be found by expressing each of the numbers in the set as a product of their prime factors.

Example

Find the LCM of 100 and 280.

$$100 = 2 \times 2 \times 5 \times 5 = 2^2 \times 5^2$$

$$280 = 2 \times 2 \times 2 \times 5 \times 7 = 2^3 \times 5 \times 7$$

The LCM is the product of the highest powers of the prime factors. Thus the LCM of 100 and 280 is

$$2^3 \times 5^2 \times 7 = 8 \times 25 \times 7 = 1400$$

- The HIGHEST COMMON FACTOR (HCF) of a set of numbers is the largest number which is a factor of each of the numbers. Thus the HCF of 24, 36 and 48 is 12 because this is the largest factor of 24, 36 and 48. The HCF may be found by expressing each of the numbers in the set as a product of its prime factors.

Example

Find the prime factors of 42, 98 and 112 and hence find their HCF.

$$42 = 2 \times 3 \times 7$$
$$98 = 2 \times 7^2$$
$$112 = 2^4 \times 7$$

Since 2 and 7 are factors of each of the three numbers

$$\text{HCF} = 2 \times 7 = 14$$

Exercise 1.5

1. What numbers are factors of:
   (a) 15 (b) 32 (c) 84?
2. Which of the following numbers are factors of 36?
   1, 2, 3, 4, 5, 6, 7, 8, 9, 10, 12, 14 and 18
3. Write down the multiples of 5 between 19 and 59.
4. Express each of the following numbers as a product of their prime factors:
   (a) 12 (b) 36 (c) 90 (d) 525
   (e) 1320.
5. Write down the three prime numbers next larger than 17.
6. Find the LCM of the following sets of numbers:
   (a) 6 and 15 (b) 12 and 16 (c) 3, 5 and 6
   (d) 4, 10 and 12 (e) 30, 36 and 42 (f) 60 and 378
   (g) 180 and 1323.
7. Find the HCF of the following sets of numbers:
   (a) 10 and 30 (b) 26, 39 and 65 (c) 28, 84 and 112
   (d) 756 and 882.
8. Write down all the multiples of 4 between 10 and 39.
9. (a) Express 132 as a product of prime factors.
   (b) Find the value of 0×143.
   (c) What is the value of $0 - 42$?

Exercise 1.6 (All of the type found in CSE examination papers)

1. Write down the next three terms of the following sequences:
   (a) 21, 17, 13, ... (b) 1, 3, 6, 10, 15, ... (EM)
2. 3^4 or 4^3. Which is the greater and by how much? (EM)

3. Given the numbers 2, 3, 4, 5, 6, 7, 8 and 9 write down:
   (a) The even numbers (b) The multiples of 3
   (c) The prime numbers. (EM)

4. From the numbers 16, 32, 64, 81 and 139 write down:
   (a) The odd numbers (b) Three square numbers
   (c) The multiples of 2.

5. Find two prime numbers whose sum is 12 and whose product is 35.

6. Write down:
   (a) All the positive whole numbers which are factors of 9
   (b) The two consecutive triangular numbers which add up to 25. (SW)

7. Find the missing terms of the sequence ?, 34, 39, 44, ?.

8. Write down the next two prime numbers after 35. (SW)

9. Subtract one hundred thousand from one million. (EM)

10. A boy has to move 27 boxes from one room to another. The most he can carry at once is 6. How many journeys will he have to make? (EM)

11. Write down the sum of the numbers 2, 5, 8 and 11.

12. $63+49=87+?$. What is the missing number?

13. A family use 15 litres of milk per week. How much do they use in a year of 52 weeks?

14. Calculate:
    (a) $(7+4)\times(8-3)$ (b) $6+3\times(8-5)$.

15. Consider the following sequence: 2, 5, 7, 12, 19, 31, 50, ?.
    (a) List the odd numbers given in the sequence.
    (b) Write down the prime numbers given in the sequence.
    (c) Find the sum of the first three numbers.
    (d) Write down the next number in the sequence.

16. How many 42-seater coaches are needed to carry 672 people at the same time? (EA)

17. The following is a sequence of triangular numbers: ?, ?, 6, 10, 15, ?. What are the missing numbers?

18. List the following numbers:
    (a) Multiples of 4 less than 25
    (b) The prime factors of 84
    (c) Triangular numbers less than 23.

19. Three prime numbers are 7, 13 and 29.

(a) Add them together and write down the result. Is their sum a prime number?

(b) Multiply them together and write down the result. Is their product a prime number?

(c) Square 13 and write your answer to the nearest 10.

(d) Square 39 and write the answer to the nearest 100.

20.

A	1	2	3	4	5	6	7
B	2	4	6	8	10	12	14
C	3	6	9	12	15	18	21
D	4	8	12	16	20	24	28
E	5	10	15	20	25	30	35
F	6	12	18	24	30	36	42
G	7	14	21	28	35	42	49

Use the table shown above to answer the following questions:

(a) Write down the next two numbers in the sequence 3, 8, 15, . . .

(b) Write down the two largest perfect square numbers in the table

(c) List the prime numbers in the table

(d) Write down the sum of the numbers in: (i) row B, (ii) row G.

21. Fill in the blanks in the sequence ?, 10, 20, 40, ?, ?.

22. Write down the missing terms in each of the following sequences:

(a) 2, 5, 8, ?, 14, ?, 20, 23, ?

(b) 1, 3, 6, 10, 15, ?, ?, 36, ?, 55

(c) 1, 4, ?, 16, ?, ?, 49, 64, 81, 100

(d) 1, 3, 9, 27, ?, 243, ?, ?

(e) 160, 80, ?, 20, 10, ?.

23. Consider the numbers 11, 27, 41, 63, 72 and 114.

(a) There are two prime numbers. Write down their sum.

(b) Three of the numbers have a common factor. What is it?

24. Write down the next two numbers in each of the following sequences:

(a) 11, 15, 23, 35, . . . (b) 256, 128, 64, . . .

(c) 5, 15, 45, ...

25. Consider the numbers 13, 24, 31, 65, 75 and 125.

(a) Two of the numbers are prime. What is the number lying exactly half way between these prime numbers?

(b) Three of these numbers have a common factor. What is it? **(AL)**

26. Consider the list of numbers: 3, 6, 10, 15 and 21.

(a) The numbers form a sequence. Write down the next number.

(b) Write down the prime number.

(c) Write down the number which is a multiple of 7.

(d) Three of the numbers add up to 24. Write down these three numbers. (S)

27. Write down the next two numbers in each of the following sequences:

(a) 23, 27, 35, 47, ... (b) 128, 64, 16, ... (S)

Multi-choice questions 1

1. The sum of all the prime numbers between 10 and 20 is

A 41 **B** 43 **C** 47 **D** 49
E 60

2. What is the difference in value between the two 6's in the number 6060?

A 0 **B** 540 **C** 594 **D** 5400
E 5940

3. What is the value of 2^5?

A 8 **B** 10 **C** 16 **D** 32
E 64

4. The smallest whole number which can be divided exactly by 3, 4, 6 and 9 is

A 12 **B** 18 **C** 24 **D** 36
E 72

5. The value of 26 390 ÷ 13 is

A 23 **B** 203 **C** 230 **D** 2030
E 2300

6. The next number in the sequence 1, 2, 6, 24, ... is

A 36 **B** 48 **C** 60 **D** 96
E 120

7. The difference between 3^4 and 4^3 is

A 0 **B** 1 **C** 7 **D** 17
E 145

8. 243 expressed as a power of 3 is

A 81 **B** 3^5 **C** 729 **D** 3^{81}

9. 2040 ÷ 20 equals

A 82 **B** 102 **C** 120 **D** 2000

10. How many prime numbers are there between 9 and 30?

A 2 **B** 6 **C** 7 **D** 9

11. Which is the smallest of these four numbers?

A 20 002 **B** 22 000 **C** 20 200 **D** 20 020

12. Consider the sequence of numbers 1, 3, 6, 10, 15, 21, ... The sum of the next two numbers in the series is

A 51 **B** 53 **C** 55 **D** 64

13. For the sequence of numbers given in question 12, the next square number in the series is

A 25 **B** 28 **C** 36 **D** 49

14. Consider the four prime numbers 53, 59, 73 and 79. What is the other prime number between 50 and 80?

A 57 **B** 61 **C** 74 **D** 77

15. What is the difference between the two numbers 90 and 144?

A 46 **B** 54 **C** 66 **D** 234

16. Consider the numbers 13, 24, 31, 65, 75 and 125. Three of the numbers have a common factor. It is

A 3 **B** 5 **C** 13 **D** 25 (AL)

17. Consider the pattern of numbers 1×2, 2×3, 3×4, 4×5, ... What is the value of the ninth term in the pattern when it is multiplied out?

A 81 **B** 90 **C** 99 **D** 110 (AL)

18. Consider the sequence of numbers 1, 3, 6, 10, 15, 21, ... The next number in the sequence is

A 25 **B** 26 **C** 27 **D** 28 (AL)

19. For the sequence of numbers given in question 18, the sum of the 15th and 16th numbers in the sequence is

A 120 **B** 225 **C** 240 **D** 256 (AL)

20. The sum of any two consecutive numbers in the sequence of numbers given in question 18 is always

A an even number **B** an odd number
C a square number **D** a prime number (AL)

21. Consider the sequence of numbers 2, 5, 8, 11, ... The difference between the next two numbers in the series is

A 31 **B** 17 **C** 14 **D** 3

22. What is the value of 61^2?

A 122 **B** 361 **C** 3601 **D** 3721 (EA)

23. Consider the two numbers 45 and 72. What is the difference between them?

A 23 **B** 27 **C** 33 **D** 117 (EA)

24. What is the largest of these four numbers?

A 40 003 **B** 43 000 **C** 40 300 **D** 40 030

25. The LCM of 3, 4 and 6 is

A 12 **B** 13 **C** 24 **D** 72

26. The HCF of 16 and 20 is

A 2 **B** 4 **C** 80 **D** 320

27. What are the next two numbers in the sequence 45, 42, 37, 30?

A 21 and 12 **B** 21 and 11 **C** 21 and 10
D 21 and 9

2 FRACTIONS

- In a fraction, the top number is called the NUMERATOR and the bottom number is called the DENOMINATOR. Thus in the fraction $\frac{3}{4}$, the numerator is 3 and the denominator is 4.
- In a PROPER FRACTION the top number is always less than the bottom number. Thus $\frac{1}{4}$ and $\frac{3}{8}$ are both proper fractions.
- In an IMPROPER FRACTION the top number is greater than the bottom number. Thus $\frac{9}{2}$ and $\frac{15}{7}$ are both improper fractions.

  Note that the value of an improper fraction is always greater than 1.
- Both proper and improper fractions are called COMMON FRACTIONS.
- A MIXED NUMBER is the sum of a whole number and a fraction. Thus $5+\frac{7}{8}$, usually written $5\frac{7}{8}$, is a mixed number. A mixed number can be expressed as an improper fraction and vice versa.

$$7\tfrac{3}{4} = \frac{(7\times 4)+3}{4} = \frac{28+3}{4} = \frac{31}{4}$$

$$\tfrac{9}{2} = 4\tfrac{1}{2} \quad (\text{since } 9\div 2 = 4 \text{ remainder } 1)$$

- Two fractions are EQUIVALENT if they have the same value. The value of a fraction remains the same if both its numerator and its denominator are multiplied or divided by the same number, provided that the number is not zero. Thus

$$\frac{3}{8} \text{ is equivalent to } \frac{3\times 2}{8\times 2} = \frac{6}{16}$$

$$\frac{5}{15} \text{ is equivalent to } \frac{5\div 5}{15\div 5} = \frac{1}{3}$$

- A fraction is said to be in its LOWEST TERMS when it is impossible to find a number which will divide exactly into both its numerator and its denominator. Thus the fractions $\frac{7}{16}$ and $\frac{9}{64}$ are in their lowest terms, but the fraction $\frac{6}{8}$ is not in its lowest terms because it can be reduced to $\frac{3}{4}$.

Example

Reduce $\frac{42}{48}$ to its lowest terms.

$$\frac{42}{48} \text{ is equivalent to } \frac{42\div 6}{48\div 6} = \frac{7}{8}$$

Exercise 2.1

Express each of the following as mixed numbers:

1. $\frac{7}{2}$ **2.** $\frac{17}{8}$ **3.** $\frac{26}{5}$ **4.** $\frac{18}{7}$ **5.** $\frac{23}{9}$.

Express each of the following as improper fractions:

6. $3\frac{1}{5}$ **7.** $1\frac{3}{4}$ **8.** $5\frac{6}{7}$ **9.** $2\frac{3}{8}$ **10.** $3\frac{9}{20}$

Reduce the following fractions to their lowest terms:

11. $\frac{9}{18}$ **12.** $\frac{25}{35}$ **13.** $\frac{3}{9}$ **14.** $\frac{18}{24}$ **15.** $\frac{6}{15}$.

- When the values of two or more fractions are to be compared, express each of the fractions with the same denominator. This common denominator should be the LCM of the denominators of the fractions to be compared. It is called the LOWEST COMMON DENOMINATOR.

Example

Arrange the fractions $\frac{11}{16}$, $\frac{7}{10}$, $\frac{9}{14}$ and $\frac{3}{4}$ in order of size beginning with the smallest.

The lowest common denominator is 560.

$$\frac{11}{16} = \frac{11 \times 35}{16 \times 35} = \frac{385}{560} \qquad \frac{7}{10} = \frac{7 \times 56}{10 \times 56} = \frac{392}{560}$$

$$\frac{9}{14} = \frac{9 \times 40}{14 \times 40} = \frac{360}{560} \qquad \frac{3}{4} = \frac{3 \times 140}{4 \times 140} = \frac{420}{560}$$

Since all the fractions have been expressed with the same denominator, all we have to do is to compare their numerators.

Thus the order of size is $\frac{360}{560}$, $\frac{385}{560}$, $\frac{392}{560}$ and $\frac{420}{560}$ or $\frac{9}{14}$, $\frac{11}{16}$, $\frac{7}{10}$ and $\frac{3}{4}$.

- The procedure when ADDING OR SUBTRACTING FRACTIONS is
  - (i) Find the lowest common denominator of the fractions to be added or subtracted
  - (ii) Express each of the fractions with this common denominator
  - (iii) Add or subtract the numerators of the new fractions to give the numerator of the answer. The denominator of the answer is the lowest common denominator found in (ii).

Example

(a) $$\frac{1}{8} + \frac{2}{3} + \frac{3}{5} = \frac{(1 \times 15) + (2 \times 40) + (3 \times 24)}{120}$$

$$= \frac{15 + 80 + 72}{120} = \frac{167}{120} = 1\frac{47}{120}$$

(b) $3\frac{3}{8}+5\frac{2}{7}+4\frac{3}{4} = (3+5+4)+\frac{3}{8}+\frac{2}{7}+\frac{3}{4}$

$$= 12+\frac{(3\times7)+(2\times8)+(3\times14)}{56}$$

$$= 12+\frac{21+16+42}{56}$$

$$= 12+\frac{79}{56} = 12+1\frac{23}{56} = 13\frac{23}{56}$$

(c) $5\frac{3}{8}-2\frac{9}{10} = \frac{43}{8}-\frac{29}{10} = \frac{(43\times5)-(29\times4)}{40}$

$$= \frac{215-116}{40} = \frac{99}{40} = 2\frac{19}{40}$$

- To MULTIPLY FRACTIONS, multiply their numerators together and then multiply their denominators together. Thus

$$\frac{7}{8}\times\frac{3}{5} = \frac{7\times3}{8\times5} = \frac{21}{40}$$

Mixed numbers must be converted into improper fractions before multiplying. Thus

$$1\frac{4}{9}\times2\frac{3}{8} = \frac{13}{9}\times\frac{19}{8} = \frac{13\times19}{8\times9} = \frac{247}{72} = 3\frac{31}{72}$$

Sometimes in calculations the word 'of' appears. It should always be taken to mean multiply. Thus

$$\tfrac{3}{4}\text{ of }24 = \frac{3}{4}\times\frac{24}{1} = \frac{3\times24}{4\times1} = \frac{72}{4} = 18$$

- To DIVIDE BY A FRACTION, invert it (i.e. turn it upside down) and proceed as in multiplication. Thus

$$\frac{3}{5}\div\frac{7}{8} = \frac{3}{5}\times\frac{8}{7} = \frac{3\times8}{5\times7} = \frac{24}{35}$$

Exercise 2.2

Arrange the following fractions in order of size, beginning with the smallest:

1. $\frac{3}{5}$ and $\frac{7}{10}$
2. $\frac{1}{2}$, $\frac{5}{6}$ and $\frac{2}{3}$
3. $\frac{3}{4}$, $\frac{5}{8}$, $\frac{9}{16}$ and $\frac{17}{32}$
4. $\frac{3}{8}$, $\frac{4}{7}$, $\frac{5}{9}$ and $\frac{3}{5}$.

Add together:

5. $\frac{1}{3}+\frac{1}{5}$
6. $\frac{3}{4}+\frac{1}{8}$

7. $\frac{3}{8}+\frac{2}{9}$

8. $\frac{1}{2}+\frac{2}{3}+\frac{3}{4}$

9. $\frac{5}{8}+\frac{3}{5}+\frac{2}{3}$

10. $1\frac{5}{8}+2\frac{5}{16}$

11. $4\frac{2}{3}+3\frac{4}{5}$

12. $2\frac{5}{8}+3\frac{3}{7}+4\frac{1}{4}$

13. $3\frac{1}{2}+2\frac{1}{3}+5\frac{5}{6}$

14. $2\frac{3}{8}+3\frac{1}{4}+\frac{7}{10}+3\frac{1}{2}$.

Subtract the following:

15. $\frac{2}{3}-\frac{1}{2}$

16. $\frac{7}{8}-\frac{1}{3}$

17. $\frac{2}{3}-\frac{3}{5}$

18. $3\frac{3}{4}-2\frac{2}{3}$

19. $4\frac{15}{32}-2\frac{7}{10}$

20. $5-2\frac{15}{16}$.

Multiply the following:

21. $\frac{3}{8}\times\frac{2}{7}$

22. $\frac{3}{4}\times\frac{6}{7}$

23. $\frac{5}{9}\times 1\frac{2}{3}$

24. $2\frac{3}{4}\times 1\frac{2}{5}$

25. $\frac{5}{8}\times\frac{4}{7}\times\frac{3}{5}$

26. $\frac{3}{4}\times 3\frac{1}{5}$

27. $5\frac{1}{4}\times 1\frac{1}{7}$

28. $\frac{1}{2}\times\frac{8}{9}\times\frac{3}{4}$

29. $3\frac{1}{3}\times 4\frac{1}{2}\times 7\frac{1}{5}$

30. $3\frac{3}{5}\times 1\frac{1}{9}\times 1\frac{1}{16}$.

Find the values of each of the following:

31. $\frac{5}{7}$ of 280

32. $\frac{3}{8}$ of 32

33. $\frac{4}{5}$ of $3\frac{1}{3}$

34. $\frac{2}{3}$ of $1\frac{1}{8}$.

35. $\frac{5}{9}$ of $6\frac{3}{10}$

Divide each of the following:

36. $\frac{4}{5}\div 3\frac{1}{3}$

37. $\frac{3}{8}\div 2\frac{1}{4}$

38. $1\frac{3}{5}\div\frac{3}{10}$

39. $\frac{2}{7}\div\frac{8}{21}$.

40. $2\frac{1}{7}\div\frac{5}{14}$.

- The SEQUENCE OF OPERATIONS when dealing with fractions is
  - (i) Work out brackets
  - (ii) Multiply and/or divide
  - (iii) Add and/or subtract.

Example

Simplify $\frac{2}{5}\times\left(\frac{2}{3}-\frac{1}{4}\right)+\frac{1}{2}$.

$$\frac{2}{5}\times\left(\frac{2}{3}-\frac{1}{4}\right)+\frac{1}{2} = \frac{2}{5}\times\left(\frac{8-3}{12}\right)+\frac{1}{2}$$

$$= \frac{2}{5}\times\frac{5}{12}+\frac{1}{2}$$

$$= \frac{1}{6}+\frac{1}{2} = \frac{1+3}{6} = \frac{4}{6} = \frac{2}{3}$$

Exercise 2.3

Find values for each of the following:

1. $\frac{9}{10}-(\frac{2}{3}\times 1\frac{1}{5})$
2. $\frac{1}{2}+(\frac{2}{5}\div\frac{7}{10})$
3. $1\frac{5}{16}\div(\frac{1}{4}+\frac{5}{8})$
4. $\frac{3}{5}\times(\frac{2}{3}-\frac{1}{4})+\frac{3}{10}$
5. $\frac{9}{10}\div(\frac{2}{5}\div\frac{4}{15})+\frac{1}{2}$
6. $(2\frac{2}{3}+1\frac{1}{5})\div 5\frac{4}{5}$
7. $6\times(\frac{3}{4}+\frac{2}{3})$
8. $7\frac{1}{3}\div(1\frac{1}{3}\times 1\frac{3}{5})$.

Exercise 2.4 (All of the type found in CSE examination papers)

1. Consider the fractions $\frac{1}{4}$, $\frac{1}{5}$, $\frac{1}{6}$ and $\frac{1}{7}$.
   (a) Which fraction is the smallest?
   (b) What is the difference between $\frac{1}{4}$ and $\frac{1}{5}$?
   (c) What is the sum of $\frac{1}{5}$ and $\frac{1}{6}$?
   (d) What is the product of $\frac{1}{5}$ and $\frac{1}{7}$?
   (e) What is the value of $\frac{1}{4}\div\frac{1}{5}$?
2. Give answers to the following in their lowest terms:
   (a) $\frac{3}{5}\times\frac{10}{21}$ (b) $\frac{5}{8}\div 2\frac{3}{4}$ (c) $\frac{7}{8}-\frac{2}{3}$ (d) $1\frac{1}{3}+2\frac{3}{4}$ (WY)
3. Find $\frac{3}{4}$ of $\frac{1}{3}$.
4. (a) Find the sum of $\frac{1}{3}$ and $\frac{1}{4}$.
   (b) Calculate the product of $\frac{3}{4}$ and $\frac{8}{9}$.
   (c) Find the value of $2\frac{3}{4}\div\frac{1}{2}$.
5. Arrange the fractions $\frac{5}{22}$, $\frac{5}{21}$ and $\frac{5}{23}$ in order of size with the smallest first. (EM)
6. Multiply $1\frac{1}{2}$ by:
   (a) 2 (b) 10 (c) 7. (EM)
7. Continue the sequence $1\frac{3}{4}$, 2, $2\frac{1}{4}$, ... for three more terms. (EM)
8. Find the value of $\frac{4}{5}\times\frac{5}{8}-\frac{3}{8}\times\frac{4}{9}$.
9. Arrange the following fractions in order of size starting with the fraction having the greatest value: $\frac{5}{8}$, $\frac{2}{3}$, $\frac{7}{12}$.
10. Find $\frac{5}{8}$ of 32.
11. Rearrange the following fractions in order of size starting with the the fraction with the greatest value: $\frac{5}{8}$, $\frac{3}{4}$, $\frac{11}{16}$.
12. Calculate:
    (a) $7-1\frac{5}{8}$ (b) $6\frac{2}{3}\times 1\frac{4}{5}$.

13. Write down the next two fractions in the sequence $\frac{1}{4}, \frac{1}{9}, \frac{1}{16}, \frac{1}{25}, \ldots$. (EA)

14. Express $\frac{1}{2}-\frac{1}{4}+\frac{1}{8}-\frac{1}{16}+\frac{1}{32}$ as a single fraction. (Y)

15. Find the value of $\frac{1}{4}+1\frac{1}{2}-\frac{1}{16}$. (Y)

Multi-choice questions 2

1. The value of $(\frac{3}{8}\times 1\frac{1}{3})-\frac{4}{15}$ is

   A $\frac{17}{120}$ B $\frac{7}{30}$ C $\frac{11}{40}$ D $\frac{3}{10}$

2. $\frac{5}{12}+\frac{3}{5}=$

   A $\frac{1}{4}$ B $\frac{8}{17}$ C $\frac{15}{17}$ D $1\frac{1}{60}$ (EA)

3. $\frac{3}{5}-\frac{5}{12}=$

   A $\frac{2}{17}$ B $\frac{11}{60}$ C $\frac{2}{7}$ D $\frac{61}{60}$ (EA)

4. Find, in its simplest terms, $\frac{3}{5}\times\frac{5}{12}$.

   A $\frac{9}{60}$ B $\frac{1}{4}$ C $\frac{15}{17}$ D $\frac{15}{12}$ (EA)

5. Which of the following fractions has the greatest value?

   A $\frac{7}{10}$ B $\frac{17}{20}$ C $\frac{7}{12}$ D $\frac{2}{3}$
   E $\frac{4}{5}$

6. $(\frac{1}{6})^2$ is equal to

   A $\frac{1}{36}$ B $\frac{1}{12}$ C $\frac{1}{6}$ D $\frac{1}{3}$

7. The value of $6\times(\frac{1}{2}+\frac{1}{4})$ is

   A 2 B $3\frac{1}{2}$ C $4\frac{1}{2}$ D $7\frac{1}{2}$

8. Consider the fractions $\frac{1}{2}$, $\frac{5}{6}$ and $\frac{2}{3}$. When arranged in order of size, beginning with the smallest, the order is

   A $\frac{1}{2}, \frac{5}{6}, \frac{2}{3}$ B $\frac{1}{2}, \frac{2}{3}, \frac{5}{6}$ C $\frac{2}{3}, \frac{1}{2}, \frac{5}{6}$ D $\frac{2}{3}, \frac{5}{6}, \frac{1}{2}$

9. Find the value of $\frac{3}{4}\div\frac{8}{9}$.

   A $\frac{24}{36}$ B $\frac{2}{3}$ C $\frac{27}{32}$ D $\frac{3}{2}$

10. $\frac{1}{4}+\frac{2}{3}+\frac{3}{5}$ is equal to

   A $\frac{1}{10}$ B $\frac{5}{12}$ C $\frac{1}{2}$ D $1\frac{31}{60}$

3 THE DECIMAL SYSTEM

- The number 3768.259 means

$$(3 \times 1000) + (7 \times 100) + (6 \times 10) + (8 \times 1) + \tfrac{2}{10} + \tfrac{5}{100} + \tfrac{9}{1000}$$

The DECIMAL POINT separates the whole numbers from the fractional parts.

Examples

$\tfrac{7}{10} = 0.7$

$\tfrac{5}{10} + \tfrac{9}{100} = 0.59$

$36 + \tfrac{7}{10} + \tfrac{8}{1000} = 36.708$

Exercise 3.1

Read off as decimals:

1. $\tfrac{9}{10}$
2. $\tfrac{3}{10} + \tfrac{8}{100}$
3. $\tfrac{6}{10} + \tfrac{7}{100} + \tfrac{2}{1000}$
4. $\tfrac{6}{100}$
5. $\tfrac{2}{10} + \tfrac{9}{1000}$
6. $\tfrac{3}{1000}$
7. $498 + \tfrac{2}{10}$
8. $54 + \tfrac{7}{100} + \tfrac{9}{1000}$.

Read off the following with denominators of 10, 100, 1000, etc.:

9. 0.3
10. 3.8
11. 7.98
12. 567.234
13. 0.002
14. 0.059
15. 0.603
16. 0.1703.

- ADDITION AND SUBTRACTION OF DECIMALS is performed in the same way as for whole numbers. Remember that the decimal points must be placed directly underneath one another.

Example

Find the sum of 127.35, 28.732 and 9027.1.

```
  127.35
   28.732
 9027.1
 --------
 9183.182
```

$127.35 + 28.732 + 9027.1 = 9183.182$

Exercise 3.2

Write down the values of:

1. $2.389 + 0.325$

2. $5.16 + 12.2$

3. $5.189 + 17.23 + 873.2$

4. $0.362 + 0.075 + 0.009 + 0.169$

5. $13.18 - 12.06$

6. $18.153 - 6.007$

7. $0.036 - 0.009$

8. $53.109 - 27.076$.

- To MULTIPLY A DECIMAL NUMBER BY 10, shift the decimal point one place to the *right*; to multiply by 100 shift the decimal point two places to the right and so on. Thus

$$27.18 \times 10 = 271.8$$

$$9.782 \times 100 = 978.2$$

$$15.3 \times 1000 = 15\,300$$

- To DIVIDE A DECIMAL NUMBER BY 10, shift the decimal point one place to the *left*; to divide by 100 shift the decimal point two places to the left and so on. Thus

$$17.6 \div 10 = 1.76$$

$$5.32 \div 100 = 0.0532$$

$$4.6 \div 1000 = 0.0046$$

Exercise 3.3

Multiply each of the following numbers by 10, 100 and 1000:

1. 0.35

2. 5.983

3. 0.038

4. 98.2345

5. 8.1624.

Divide each of the following numbers by 10, 100 and 1000:

6. 189

7. 18.13

8. 527.31

9. 0.03

10. 0.325.

- The NUMBER OF DECIMAL PLACES in a number is the number of figures which follow the decimal point. A number may be approximated by stating the number correct to so many decimal places. The rule is: If the first figure to be discarded is 5 or more, then increase the previous figure by 1. Thus

93.7257 = 93.726 correct to 3 decimal places
= 93.73 correct to 2 decimal places
= 93.7 correct to 1 decimal place

0.007 682 = 0.007 68 correct to 5 decimal places
= 0.0077 correct to 4 decimal places

8.703 = 8.70 correct to 2 decimal places
= 8.7 correct to 1 decimal place

- A second way of approximating a number is to use SIGNIFICANT FIGURES. In the number 9873, 9 is the most significant figure because it has the greatest value; 98 are the two most significant figures, while 987 are the three most significant figures. The rules regarding the use of significant figures are as follows:

  (i) If the first figure to be discarded is 5 or more, increase the previous figure by 1.

7.192 53 = 7.1925 correct to 5 significant figures
= 7.193 correct to 4 significant figures
= 7.19 correct to 3 significant figures
= 7.2 correct to 2 significant figures

  (ii) Zeros must be kept to show the position of the decimal point or to indicate that zero is a significant figure.

35 291 = 35 290 correct to 4 significant figures
= 35 300 correct to 3 significant figures
= 35 000 correct to 2 significant figures
0.0739 = 0.074 correct to 2 significant figures
= 0.07 correct to 1 significant figure
18.403 = 18.40 correct to 4 significant figures
= 18.4 correct to 3 significant figures

Exercise 3.4

Write down the following numbers correct to the number of decimal places stated:

1. 19.372: (a) to 2, (b) to 1
2. 0.007 519: (a) to 5, (b) to 3, (c) to 2
3. 4.9703: (a) to 3, (b) to 2
4. 153.2617: (a) to 3, (b) to 2, (c) to 1.

Write down the following numbers correct to the number of significant figures stated:

5. 24.935 82: (a) to 6, (b) to 4, (c) to 2

6. 0.007 326: (a) to 4, (b) to 3, (d) to 2

7. 35.604: (a) to 4, (b) to 3

8. 35 682: (a) to 4, (b) to 3, (c) to 2

9. 17 359 285: (a) to 5, (b) to 2

10. 0.007 803: (a) to 2, (b) to 1.

- The worst mistake you can make in a calculation is to misplace the decimal point. To prevent this occurring a ROUGH CHECK should always be made before actually performing the calculation. In the rough check try to select numbers which are easy to multiply or that will cancel.

Example

(a) 32.7×0.259.

For a rough check we will take $32 \times 0.25 = 8$.

Correct product $= 8.4693$.

(The rough check shows that the answer is not 84.693 nor 0.846 93.)

(b) $79.32 \div 21.46$.

For a rough check we will take $80 \div 20 = 4$.

Correct answer $= 3.696$ correct to 4 significant figures.

(The rough check shows that the answer is not 39.96 nor 0.3696.)

(c) $\dfrac{47.5 \times 36.52}{11.3 \times 2.75}$.

For a rough check we will take $\dfrac{50 \times 36}{10 \times 3} = 60$.

The answer is 55.82 correct to 4 significant figures.

Exercise 3.5

Perform rough checks for each of the following:

1. 22×0.57

2. 41.35×0.26

3. $0.732 \times 0.098 \times 2.17$

4. $92.17 \div 31.45$

5. $0.092 \div 0.035$

6. $27.18 \times 29.19 \times 0.031$

7. $\dfrac{1.456 \times 0.0125}{0.0532}$

8. $\dfrac{29.92 \times 31.32}{10.89 \times 2.95}$.

- To CONVERT A FRACTION TO A DECIMAL remember that the line separating the numerator and the denominator acts as a division sign. Thus

$$\tfrac{2}{5} = 2 \div 5 = 0.4$$

$$2\tfrac{3}{8} = 2+(3\div 8) = 2+0.375 = 2.375$$

- Converting $\frac{2}{3}$ into a decimal gives $2\div 3 = 0.666\,666\ldots$ (all we get is a succession of 6's). This is an example of a RECURRING DECIMAL. In order to prevent endless repetition the result is written $\dot{6}$. Here are some more examples

$$\tfrac{1}{3} = 0.\dot{3} \quad \text{(meaning } 0.333\,333\ldots\text{)}$$

$$\tfrac{1}{6} = 0.1\dot{6} \quad \text{(meaning } 0.166\,666\ldots\text{)}$$

$$\tfrac{7}{11} = 0.\dot{6}\dot{3} \quad \text{(meaning } 0.636\,363\ldots\text{)}$$

For all practical purposes recurring decimals are not needed; what is needed is the answer given to so many decimal places or significant figures.

- Decimals are fractions with denominators 10, 100, 1000, etc. The CONVERSION OF A DECIMAL NUMBER INTO A FRACTION can therefore be made. Thus

$$0.53 = \tfrac{53}{100}$$

$$0.625 = \tfrac{625}{1000} = \tfrac{5}{8}$$

Example

Find the difference between $2\frac{5}{16}$ and 2.3214.

$$2\tfrac{5}{16} = 2+(5\div 16) = 2+0.3125 = 2.3125$$

$$2.3214-2.3125 = 0.0089$$

Exercise 3.6

Convert the following fractions into decimals stating, where necessary, the answer to 4 significant figures:

1. $\frac{3}{4}$

2. $\frac{1}{5}$

3. $\frac{7}{8}$

4. $\frac{13}{16}$

5. $\frac{3}{7}$

6. $\frac{5}{6}$

7. $1\frac{19}{32}$

8. $3\frac{15}{64}$.

Write down the following recurring decimals correct to 4 significant figures:

9. $0.\dot{5}$

10. $0.\dot{8}$

11. $0.1\dot{7}$

12. $0.4\dot{5}$

13. $0.\dot{3}\dot{5}$

14. $0.\dot{2}\dot{1}$

15. $0.\dot{5}6\dot{2}$

16. $0.\dot{7}3\dot{1}$.

Convert the following decimals to fractions in their lowest terms:

17. 0.3 **19.** 0.4375 **21.** 1.75

18. 0.65 **20.** 2.62 **22.** 9.185.

23. Find the difference between $\frac{3}{16}$ and 0.17.

24. What is the difference between $5\frac{3}{8}$ and 3.627?

25. Calculate $\frac{9}{11}-0.723$, stating the answer correct to 3 decimal places.

Exercise 3.7 (All of the type found in CSE examination papers)

1. Calculate:
(a) 4.8×6 (b) $4.8\div 6$ (c) 4.8×0.6 (d) $4.8\div 0.6$.

2. Calculate:
(a) 37.9×9 (b) 3.4×1.1 (c) 41.32×20.

3. Write $\frac{7}{8}$ as a decimal number.

4. Express $81\div 108$ as:
(a) a vulgar fraction in its lowest terms (b) a decimal number.
(WY)

5. Calculate:
(a) 28.73×100 (b) $28.73\div 100$. (EM)

6. Find the value of $\dfrac{0.8\times 7\times 1.3}{2.1\times 4}$. (EM)

7. Work out:
(a) 73.8×1000 (b) $\frac{1}{4}$ of 3.236. (EM)

8. Which is the larger, 23 or 5.6 and by how much? (EM)

9. (a) Multiply 0.06 by 0.03. (b) Divide 174.3 by 1000.

10. Work out $5+0.73-0.48$.

11. In the number 52.058:
(a) Write down the actual value of the digit 8
(b) Find the difference between the actual values of the digits 5.

12. Write 7.379 correct to one place of decimals and 2938 correct to 2 significant figures.

13. A football match attracted 48 359 spectators. Express this attendance correct to 3 significant figures.

14. Arrange the numbers 0.5, 0.05 and 0.22 in order of size, beginning with the smallest.

15. (a) Find the sum of 3.7, 167.09 and 0.516.
(b) Write the number 167.09 correct to: (i) 4 significant figures, (ii) 2 significant figures.

16. (a) Add 3.98, 0.745 and 52.073.
(b) Write the number 738.0584 correct to (i) 2 decimal places, (ii) 2 significant figures.

17. Express as decimal numbers:
(a) $\frac{5}{8}$ (b) $\frac{11}{16}$.

18. Express as common fractions in their lowest terms:
(a) 0.7 (b) 0.12. (EA)

Multi-choice questions 3

1. In the number 8.6972 the value of the digit 9 is
A 90 **B** $\frac{9}{10}$ **C** $\frac{9}{100}$ **D** $\frac{9}{1000}$

2. $0.017 \div 0.027$ is equal to (correct to 2 significant figures)
A 0.063 **B** 0.63 **C** 6.3 **D** 63

3. The number 158 861 correct to 2 significant figures is
A 15 **B** 16 **C** 150 000 **D** 160 000

4. The number 0.081 778 correct to 3 significant figures is
A 0.081 **B** 0.0817 **C** 0.0818 **D** 0.082

5. What is 0.047 85 when written to the nearest thousandth?
A 0.047 **B** 0.0478 **C** 0.0479 **D** 0.048

6. The best estimate of $\dfrac{27.5 \times 30.52}{11.3 \times 2.73}$ is
A 2.7 **B** 27 **C** 270 **D** 2700

7. Correct 7342.7 to 3 significant figures.
A 7 **B** 734 **C** 7000 **D** 7340

8. Write $\frac{5}{8}$ as a decimal number.
A 0.13 **B** 0.58 **C** 0.625 **D** 1.6

9. The value of $4.7 - 1.9 + 2.1$ is
A 0.7 **B** 1.7 **C** 4.9 **D** 5.9

10. Given that $225 \times 35 = 7875$ then 22.5×0.35 is
A 787.5 **B** 7.875 **C** 0.7875 **D** 0.078 75

4 DECIMAL CURRENCY

- The BRITISH SYSTEM uses the pound sterling as the basic unit. The only sub-unit used is the penny such that

$$100\,\text{p} = £1$$

- The ADDITION, SUBTRACTION, MULTIPLICATION AND DIVISION of sums of money is performed in a similar way to that used for decimals.

Example

(a) Add together £3.72, £15.29 and £16.34.

£ 3.72 £3.72 + £15.29 + £16.34 = £35.35
£15.29
£16.34
£35.35

(b) Find the difference between £8.73 and £4.38

£8.73 Difference between £8.73 and £4.38
£4.38 = £8.73 − £4.38 = £4.35
£4.35

(c) Find the cost of 28 articles whose price is 31 p each.

Total cost = 28 x 31 p = 868 p = £8.68

(d) If 241 identical articles cost £50.61, how much does each article cost?

Cost of each article = £50.61 ÷ 241
= £0.21
= 21 p

Exercise 4.1

Add together the following sums of money:

1. £17.68, £2.58 and £133.26
2. £15.24, £600.98, £7.58 and £98.63
3. £5.06, £7.38 and £83.98
4. 73 p, 41 p and 86 p
5. 16 p, 29 p, 81 p and 93 p.

Subtract the following:

6. £18.34 from £19.48
7. £3.49 from £50.00
8. 54 p from £5
9. 39 p from £7.34
10. £19.69 from £20.63.

Find the total cost of the following:

11. 12 articles at £13 each
12. 89 articles at 27 p each
13. 15 articles at 39 p each
14. 24 articles at £7.03 each.
15. If 12 identical articles cost £4.44, what is the cost of each article?
16. 35 articles cost a total of £8.05. How much does each article cost?

- FOREIGN EXCHANGE is used to convert the currency of one country into the currency of another country.

Foreign Monetary Systems and Exchange Rate at January 1983

Country	*Monetary Unit*			*Rate of Exchange*		
Belgium	100 centimes	=	1 franc	74.50 B fr.	=	£1
France	100 centimes	=	1 franc	10.47 Fr fr.	=	£1
West Germany	100 pfennig	=	1 mark	DM 3.68	=	£1
Greece	100 lepta	=	1 drachma	dr. 138	=	£1
Italy	100 centesimos	=	1 lira	L. 2100	=	£1
Spain	100 centimos	=	1 peseta	pta. 194	=	£1
Switzerland	100 centimes	=	1 franc	3.01 Sw fr.	=	£1
United States	100 cents	=	1 dollar	$1.52	=	£1

Example

(a) If £1 = 194 pesetas, find to the nearest penny the value in British money of 20 000 pesetas.

$$194 \text{ pesetas} = £1$$

$$1 \text{ peseta} = £\frac{1}{194}$$

$$20\,000 \text{ pesetas} = £\frac{1}{194} \times \frac{20\,000}{1}$$

$$= £\frac{20\,000}{194}$$

$$= £103.09$$

(b) A tourist changes traveller's cheques for £50 into US dollars. If the exchange rate is \$1.52 = £1, how many dollars does he get?

$$£50 = \$1.52 \times 50 = \$76$$

Exercise 4.2

Where necessary give the answers correct to 2 decimal places.

Using the exchange rates given opposite find:

1. The number of French francs equivalent to £40
2. The number of Spanish pesetas equivalent to £85
3. The number of Italian lira equivalent to £120
4. The number of pounds equivalent to 345 US dollars
5. The amount of sterling equivalent to 220 Swiss francs
6. The number of pounds equivalent to 7000 Belgian francs.

Exercise 4.3 (All of the type found in CSE examination papers)

1. Write the amount 'fifteen pounds and eight pence' in figures.
2. A blouse costs £8.53. How much change will there be if it is paid for with a £10 note?
3. Find the total cost of renting a television set for six months at £9.53 per month.
4. If there are 2 US dollars to £1, find the value in British money of \$300.
5. Find the total cost of a skirt costing £11.45 and three pairs of socks costing £1.21 per pair.
6. A tourist has 54 French francs left after his holiday. If the exchange rate is 10.8 francs to £1, how much in sterling will he get for his 54 francs?
7. Calculate $\frac{3}{4}$ of £8.32.
8. A person earns £81 and spends three-quarters of it. Calculate the amount *not* spent.

9. A holiday maker changes 74 French francs into sterling. If the exchange rate is 9.25 French francs to £1, how much in British money will she get?

10. How many Italian lire should I receive for £45 when the exchange rate is 1810 lire to £1. (EA)

11. Find two-thirds of £5.70. (EA)

12. A person earns £81 and spends two-thirds of it. Calculate the amount of money spent. (EA)

13. Find $\frac{3}{5}$ of £70.

14. The rate of exchange when Roma went to France on holiday was £1 = 8.80 francs.
    (a) How many francs did she receive for her £30 pocket money?
    (b) Find the equivalent cost in sterling of some flowers which cost 22 francs. (SW)

15. How much change should I receive from £1 when I buy four packets of tea at 23 p per packet. (SW)

16. Find the cost of 30 eggs when six eggs cost 27 p. (SW)

Multi-choice questions 4

1. The cost of 4 articles at 72 p each is

   A £2.88 **B** £2.89 **C** £2.90 **D** £2.95

Questions 2, 3 and 4 are about the amount of money spent by three girls in a shop. Angela spent 65 p. Betty spent 91 p and Celia spent £1.04.

2. How much change should Angela have from £1?

   A 15 p **B** 25 p **C** 35 p **D** 45 p (EA)

3. How much more did Celia spend than Angela?

   A 29 p **B** 39 p **C** 41 p **D** £1.69 (EA)

4. What fraction, in its lowest terms, of the total amount was spent by Celia?

   A $\frac{2}{5}$ **B** $\frac{1}{2}$ **C** $\frac{5}{11}$ **D** $\frac{104}{260}$ (EA)

5. The total cost for a family of four for a package holiday in Spain is 94 500 pesetas. If there are 140 pesetas to the £1, how much does their holiday cost them?

   A £67.50 **B** £132.30 **C** £675 **D** £1323
   E £6750

6. The value of £6.92 − 75 p + £0.96 is

   A £5.21 **B** £5.23 **C** £7.13 **D** £7.15
   E £7.27

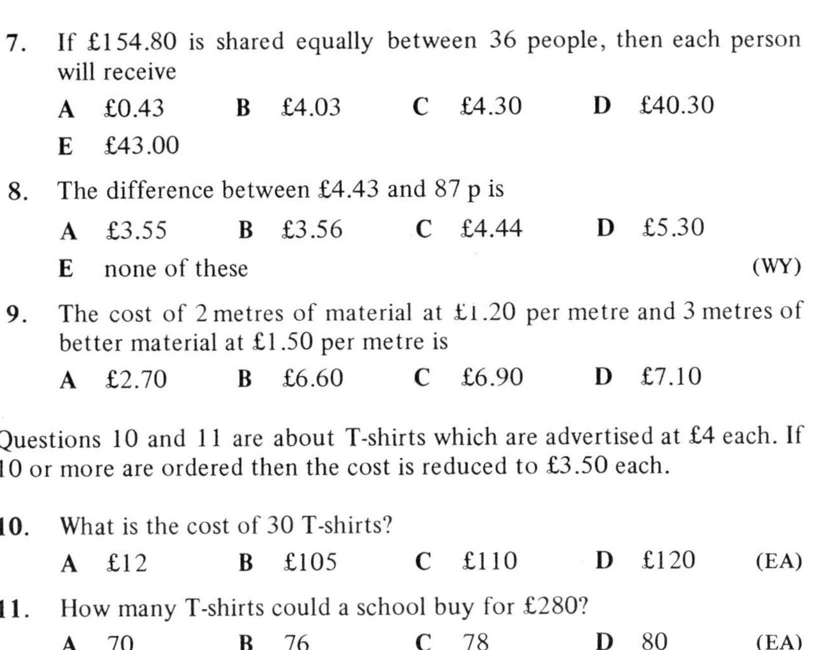

7. If £154.80 is shared equally between 36 people, then each person will receive

   A £0.43 **B** £4.03 **C** £4.30 **D** £40.30
   E £43.00

8. The difference between £4.43 and 87 p is

   A £3.55 **B** £3.56 **C** £4.44 **D** £5.30
   E none of these (WY)

9. The cost of 2 metres of material at £1.20 per metre and 3 metres of better material at £1.50 per metre is

   A £2.70 **B** £6.60 **C** £6.90 **D** £7.10

Questions 10 and 11 are about T-shirts which are advertised at £4 each. If 10 or more are ordered then the cost is reduced to £3.50 each.

10. What is the cost of 30 T-shirts?

    A £12 **B** £105 **C** £110 **D** £120 (EA)

11. How many T-shirts could a school buy for £280?

    A 70 **B** 76 **C** 78 **D** 80 (EA)

12. The change from £1 after buying 18 buttons at 2 p each is

    A 36 p **B** 50 p **C** 54 p **D** 64 p

5 THE METRIC SYSTEM

- The METRIC UNIT OF LENGTH is

$$\begin{aligned} 1 \text{ metre (m)} &= 10 \text{ decimetres (dm)} \\ &= 100 \text{ centimetres (cm)} \\ &= 1000 \text{ millimetres (mm)} \end{aligned}$$

and $$1 \text{ kilometre (km)} = 1000 \text{ metres}$$

Example

(a) Convert 15.36 m into millimetres.

$$15.36\text{ m} = 15.36 \times 1000\text{ mm} = 15\,360\text{ mm}$$

(b) Convert 963 cm into metres.

$$963\text{ cm} = \frac{963}{100}\text{m} = 9.63\text{ m}$$

- The METRIC UNIT OF MASS is

$$\begin{aligned} 1 \text{ kilogram (kg)} &= 1000 \text{ grams (g)} \\ 1 \text{ gram} &= 1000 \text{ milligrams (mg)} \end{aligned}$$

For very large masses the tonne is used such that

$$1 \text{ tonne (t)} = 1000 \text{ kilograms}$$

Example

(a) How many grams are equivalent to 8945 mg?

$$8945\text{ mg} = \frac{8945}{1000} = 8.945\text{ g}$$

(b) How many grams are there in 19 kg?

$$19\text{ kg} = 19 \times 1000\text{ g} = 19\,000\text{ g}$$

(c) Convert 15 t into kilograms.

$$15\text{ t} = 15 \times 1000\text{ kg} = 15\,000\text{ kg}$$

- The ADDITION AND SUBTRACTION of metric quantities is done in the same way as for decimal numbers. However, it is important that all the quantities should be in the same units.

Example

Add together 15.2 m, 39.2 cm and 150.2 mm and state the answer in metres.

$$39.2\text{ cm} = 0.392\text{ m} \quad \text{and} \quad 150.2\text{ mm} = 0.1502\text{ m}$$

$$\text{Total length} = 15.2 + 0.392 + 0.1502 = 15.7422\text{ m}$$

- The MULTIPLICATION AND DIVISION of metric quantities is done in exactly the same way as for decimal numbers.

Example

(a) 57 lengths of wood each 95 cm long are required by a builder. What total length of wood, in metres, is needed?

$$95\text{ cm} = 0.95\text{ m}$$

$$\text{Total length required} = 0.95 \times 57 = 54.15\text{ m}$$

(b) Frozen peas are packed in bags containing 450 grams. How many packets can be filled from 2000 kg of peas?

$$450\text{ g} = 0.45\text{ kg}$$

$$\text{Number of bags} = \frac{2000}{0.45} = 4444$$

(c) A certain cloth costs £3.68 per metre. How much will 12 km of this cloth cost?

$$12\text{ km} = 12\,000\text{ m}$$

$$\text{Total cost} = 12\,000 \times £3.68 = £44\,160$$

Exercise 5.1

1. Convert to metres:

(a) 6.78 km	(b) 0.79 km	(c) 693 cm
(d) 5.3 cm	(e) 7395 mm	(f) 7 mm.

2. Convert to kilometres:

(a) 9375 m	(b) 368 m	(c) 2.75 m
(d) 3941 cm	(e) 735 682 cm.	

3. Convert to centimetres:

(a) 17.15 m	(b) 0.395 m	(c) 1.78 km
(d) 864 mm	(e) 5.2 mm.	

4. Convert to millimetres:

(a) 58 m	(b) 0.235 m	(c) 0.16 km
(d) 39.2 cm	(e) 0.59 cm.	

5. Convert to kilograms:

(a) 680 g (b) 37 800 g

(c) 2987 mg (d) 459 000 mg.

6. Convert to grams:

(a) 78 000 mg (b) 45 mg

(c) 19.1 kg (d) 0.59 kg.

7. Convert 27 000 kg into tonnes.

8. Convert 7.32 t into kilograms.

9. Add together the following lengths, stating the answers in metres:

(a) 47 cm, 5.83 m and 15 mm (b) 93 km, 462 m and 536 cm

(c) 0.185 m, 7.36 cm and 8.2 mm.

10. A length of rope 2.5 m long has the following lengths cut from it: 25 cm, 863 mm and 0.7 m. What length of rope remains?

11. Add together the following masses and state the answers in kilograms:

(a) 792 g, 15 000 mg and 1.265 kg

(b) 450 g, 25 kg and 793 000 mg.

12. A greengrocer starts the day with 127 kg of carrots. He sells $3\frac{1}{2}$ kg, 450 g and 25 kg. What mass of carrots remains?

13. Calculate the amount of ribbon left on a reel containing 25 m when the following lengths are cut: $\frac{1}{4}$ m, 580 cm, 3 m 45 cm and $9\frac{3}{4}$ m.

14. 95 lengths of wood 127 cm long are needed. What total length of wood, in metres, is required?

15. 209 lengths of cloth each 135 cm long have to be cut from a bale containing 300 m. What length remains?

16. How many lengths of string 53 cm long can be cut from a ball containing 25 m and how much string remains?

17. How many lengths of cloth each 36 cm long can be cut from a roll containing 30 m?

18. Butter is packed in 250 g packets. How many packets can be obtained from 8 t of butter?

19. A certain spice is packed in jars containing 32 g. How many jars can be filled from 15 kg of spice?

20. A certain type of tablet has a mass of 8.2 mg. What is the mass, in kilograms, of 5 000 000 of these tablets?

- Problems are sometimes set involving IMPERIAL UNITS and their metric equivalents. The problems usually involve the conversion of quantities using approximate conversion factors.

Example

If 1 pound = 2.2 kg, find in pounds the mass of a sack of potatoes having a mass of 25 kg.

$$\text{Mass of potatoes} = 25 \times 2.2 = 55\,\text{lb}$$

Exercise 5.2

Using the conversion 1 km = $\frac{5}{8}$ mile, answer the following questions:

1. Find in miles the distance equivalent to 32 km.
2. Find in kilometres the distance equivalent to 25 miles.
3. How many centimetres are there in 1 inch?
4. A distance is measured as being 532 m. How many yards is this?

Using 1 kg = 2.2 pounds, answer the following questions:

5. A woman buys 2 kg of tomatoes. How many pounds of tomatoes does she buy?
6. Butter is packed in 250 g packets. What is the mass in pounds of a packet of butter?
7. A shopper buys 5 lb of potatoes. How many kilograms of potatoes are purchased?
8. A consignment of steel having a mass of 5 t is delivered to a factory. How many pounds of steel are delivered?

Exercise 5.3 (All of the type found in CSE examination papers)

1. (a) Change 15.32 kg into grams.
   (b) Change 293.5 cm into metres.
   (c) Change 2895 m into kilometres.
2. (a) Change 8539 g into kilograms.
   (b) Express 946 m in centimetres.
   (c) Change 84.6 km into metres.
3. How many:
   (a) millimetres in 1 metre (b) milligrams in 3.4 g? (WY)
4. Calculate three-quarters of 7.6 m giving your answer in centimetres. (EM)
5. (a) How many pieces of string 30 cm long can be cut from a piece 4 m long?
   (b) A line is 70 cm long. Write this length: (i) in metres, (ii) in millimetres. (EM)
6. How many grams are there in $\frac{3}{4}$ kg?
7. A 1500 g packet of salt has a mass of 3.30 lb. What will be the mass, in pounds, of a 500 g packet of salt?

8. 1 kilogram of butter costs £1.80. A cook uses 100 grams of this butter in making a cake. How much did this butter cost?

9. A coal merchant delivered 28 bags of coal each containing 25 kg. How much coal was delivered altogether?

10. In order to make a fitted sheet, four pieces of elastic each measuring 34 cm are required. How much elastic, in metres, is required altogether? (EA)

11. How many 250 g packets of cheese have a total mass of $3\frac{1}{2}$ kg?

12. To hire a coach I pay an initial charge of £18 and then 43 p for each kilometre travelled. How much should I pay for a journey of 160 km?

13. When six pieces each 300 mm long are cut from a bar of steel 2 m long, how many millimetres are left? (EA)

14. A large piece of cheese has a mass of 2.25 kg. Smaller pieces having masses of 525 g, 485 g and 370 g are cut from it. What mass of cheese is left? (EA)

15. To make 5 kg of blackcurrant jam you need 3 kg of blackcurrants and 3 kg of sugar. Blackcurrants cost 52 p per kilogram and sugar costs 38 p per kilogram. Ignoring any other costs involved, find:

    (a) the total cost of making 5 kg of jam

    (b) the cost of 1 kg of jam. (EA)

16. If 1 inch = 2.5 cm, find, in centimetres, the length of a 12 inch ruler. (EA)

17. Sugar is sold in bags containing 1 kilogram. Taking 1 kg = 2.2 lb, find the mass in pounds of 25 such bags. (EA)

18. Two men have to load a lorry with coal. The coal is in 50 kg bags and it takes each man $1\frac{1}{2}$ minutes to carry and load each bag on the lorry. How long will it take to load the lorry with 3 t of coal? (Y)

19. (a) How many grams are there in 5.2 kg?

    (b) Divide 20 m into 40 equal lengths giving your answer in centimetres. (SW)

20. How many whole lengths of wallpaper, each 230 cm in length, can be cut from a roll containing 12 m and what length remains?

Multi-choice questions 5

1. A table is 1.45 m wide. How many millimetres is this?

   A 14.5 **B** 145 **C** 1450 **D** 14 500

2. The difference between 2 kg and 1.8 kg, in grams is

   A 1.6 **B** 2 **C** 20 **D** 200

3. A piece of string is 4 m long. What is one-fifth of the string's length?

   A 0.4 cm **B** 0.8 cm **C** 40 cm **D** 80 cm

4. A piece of tape is 2 m long. It is cut into 16 pieces of equal length. What is the length of each piece?

A 6.25 cm **B** 8 cm **C** 12 cm **D** 12.5 cm

5. If 1 km = 0.6 miles, then 120 miles is equivalent to

A 72 km **B** 200 km **C** 720 km **D** 2000 km

6. 870 mm is equal to

A 0.87 m **B** 8.7 m **C** 87 m **D** 870 m

7. The sum of 7.5 m, 315 cm and 18 750 mm is

A 29.4 m **B** 30.75 m **C** 307.5 m **D** 341.25 m

8. What is the cost of 1 kg of tea if a 125 g packet costs 32 p?

A £1.28 **B** £2.50 **C** £2.52 **D** £2.56

E none of these **(WY)**

9. Calculate the cost of 400 g of bacon if 1 kg costs £3.05.

A 35 p **B** 61 p **C** £1.22 **D** £12.20

E none of these **(WY)**

10. Calculate the number of packets of chemical each containing 240 g that can be filled from a sack containing 120 kg.

A 5 **B** 50 **C** 500 **D** 5000

6 RATIO AND PROPORTION

- A RATIO is a comparison between two similar quantities.

Example

Find the ratio of 5 cm to 2 m.

$$\frac{5\text{ cm}}{2\text{ m}} = \frac{5\text{ cm}}{200\text{ cm}} = \frac{1}{40}$$

This ratio can also be written as 1:40 which reads 'one to forty'. This tells us that the length of 2 m is 40 times the length of 5 cm.

- A ratio can therefore be written in two ways, either in the form $a:b$ or as a fraction $\frac{a}{b}$.

When converting from the form $a:b$ to a fraction, note that

(i) The first quantity mentioned becomes the numerator of the fraction

(ii) The two quantities must be stated in the same units; the units then cancel leaving the ratio as a number

(iii) A ratio is always expressed in its lowest terms.

Example

Express the ratio $5\frac{1}{2}:3$ as a fraction in its lowest terms.

$$5\tfrac{1}{2}:3 \text{ is equivalent to } \frac{5\frac{1}{2}}{3} = \frac{5\frac{1}{2}\times 2}{3\times 2} = \frac{11}{6}$$

Exercise 6.1

Express the following ratios as fractions in their lowest terms:

1. 9:7
2. 5:10
3. 14:7
4. 12:15
5. 16:24
6. 50 p:£4
7. 5 cm:8 m
8. 80 m:2 km
9. 5 kg:250 g
10. $5:\frac{1}{2}$
11. $\frac{1}{4}:4$
12. $\frac{3}{4}:7$
13. $9:\frac{5}{8}$.

- The line AB (Fig. 6.1) whose length is 15 cm has been divided into two parts in the ratio 2:3. The line has been divided into its PROPORTIONAL PARTS, and as can be seen from the diagram the line has been divided into a total of 5 parts. The length AC contains 2 of those parts and the length BC contains 3 of them. Each part is $\frac{15\text{ cm}}{5} = 3$ cm long.

  Hence AC = 2 x 3 = 6 cm and BC = 3 x 3 = 9 cm.

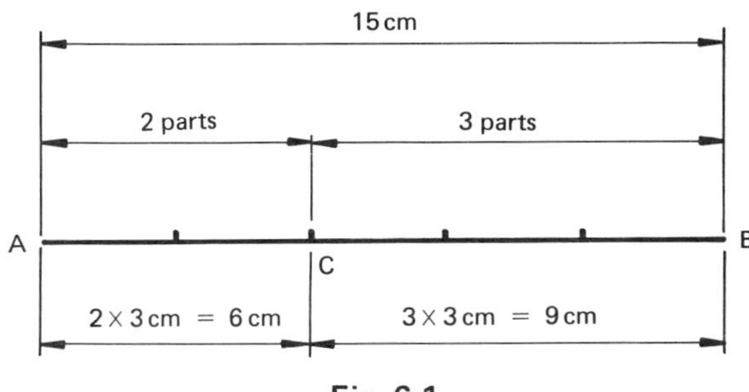

Fig. 6.1

The problem could be tackled in the following way.

$$\text{Total number of parts} = 2 + 3 = 5$$

$$\text{Length of each part} = \frac{15\text{ cm}}{5} = 3\text{ cm}$$

$$\text{Length of AC} = 2 \times 3\text{ cm} = 6\text{ cm}$$

$$\text{Length of BC} = 3 \times 3\text{ cm} = 9\text{ cm}$$

Example

Divide £240 in the ratio 5:4:3.

$$\text{Total number of parts} = 5 + 4 + 3 = 12$$

$$\text{Value of each part} = \frac{£240}{12} = £20$$

$$\text{Value of the first part} = 5 \times £20 = £100$$

$$\text{Value of the second part} = 4 \times £20 = £80$$

$$\text{Value of the third part} = 3 \times £20 = £60$$

Exercise 6.2

1. Divide £1600 in the ratio 5:3.
2. Divide 80 kg in the ratio 7:3.
3. Divide 120 m in the ratio 2:3:5.

4. A line 1.68 m long is to be divided into 3 parts in the ratio 2:7:11. Find, in millimetres, the length of each part.

5. Two lengths are in the ratio of 7:5. If the first length is 35 cm, find: (a) the length of the second part, (b) the total length.

6. Three amounts of money are in the ratio 2:4:5. If the largest amount is £40, what are the other two amounts?

- Two quantities are in DIRECT PROPORTION if corresponding pairs of values are in the same ratio. If we buy butter at £2 per kilogram, we expect to pay £4 for 2 kg. That is we have doubled the amount bought and paid twice as much. We expect to pay £1 for $\frac{1}{2}$ kg. That is we have halved the amount and paid half the price. Thus the amount of butter bought and the price paid are in direct proportion.

Example

If 74 exercise books cost £17.02, how much do 27 cost?

$$\text{74 books cost £17.02} = 1702\text{ p}$$

$$\text{1 book costs } \frac{1702}{74} = 23\text{ p}$$

$$\text{27 books cost } 23\text{ p} \times 27 = 621\text{ p} = \text{£6.21}$$

- Two quantities are in INVERSE PROPORTION if an increase in one quantity produces a decrease in the second quantity. Suppose 4 men working on a certain job take 6 days to complete it. If we double the number of men to 8, we expect the job to take half the time, i.e. 3 days. If we halve the number of men to 2, then we expect the job to take twice as long, i.e. 12 days. This is an example of inverse proportion.

Example

A bag contains sweets. When divided among 7 children each receives 8 sweets. If the sweets were divided among 4 children, how many sweets would each get?

The two quantities, the number of sweets each child gets and the number of children, are in inverse proportion.

$$\text{Total number of sweets} = 7 \times 8 = 56$$

$$\text{Number of sweets for each child} = \frac{56}{4} = 14$$

Exercise 6.3

1. If 14 kg of pears cost £5.60, how much do 3 kg cost?

2. If 27 textbooks cost £67.50, how much do 48 cost?

3. 50 articles cost £15. How much do 35 identical articles cost?
4. A car travels 410 km on 40 litres of petrol. How much petrol will be needed for a journey of 325 km?
5. 18 m of stair carpet cost £42. How much will 12 m cost?
6. A farmer employs 12 people to harvest his fruit crop and they take 18 days to complete the work. If he had employed 27 people, how long would they have taken?
7. 20 men produce 1200 components in 8 working days. How long would 8 men take to produce the 1200 components?
8. 8 women can do a piece of work in 60 hours. How many women would it take to complete the work in 12 hours?

Exercise 6.4 (All of the type found in CSE examination papers)

1. Divide £10 between two people in the ratio 1:3. (EM)
2. If five pencils cost 25 p, work out the cost of twelve pencils. (EM)
3. A car does 7 km per litre of petrol. How far will it go on $4\frac{1}{2}$ litres of petrol? (AL)
4. If a man walks 3 km in 40 minutes, how many kilometres will he have walked in 60 minutes if he keeps going at the same rate? (AL)
5. If 4 loaves cost £1.28, how much will 7 cost? (AL)
6. Find the larger share when £650 is divided in the ratio 7:3. (EA)
7. A sum of money was divided between Julie and Joan. Julie received £3.20 and Joan received £4. In what ratio was the money divided? Express your answer in the simplest possible form. (Y)
8. £45 is divided between John and Janet in the ratio 2:3.
   (a) Find John's share.
   (b) Calculate how much more Janet gets than John. (SW)
9. Share £350 in the ratio 3:4. (S)
10. James and John divided £6 between them in the ratio of 5:3 respectively. How much did James receive? (Y)
11. At a wedding reception for 120 people it was estimated that one bottle of wine would be sufficient for eight people.
    (a) How many bottles would supply the 120 guests?
    (b) At a cost of £1.90 per bottle, what would be the total bill for wine? (Y)
12. Express the ratio 60 cm to 1 km as a fraction in its lowest terms. (Y)

Multi-choice questions 6

1. When the number 720 is divided in the ratio 3 : 5, the smaller part is

   A 144 **B** 240 **C** 270 **D** 432
   E 450 (AL)

2. If £90 is divided in the ratio 4 : 5 : 6, the value of the largest share is

   A £22.50 **B** £33.75 **C** £36 **D** £45 (WM)

3. The ratio of *A*'s share to *B*'s share in the profits of a business is 5 : 4. If the total profit is £450, then *A*'s share is

   A £90 **B** £200 **C** £225 **D** £250

4. If £1 is shared in the ratio 7 : 13, the larger share is

   A 65 p **B** 35 p **C** 30 p **D** 13 p
   E 5 p

5. A mortar mixture is made of cement, sand and water mixed by weight in the ratio 1 : 3 : 5 respectively. What weight of sand is contained in 63 kg of the mixture?

   A 7 kg **B** 21 kg **C** 35 kg **D** 42 kg

6. Which one of the following is *not* in the same ratio as 45 : 72?

   A 5 : 9 **B** 10 : 16 **C** 25 : 40 **D** 65 : 104 (EA)

7. Three girls, Angela, Betty and Celia, spent some money in the same proportion as 65 : 91 : 104. If Angela spent £1, how much did Celia spend?

   A £1.39 **B** £1.60 **C** £1.78 **D** £2.04 (EA)

8. If £240 is divided in the ratio 2 to 3, then the smaller share is

   A £80 **B** £96 **C** £144 **D** £160

7 PERCENTAGES

- A PERCENTAGE is a fraction with a denominator of 100. Thus

$$\tfrac{1}{5} = \tfrac{20}{100} = 20\%$$

$$\tfrac{3}{4} = \tfrac{75}{100} = 75\%$$

- To CONVERT A FRACTION OR A DECIMAL INTO A PERCENTAGE, multiply it by 100. Thus

$$\tfrac{7}{10} = \tfrac{7}{10} \times 100\% = 70\%$$

$$\tfrac{2}{3} = \tfrac{2}{3} \times 100\% = 66.7\%$$

$$0.49 = 0.49 \times 100\% = 49\%$$

- To CONVERT A PERCENTAGE INTO A DECIMAL NUMBER divide it by 100. Thus

$$89\% = \tfrac{89}{100} = 0.89$$

Exercise 7.1

Convert the following fractions into percentages:

1. $\tfrac{9}{10}$ **2.** $\tfrac{13}{20}$ **3.** $\tfrac{14}{25}$ **4.** $\tfrac{29}{50}$ **5.** $\tfrac{1}{4}$.

Convert the following decimal numbers into percentages:

6. 0.8
7. 0.94
8. 0.05
9. 0.562
10. 0.752.

Convert the following percentages into decimal numbers:

11. 44%
12. 15%
13. 9%
14. 8.3%
15. 95.2%.

- To find the PERCENTAGE OF A QUANTITY it helps to express the percentage as a fraction.

Example

(a) Find 20% of 80.

$$20\% \text{ of } 80 = \tfrac{20}{100} \text{ of } 80 = \tfrac{20}{100} \times 80 = 16$$

(b) 13.3 cm is 15% of a certain length. What is the complete length?

$$15\% \text{ of the length} = 13.3 \text{ cm}$$

$$1\% \text{ of the length} = \frac{13.3}{15} \text{ cm}$$

$$\text{Now the complete length} = 100\%$$

$$= \frac{13.3}{15} \times 100 = 88.7 \text{ cm}$$

Exercise 7.2

1. What is:
   (a) 30% of 70 (b) 15% of 80 (c) 12% of 50
   (d) 9% of 36?
2. What percentage is:
   (a) 25 of 400 (b) 30 of 150 (c) 15 of 75
   (d) 18 of 54?
3. A student scores 45 marks out of 60 in an examination. What was the percentage mark? If the percentage needed to pass the examination is 40%, what is the pass mark?
4. If 35% of a length is 17.5 cm, what is the complete length?
5. What is:
   (a) 12% of £80 (b) 15% of £120 (c) 70% of £580?
6. 27% of a consignment of fruit is bad. If the consignment weighs 8 t, how many kilograms were bad?
7. A retailer buys 8000 ball point pens and finds that 5% of them were faulty. How many faulty pens were there?
8. A manufacturer of light bulbs states that 6% of them will be faulty. If the shopkeeper buys 5000 of these bulbs, how many of them will be usable?

- PROFIT is the difference between the selling price and the cost price. That is

$$\text{Profit} = \text{selling price} - \text{cost price}$$

The profit per cent is always calculated on the cost price.

$$\text{Profit } \% = \frac{\text{selling price} - \text{cost price}}{\text{cost price}} \times 100$$

- If a LOSS is made

$$\text{Loss } \% = \frac{\text{cost price} - \text{selling price}}{\text{cost price}} \times 100$$

Example

(a) A dealer buys an article for £10 and sells it for £12. Calculate the profit per cent.

We are given that selling price = £12 and the cost price = £10. Hence

$$\text{Profit \%} = \frac{12-10}{10} \times 100 = 20\%$$

(b) A woman buys a car for £3200 and sells it for £2400. Calculate her percentage loss.

We are given that the cost price = £3200 and the selling price = £2400. Hence

$$\text{Loss \%} = \frac{3200-2400}{3200} \times 100 = 25\%$$

Exercise 7.3

1. A shopkeeper buys an article for 40 p and sells it for 50 p. Calculate the percentage profit.
2. An article is bought for £5 and sold for £4. What is the loss per cent?
3. A greengrocer buys a box of grapefruit containing 150 for £12. He sells grapefruit at 12 p each. What is his profit per cent?
4. A car is bought for £6200 and sold for £1550. What is the percentage loss?
5. A dealer buys 200 identical articles for £40 and he sells them for 25 p each. What is his profit per cent?

● DISCOUNT is the amount a dealer takes off the selling price of an article when the customer pays in cash. For instance, a discount of 8% means that the customer pays only 92% of the selling price.

Example

A wardrobe is offered for sale at £270. A customer is offered a discount of 10% for cash. How much does the customer actually pay?

$$\text{Discount} = 10\% \text{ of } £270 = £27$$

$$\text{Amount actually paid} = £270 - £27 = £243$$

Exercise 7.4

1. An armchair is offered for sale at £200. If the shopkeeper offers a 10% discount for cash, how much will a cash-paying customer pay for the chair?
2. During a sale a clothing shop offers a discount of 12%. If the price of a suit is £60, how much will it sell for in the sale?

3. A furniture shop offers a 7% discount for cash. How much discount will be allowed on furniture costing £1100?

4. An electrical shop offers a $12\frac{1}{2}$% discount on goods purchased for cash. A customer bought a refrigerator priced at £250 and paid cash. How much did she actually pay?

Exercise 7.5 (All of the type found in CSE examination papers)

1. Calculate 12% of 60.

2. A table is bought for £100 and sold for £150. What is the percentage profit?

3. In a local election there were three candidates for one seat. Brown received 952 votes, Black 789 votes and Green 596 votes. If 5000 people could have voted, work out the percentage of people who actually voted. (EM)

4. What is $35\frac{1}{2}$% of £800.

5. (a) A dealer buys an article for £1.60. He sells it to make a profit of 30%. Find the dealer's selling price.
   (b) If the dealer takes 8 p off his selling price, calculate his percentage profit now.

6. (a) Find 40% of 40. (b) Increase £5000 by 5%. (SW)

7. A motor bike is bought for £168 and sold later for £126.
   (a) What is the loss?
   (b) What is the loss per cent?

8. What percentage is 350 m of 10 km?

9. Below is a table of corresponding fractions, decimals and percentages.

Fraction	*Decimal*	*Percentage*
$\frac{1}{2}$	0.5	50
a	0.75	b
$\frac{3}{5}$	c	d
e	f	$12\frac{1}{2}$

Write down the figures which should be placed in each of the spaces marked a, b, c, d, e and f, expressing fractions in their lowest terms. (EA)

10. Change the following to fractions in their lowest terms:
    (a) 48% (b) $6\frac{1}{4}$%. (EA)

11. A has 25% more money than B.
    (a) If B has £12, how much has A?
    (b) If A has £20, how much has B?

12. In a class of 30 children, 40% are boys.

(a) What percentage of the class are girls?

(b) How many boys are there in the class?

13. A suit of clothes is advertised at £68.80 with a discount of 10% for cash. What is the discounted price of the suit?

14. In a sale a television set marked at £250 is sold at a discount of 1% for cash. What is the sale price?

15. 30% of a sum of money is £150. Calculate:

(a) the sum of money (b) 40% of the sum of money.

Multi-choice questions 7

1. 28 expressed as a percentage of 112 is

A $\frac{1}{4}$% **B** 4% **C** 25% **D** 28%

E none of these (WY)

2. Written as a vulgar fraction, $12\frac{1}{2}$% is

A $\frac{1}{25}$ **B** $\frac{2}{25}$ **C** $\frac{1}{8}$ **D** $\frac{1}{3}$

E none of these (WY)

3. The price of a machine was £2000 but it is increased in price by 10%. What is the new price of the machine?

A £2010 **B** £2020 **C** £2100 **D** £2200

4. £1440 was divided between two people so that one person was given 45% of the total. How much did the other person receive?

A £628 **B** £696 **C** £744 **D** £792

E £990

5. T-shirts are advertised at £4 each. If ten or more are ordered the cost is reduced to £3.50 each. What percentage is saved by buying ten shirts at the reduced rate?

A 5% **B** $12\frac{1}{2}$% **C** $14\frac{2}{7}$% **D** 50% (EA)

6. 30% of £120 is

A £4 **B** £36 **C** £360 **D** £400

E £3600

7. A shopkeeper buys an article for £3.60. His marked selling price is 20% more than the cost price. What is the marked selling price?

A £3.78 **B** £4.20 **C** £4.32 **D** £4.80

E £5.40

8. A piece of wood is 6 m long. 15% of the wood's length is

A 0.9 cm **B** 22.5 cm **C** 90 cm **D** 225 cm

E 900 cm

9. If 40% of a sum of money is £400 then the whole sum of money is

A £40 **B** £1000 **C** £4000 **D** £20 000

10. What is 72 as a percentage of 45?

A 27% **B** $62\frac{1}{2}$% **C** 144% **D** 160% (EA)

8 DIRECTED NUMBERS

- DIRECTED NUMBERS are numbers which have either a plus or minus sign attached to them. Thus $+7$ is a positive number and -3 is a negative number.

- To ADD several numbers together whose signs are the same, add the numbers together. The sign of the sum is the same as the sign of each of the numbers. Thus

$$+7+(+3) = +10$$

$$-2+(-5)+(-7) = -14$$

When a plus sign means 'add' it is usually omitted and when a calculation begins with a + sign this too is also omitted.

$$+7+(+5) = +12$$

More often this is written

$$7+5 = 12$$

$$-7+(-9) = -16$$

More often this is written

$$-7-9 = -16$$

- To ADD two numbers having different signs, subtract the numerically smaller from the numerically larger. The sign of the result will be the same as the sign of the numerically larger number. Thus

$$-12+6 = -6$$

$$11-3 = 8$$

When dealing with several numbers having mixed signs add the positive numbers together and then add the negative numbers together. The set of numbers is then reduced to two numbers, one positive and the other negative, which can then be dealt with as shown above.

$$-3+7-8-9+5+6+3 = -20+21 = 1$$

- To SUBTRACT a directed number, change its sign and add the resulting number. Thus

$$5-(-4) = 5+4 = 9$$

$$-7-(+8) = -7-8 = -15$$

- The PRODUCT of two numbers having like signs is positive while the product of two numbers with unlike signs is negative. That is

$$\text{positive} \times \text{positive} = \text{positive}$$
$$\text{positive} \times \text{negative} = \text{negative}$$
$$\text{negative} \times \text{positive} = \text{negative}$$
$$\text{negative} \times \text{negative} = \text{positive}$$
$$3 \times 4 = 12 \qquad 3 \times (-4) = -12$$
$$(-4 \times 3) = -12 \qquad (-3) \times (-4) = 12$$

- When DIVIDING, numbers with like signs give a positive answer while numbers with unlike signs give a negative answer. Thus

$$\frac{20}{4} = 5 \qquad \frac{-20}{4} = -5 \qquad \frac{20}{-4} = -5 \qquad \frac{-20}{-4} = 5$$

Exercise 8.1

Find values for each of the following:

1. $8+5$
2. $-6-3$
3. $-14-18$
4. $-7-6-3$
5. $-8-7-15$
6. $8-12$
7. $9-17$
8. $-6+10$
9. $-3-2+7$
10. $-8+11-9+17-34$
11. $-40-23+72+15-18$
12. $8-(-5)$
13. $-3-(-7)$
14. $-2-(+3)$
15. $7-(-8)$
16. $5 \times (-6)$
17. $(-3) \times 2$
18. $5 \times (-4)$
19. $(-2) \times (-4)$
20. $(-2) \times (-5) \times (-6)$
21. $4 \times (-3) \times (-7)$
22. $(-5)^2$
23. $8 \div (-2)$
24. $(-9) \div 3$
25. $(-8) \div (-2)$
26. $10 \div (-2)$
27. $(-12) \div (-4)$
28. $(-14) \div 7$
29. $\dfrac{-16}{(-2) \times (-4)}$
30. $\dfrac{2 \times (-8) \times 6}{(-3) \times (-2) \times (-4)}$.

- COUNTING NUMBERS are the numbers 1, 2, 3, 4, 5, ...
- NATURAL NUMBERS are the numbers 0, 1, 2, 3, 4, ...

- INTEGERS are the numbers . . . , $-2, -1, 0, 1, 2, 3,$ . . . .

- A RATIONAL NUMBER is one which can be expressed as a common fraction. Thus 0.875 is a rational number because it can be expressed as $\frac{7}{8}$. Note that all recurring decimals are rational. For instance $0.\dot{1} = \frac{1}{9}$ and $0.\dot{6}\dot{3} = \frac{7}{11}$.

- An IRRATIONAL NUMBER cannot be written as a common fraction. For instance, $\sqrt{2}$, $\sqrt{13}$ and π are all irrational numbers. Note that not all square roots are irrational numbers. For instance $\sqrt{0.25} = 0.5 = \frac{1}{2}$.

- Numbers like $\sqrt{-1}$, $\sqrt{-4}$, etc. have no real meaning and they are called IMAGINARY NUMBERS.

- Numbers which are not imaginary are said to be REAL NUMBERS. Thus counting numbers, natural numbers, integers, rational numbers and irrational numbers are all real numbers.

- Real numbers can be represented on a NUMBER LINE (Fig. 8.1).

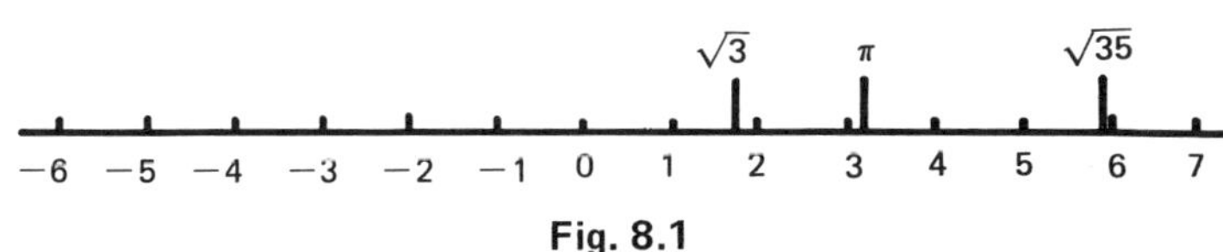

Fig. 8.1

Exercise 8.2

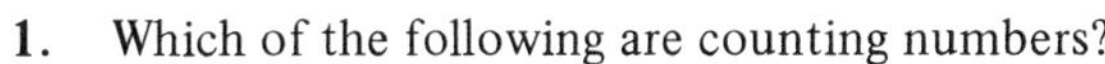

1. Which of the following are counting numbers?
   (a) $\frac{1}{2}$ (b) -3 (c) 5 (d) 8
   (e) 2.35 (f) 0 (g) 7.

2. Which of the following are natural numbers?
   (a) $\frac{3}{4}$ (b) -2 (c) 0 (d) 5
   (e) 3.1 (f) $0.\dot{3}$.

3. Which of the following are integers?
   (a) 3 (b) -5 (c) $\frac{7}{8}$ (d) $\sqrt{15}$
   (e) 1.75.

4. Which of the following are real numbers?
   (a) $\frac{1}{2}$ (b) -4 (c) $\sqrt{-4}$ (d) $\sqrt{13}$
   (e) $\sqrt{-1}$.

5. Which of the following are rational numbers?
   (a) 0.65 (b) $0.\dot{7}\dot{3}$ (c) $\sqrt{5}$ (d) -3.7
   (e) $\sqrt{17}$.

Exercise 8.3 (All of the type found in CSE examination papers)

1. Find the value of:

   (a) $(-5)\times(-8)$ (b) $-5-8$

   (c) $5-(-8)$ (d) $30-(-10)$. (WY)

2. Calculate the value of $(-7)^2$.

3. Work out the value of $(8-3)-(5-3)$.

4. Find the value of: (a) $(5-2)\times(7-2)$ (b) $8-3\times(5-2)$.

5. Work out the value of $64-7\times(8-2)$.

6. Write down in order of size, smallest first, the numbers -2, 0, -4 and 6.

7. Rearrange in order of size, largest first, the numbers -9, -42, -2 and -5.

8. Remove the brackets and work out the answer:

   (a) $(7-3)\times(7+4)$ (b) $7-3\times(7+4)$.

9. In each of the following insert a + sign or a − sign to make the statement correct:

   (a) $29+7-12=29+(7\ \ldots\ 12)$

   (b) $15-6+8=15\ \ldots\ (6-8)$

   (c) $(-17)\times 5=\ldots 85$

   (d) $(-80)\div(\ldots 4)=-20$.

10. Copy the number line shown in Fig. 8.2 and then mark and label the positions of the numbers $-1\frac{1}{2}$, 0.75 and 2.1.

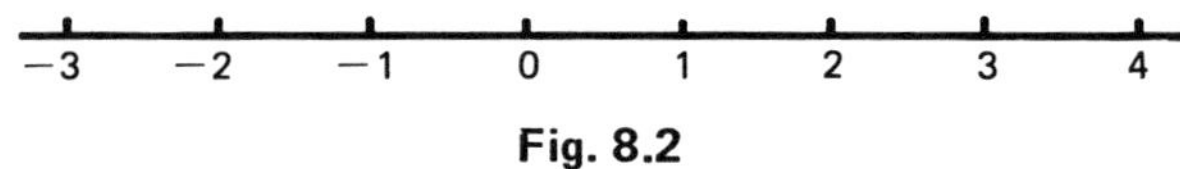

Fig. 8.2

Multi-choice questions 8

1. The value of $\dfrac{(-1)-(-1)}{(-1)}$ is

   A 2 **B** 1 **C** 0 **D** −2

2. A thermometer rises from −5°C to 15°C. What is the rise in temperature?

   A −20 **B** −10 **C** 10 **D** 20

3. $(-3)^2\times 4=$

   A −36 **B** −24 **C** 10 **D** 36

4. $(-2)\times(-4)$ equals

   A −8 **B** −6 **C** −2 **D** 8

5. What is the value of $7.08 - 12$?

   A -7.92 **B** -7.08 **C** -4.92 **D** 6.96

   E 19.08

6. Find the value of $-4-7+6$.

   A -17 **B** -9 **C** -5 **D** -3

7. Find the value of $4 \times (-5) \times (-4)$.

   A 80 **B** 22 **C** -22 **D** -80

8. Which of the following is a rational number?

   A $\sqrt{7}$ **B** $\sqrt{6}$ **C** $\sqrt{4}$ **D** $\sqrt{2}$

9 BASIC ALGEBRA

- Statements in words can be translated into ALGEBRAIC SYMBOLS. Any symbol can be chosen to represent the quantities concerned. Thus
  (i) If x and y are two numbers, the sum of the numbers is $x+y$
  (ii) The product of three numbers a, b and c is $a \times b \times c$
  (iii) Five times a number N is $5 \times N$.

Exercise 9.1

Translate each of the following into algebraic symbols:

1. Six times a number x
2. Four times a number a minus three
3. The sum of five times a number y and a number z
4. The product of the three numbers x, y and z
5. Five times the product of the two numbers m and n.

- SUBSTITUTION is the process of finding the numerical value of an algebraic expression.

Example

Find the value of $3x+5y-3z$ when $x=2$, $y=7$ and $z=4$.

$$\begin{aligned} 3x+5y-3z &= (3\times 2)+(5\times 7)-(3\times 4) \\ &= 6+35-12 \\ &= 29 \end{aligned}$$

(Note that multiplication signs are often missed out when writing algebraic terms. Thus $5y$ means $5 \times y$. These missed multiplication signs must reappear when numbers are substituted for the symbols.)

Exercise 9.2

If $a=3$, $b=6$ and $c=7$, find the values for the following:

1. $b+3$
2. $a-3$
3. $12-c$
4. $5a$
5. $3c$
6. bc

7. abc

8. $5ab$

9. $\dfrac{12}{a}$

10. $\dfrac{ab}{9}$

11. $a+3b-4c$

12. $a+3b-8c$.

- The third POWER of a is written a^3 which means $a \times a \times a$.

The number 3 which indicates the number of a's to be multiplied together is called the INDEX (plural: INDICES).

$$2^5 = 2 \times 2 \times 2 \times 2 \times 2$$

$$b^4 = b \times b \times b \times b$$

Example

Find the value of y^5 when $y = 2$.

$$y^5 = 2^5 = 2 \times 2 \times 2 \times 2 \times 2 = 32$$

When dealing with expressions like $5ab^3$, note that it is only the symbol b which is raised to the third power. Thus

$$5ab^3 = 5 \times a \times b \times b \times b$$

Example

Find the value of $5b^2c^3$ when $b = 2$ and $c = 3$.

$$5b^2c^3 = 5 \times 2^2 \times 3^3$$

$$= 5 \times 2 \times 2 \times 3 \times 3 \times 3$$

$$= 540$$

Exercise 9.3

If $p = 2$, $q = 3$ and $r = 5$, find values for the following:

1. p^3
2. q^4
3. r^2
4. pq^4
5. $3p^4$
6. $4p^2q$
7. p^3r^2
8. pqr^3
9. $4pq^2r$
10. p^2+q^2.

- LIKE TERMS are numerical multiples of the same algebraic quantity. Thus $5x$, $3x$ and $-8x$ are three like terms.

Like terms may be added together by adding their numerical coefficients. Thus

$$5x+3x-2x = (5+3-2)x = 6x$$

Only like terms can be added or subtracted. Thus $7a+3b-c$ is an expression containing three unlike terms and it cannot be simplified further.

It is possible to have several sets of like terms in an expression and each set may then be simplified by adding and/or subtracting.

$$\begin{aligned} &8p+2q-5r+6r-4p+7q-3p+3r \\ &= (8-4-3)p+(2+7)q+(-5+6+3)r \\ &= p+9q+4r \end{aligned}$$

Exercise 9.4

Simplify the following:

1. $5x+3x$
2. $7y-3y$
3. $8p+5p-3p$
4. $2q-5q$
5. $-3x-4x+x$
6. $5m+8m-15m$
7. $4x-8x-2x$
8. $3p+4q+2p+5q$
9. $5x-4y-3z+7x+5z-2y-6y+8z+x+3y$.

- The rules used for the MULTIPLICATION AND DIVISION of algebraic terms are the same as those used for directed numbers. Thus

$$\begin{aligned} (+x)(+y) &= +xy = xy \\ (-x)(+y) &= -xy \\ (+x)(-y) &= -xy \\ (-x)(-y) &= +xy = xy \\ (2x)(-3y) &= -(2\times3)\times x\times y = -6xy \\ (5a)(2b)(-3c) &= -(5\times2\times3)\times a\times b\times c = -30abc \end{aligned}$$

Exercise 9.5

Simplify each of the following:

1. $(2y)(3p)$
2. $(3m)(-2p)$
3. $(-4a)(-5b)$
4. $(3a)(2b)(3c)$
5. $(-2x)(-3y)(2z)$
6. $(7m)(-2n)(p)$
7. $(3a)(2b)(-c)(5d)$
8. $(-3)(-5y)(2z)(-4q)$.

- When REMOVING A BRACKET each term within the bracket is multiplied by the term outside the bracket.

Example

(a) $5(2m+3p) = (5\times 2m)+(5\times 3p) = 10m+15p$

(b) $-2(3a-2b) = [(-2)\times 3a]+[(-2)\times(-2b)] = -6a+4b$

- When a bracket has a minus sign outside it, all the signs inside the bracket are changed when the bracket is removed.

Example

(a) $-(p+q) = -p-q$

(b) $-(4m-3q) = -4m+3q$

(c) $-5(a-2b) = -5a+10b$

- When SIMPLIFYING EXPRESSIONS CONTAINING BRACKETS remove the brackets first and then add or subtract the like terms.

Example

$$5(3a+2b)-3(2a-6b) = 15a+10b-6a+18b = 9a+28b$$

Exercise 9.6

Remove the brackets from each of the following:

1. $3(a+2b)$
2. $5(2x-3y)$
3. $5(4p-3q)$
4. $-(p-q)$
5. $-2(3x-4y)$
6. $-5(2m-3n)$.

Remove the brackets and simplify:

7. $3(a-2b)+5(2a-3b)$
8. $-(p+2q)-(3p+5q)$
9. $2(3m+5n)-3(2m-5n)$
10. $5(2x-3y)-3(x-2y)$.

Exercise 9.7 (All of the type found in CSE examination papers)

1. Simplify:
   (a) $6x-3x$ (b) $6x \times 3x$ (c) $6x \div 3x$.
2. In each of the following remove the bracket and simplify:
   (a) $4(x-y)+3(2x+y)$ (b) $4(x-y)-3(2x-y)$
   (c) $(3x-y)-(2x-y)$.
3. Find the value of $8y^2+xy-z^2$ when $x=1$, $y=2$ and $z=-3$.
4. Given that $a=2$, $b=3$ and $c=-1$, find the value of:
   (a) $2a-5b$ (b) a^2-4c^2.
5. Simplify $4x+3y-(3x-y)$. (SW)

6. Simplify the following expressions:
(a) $8x-y-3x-5y$ (b) $5a\times 2b\times 3a$. (EA)

7. Given that $x=6$, $y=4$ and $z=-2$, work out the values of the following expressions:
(a) $x-y+z$ (b) $x-(y-z)$ (c) $\frac{x^2}{3y}$. (EA)

8. If $a=2$, $b=-3$ and $c=5$, evaluate:
(a) ab^2 (b) $\frac{ab}{c}$.

9. If $x=4$ and $y=-3$, find the values of:
(a) $(x-y)^2$ (b) $2x^2-5xy+y^2$.

10. Work out the value of:
(a) $a+2b$ when $a=1$ and $b=\frac{1}{2}$
(b) $9a-c$ when $a=1$ and $c=\frac{1}{3}$. (SW)

Multi-choice questions 9

1. What is the value of x^2+x-6 when $x=3$?
A 0 B 3 C 6 D 18 (EA)

2. Simplify $(x^2+x-6)+(x^2+4x+3)$.
A $2x^2+5x-3$ B x^4+4x^2-3
C $2x^4+4x^2-18$ D $3x+9$

3. Simplify $3(2x-y)-2(3x-2y)$.
A y B $-7y$ C $12x+3y$ D $12x+y$

4. Find the value of $\frac{p-q}{p+q}$ when $p=10$ and $q=-6$.
A -4 B -1 C $\frac{1}{4}$ D 4

5. If $F=2x^2+4x$, what is the value of F when $x=-2$?
A -16 B 0 C 8 D 16 (EA)

6. If $a=\frac{5}{16}$ and $b=\frac{2}{5}$, what is the value of ab?
A $\frac{1}{8}$ B $\frac{1}{4}$ C $\frac{1}{3}$ D $\frac{5}{8}$

7. If $x=0.2$ and $y=4.8$, calculate the value of $\frac{y}{x}$.
A 240 B 24 C 2.4 D 0.24

8. If $p=7$, $q=-3$ and $r=0$, what is the value of pqr?
A -21 B 0 C 4 D 21

9. $3x+y-x=$
A y B $3-y$ C $3+y$ D $2x+y$

10. If $y=2x^2-3x-7$, calculate the value of y when $x=3$.
A 16 B 9 C 8 D 2

10 FACTORISATION

- A FACTOR is a common part of two or more terms which make up an algebraic expression. Thus the expression $7a+7b$ has two terms, $7a$ and $7b$, which have the number 7 common to both of them. Hence

$$7a+7b = 7(a+b)$$

7 and $(a+b)$ are said to be the factors of $7a+7b$.

Example

(a) Find the factors of $ap+aq$.

ap and aq have the quantity a common to both of them.

$$\therefore \quad ap+aq = a(p+q)$$

To find the terms inside the bracket, divide each of the terms making up the original expression by the common quantity (i.e. the quantity to be placed outside the bracket). Thus

$$\frac{ap}{a} = p \quad \text{and} \quad \frac{aq}{a} = q$$

(b) Factorise $m(x-y)-n(x-y)$.

The terms $m(x-y)$ and $n(x-y)$ have $(x-y)$ as a common factor. Hence

$$m(x-y)-n(x-y) = (x-y)(m-n)$$

Exercise 10.1

Factorise each of the following:

1. $2x+2y$
2. $3p-3q$
3. $5x+15y$
4. $br-bs$
5. $4x-6y$
6. ax^2+bx
7. $x(a-b)+y(a-b)$
8. $p(x+y)-q(x+y)$.

- A BINOMIAL EXPRESSION consists of two terms. Thus $(3x+5)$, $(a+b)$ and $(3x-2y)$ are all binomial expressions. To find the product of two binomial expressions multiply the terms connected by a line

$$(a+b)(c+d) = ac+ad+bc+bd$$

Example

$$(2x+5)(3x-2) = 2x \times 3x + 2x \times (-2) + 5 \times 3x + 5 \times (-2)$$
$$= 6x^2 - 4x + 15x - 10$$
$$= 6x^2 + (-4+15)x - 10$$
$$= 6x^2 + 11x - 10$$

Exercise 10.2

Remove the brackets from each of the following:

1. $(x+2)(x+3)$
2. $(x-5)(x-1)$
3. $(x+3)(x-2)$
4. $(2x+1)(x-3)$
5. $(3x-2)(2x+3)$
6. $(5x-2)(x-3)$
7. $(x+1)^2$
8. $(x-2)^2$
9. $(x+2)(x-2)$
10. $(2x+3)(2x-3)$.

- An expression of the type ax^2+bx+c, where a, b and c are constants, is called a QUADRATIC EXPRESSION. Thus x^2+7x-3 and $3x^2-5x+3$ are both quadratic expressions.

In a simple expression with $a=1$, like x^2+5x+6, the factors will be $(x+m)(x+n)$ where m and n are two numbers which add up to 5 and which, when multiplied, make 6. These two numbers must be 2 and 3. Hence

$$x^2+5x+6 = (x+3)(x+2)$$

Example

Factorise $x^2-2x-15$.

If we let the factors be $(x+m)(x+n)$, then the sum of m and n must be -2 and the product must be -15. These numbers must be -5 and $+3$ because $-5+3=-2$ and $(-5)\times 3=-15$. Hence

$$x^2-2x-15 = (x-5)(x+3)$$

In an expression like $2x^2+13x+15$, work out the factors of 2 and 15 and proceed by trial and error. Thus the factors of 2 can only be 2 and 1 while the factors of 15 can only be 3 and 5. We must now try and find the correct combination of these factors.

$$(2x+5)(x+3) = 2x^2+11x+15$$

This is not correct so try again.

$$(2x+3)(x+5) = 2x^2+13x+15$$

This is correct and hence

$$2x^2+13x+15 = (2x+3)(x+5)$$

Exercise 10.3

Factorise the following:

1. x^2+3x+2
2. x^2+x-2
3. $x^2+8x+15$
4. $x^2+2x-15$
5. $x^2-2x-15$
6. x^2-2x-8
7. x^2-6x+8
8. $x^2-12x+32$
9. $3x^2+5x+2$
10. $2x^2+7x+6$
11. $5x^2-11x+2$
12. $3x^2-7x-6$.

- Sometimes the factors of a quadratic expression form a PERFECT SQUARE.

Consider
$$(a+b)^2 = (a+b)(a+b) = a^2+2ab+b^2$$
$$(a-b)^2 = (a-b)(a-b) = a^2-2ab+b^2$$

Therefore the square of a binomial expression consists of

$$(\text{first term})^2+2\times(\text{first term})\times(\text{second term})+(\text{second term})^2$$

Example

(a) Remove the brackets from $(2x+3)^2$.

$$(2x+3)^2 = (2x)^2+2\times 2x\times 3+3^2$$
$$= 4x^2+12x+9$$

(b) Remove the brackets from $(3x-2)^2$.

$$(3x-2)^2 = (3x)^2+2\times 3x\times(-2)+(-2)^2$$
$$= 9x^2-12x+4$$

(c) Factorise $25x^2-20x+4$.

$$25x^2 = (5x)^2 \quad \text{and} \quad 4 = (-2)^2$$
$$-20x = 2\times 5x\times(-2)$$
$$25x^2-20x+4 = (5x-2)^2$$

- The expression a^2-b^2 is called the DIFFERENCE OF THE TWO SQUARES a^2 and b^2.

$$a^2-b^2 = (a+b)(a-b)$$

Example

(a) Remove the brackets from $(3x+5)(3x-5)$.

$$(3x+5)(3x-5) = (3x)^2-5^2 = 9x^2-25$$

(b) Factorise $4x^2-9$.

Since $4x^2 = (2x)^2$ and $9 = 3^2$

$$4x^2-9 = (2x+3)(2x-3)$$

Exercise 10.4

Remove the brackets from the following:

1. $(x+2)^2$
2. $(x-3)^2$
3. $(2x+1)^2$
4. $(3x-4)^2$
5. $(2x+7)^2$
6. $(x+1)(x-1)$
7. $(2x+5)(2x-5)$
8. $(3x+4)(3x-4)$.

Factorise the following:

9. x^2+4x+4
10. x^2-4x+4
11. $9x^2-12x+4$
12. $4x^2+12x+9$
13. x^2-4
14. $9x^2-16$
15. $25x^2-49$.

Exercise 10.5 (All of the type found in CSE examination papers)

1. Factorise completely $3pq-12pq^2$.
2. Calculate the value of 108^2-92^2.
3. If $(x-4)^2=x^2+px+16$, state the value of p. (EA)
4. (a) Remove the brackets from $(3x-2)^2$.
   (b) Factorise y^2-16.
5. Find the factors of $x^2+12x+11$. (SW)
6. Expand $(x+3)(x+2)$, i.e. remove the brackets. (SW)
7. Remove the brackets and simplify:
   (a) $(7-3)(7+4)$ (b) $(x-2y)(x+2y)$ (c) $(2x-5)^2$. (EA)
8. (a) Write without brackets, as simply as possible, $(q-p)(q+p)$.
   (b) Find the exact value of $50\,001^2-49\,999^2$. (EA)
9. Factorise $5x^2-10x$.
10. Remove the brackets and simplify $(3x-1)(2x+5)$.
11. Factorise:
    (a) x^2-25 (b) $5ab^3-2a^2b$.
12. Factorise:
    (a) $5xyz-15yz$ (b) $3(a-b)-p(a-b)$
    (c) x^2-x-2 (d) $4x^2-9y^2$.
13. Remove the brackets and simplify:
    (a) $5(x+y)-3(x-y)$ (b) $(a+3b)^2$
    (c) $(a+3b)(a-3b)$.
14. Factorise:
    (a) $2x^2-4x$ (b) $9-x^2$. (SW)
15. If $(x+p)(x+7)=x^2+qx-21$, write down the values of p and q. (EA)

Multi-choice questions 10

1. The expansion of $(2x+y)(x-2y)$ is
   A $2x^2-2y^2$ B $2x^2+2y^2$
   C $2x^2-3xy-2y^2$ D $2x^2+3xy-2y^2$

2. $(x-2)$ and $(x+3)$ are the factors of
   A x^2+x-6 B $2x+1$
   C x^2-9 D x^2+x+6

3. Which one of the following is equal to $ab+bc$?
   A $a(b+c)$ B $(ab)(bc)$ C $a(bc)$ D $b(a+c)$

4. What are the factors of $2p^2+4p$?
   A $2p(p+4)$ B $p(p+4)$
   C $6p^3$ D $2p(p+2)$

5. The exact value of $3.6^2-3.5^2$ is
   A 0.1 B 0.71 C 7.1 D 71

6. Given the two algebraic expressions x^2+x-6 and x^2+4x+3, which of the following is a factor of both expressions?
   A $(x+3)$ B $(x+2)$ C $(x-1)$ D $(x-2)$ (EA)

7. Which of the following is equal to $(2x+1)(x-3)$?
   A $2x^2-x-3$ B $2x^2-5x-3$
   C $2x^2+5x-3$ D $2x^2+x-3$

8. If $a^2+b^2=41$ and $ab=9$, then $(a+b)^2$ is equal to
   A 41 B 50 C 59 D 81

9. The factors of x^2-9x are
   A $(x-3)(x+3)$ B $(x-3)^2$
   C $(x-1)(x+9)$ D $x(x-9)$

10. When $(4-x)$ is multiplied by $(3x-7)$, the term in x is
    A $5x$ B $7x$ C $12x$ D $19x$

11 ALGEBRAIC FRACTIONS

- To SIMPLIFY A FRACTION, factors which are common to both numerator and denominator may be cancelled. Thus

$$\frac{9p^2q}{5pq^3} = \frac{9 \times \not{p} \times p \times \not{q}}{5 \times \not{p} \times \not{q} \times q \times q} = \frac{9p}{5q^2}$$

- To MULTIPLY FRACTIONS first cancel common factors and then multiply the numerators and denominators separately. Thus

$$\frac{3a^2b}{5cd^2} \times \frac{4c^2d^3}{7a^3b^2} = \frac{3 \times \not{a} \times \not{a} \times \not{b} \times 4 \times \not{c} \times c \times \not{d} \times \not{d} \times d}{5 \times \not{c} \times \not{d} \times \not{d} \times 7 \times \not{a} \times \not{a} \times a \times \not{b} \times b}$$

$$= \frac{3 \times 4 \times c \times d}{5 \times 7 \times a \times b}$$

$$= \frac{12cd}{35ab}$$

- DIVISION is performed by multiplying the first fraction by the inverse of the second fraction. Thus

$$\frac{5p^2q^3}{8mn^2} \div \frac{7pq^4}{3m^3n^2} = \frac{5p^2q^3}{8mn^2} \times \frac{3m^3n^2}{7pq^4}$$

$$= \frac{5 \times \not{p} \times p \times \not{q} \times \not{q} \times \not{q} \times 3 \times \not{m} \times m \times m \times \not{n} \times \not{n}}{8 \times \not{m} \times \not{n} \times \not{n} \times 7 \times \not{p} \times \not{q} \times \not{q} \times \not{q} \times q}$$

$$= \frac{5 \times 3 \times p \times m \times m}{8 \times 7 \times q}$$

$$= \frac{15m^2p}{56q}$$

Exercise 11.1

Simplify each of the following:

1. $\frac{pq^2}{pq}$ 2. $\frac{x^2y^2}{x^3y}$ 3. $\frac{6abc^3}{3ab^2c}$ 4. $\frac{9x^2y^3z}{6xy^2z^2}$.

Multiply each of the following:

5. $\dfrac{8a^2b}{3m^2n} \times \dfrac{9n^3m}{4ab^2}$

6. $\dfrac{6pq}{4rs} \times \dfrac{8s^2}{3p}$

7. $\dfrac{6ab}{c} \times \dfrac{ad}{2b} \times \dfrac{8cd^2}{4bc}$.

Divide out each of the following:

8. $\dfrac{ab^2}{bc^2} \div \dfrac{a^2}{bc^3}$

9. $\dfrac{5xy^2}{7a^2b} \div \dfrac{10x^2y}{14ab^2}$

10. $\dfrac{m^2n^3}{pq^3} \div \dfrac{m^3n^2}{p^2q}$.

- The procedure when ADDING OR SUBTRACTING FRACTIONS is the same as in arithmetic. The procedure is:
  (i) Find the LCM of the denominators
  (ii) Express each fraction with this common denominator
  (iii) Add or subtract the fractions.

Example

(a) Simplify $\dfrac{a}{3}+\dfrac{a}{4}-\dfrac{a}{5}$.

The LCM of 3, 4 and 5 is 60.

$$\frac{a}{3}+\frac{a}{4}-\frac{a}{5} = \frac{20a}{60}+\frac{15a}{60}-\frac{12a}{60}$$
$$= \frac{20a+15a-12a}{60} = \frac{23a}{60}$$

(b) Simplify $\dfrac{x}{2}-\dfrac{x-1}{3}$.

The LCM of 2 and 3 is 6.

$$\frac{x}{2}-\frac{x-1}{3} = \frac{3x-2(x-1)}{6}$$
$$= \frac{3x-2x+2}{6}$$
$$= \frac{x+2}{6}$$

Note that the sign in front of a fraction applies to the fraction as a whole. The line which separates the numerator and the denominator acts as a bracket.

Exercise 11.2

Simplify each of the following:

1. $\frac{a}{4}+\frac{b}{5}$
2. $\frac{p}{2}-\frac{q}{3}$
3. $\frac{3a}{4}-\frac{a}{5}$
4. $\frac{x}{2}+\frac{x}{3}+\frac{x}{4}$
5. $\frac{3}{2x}+\frac{2}{3x}$
6. $\frac{m}{x}-\frac{3m}{2x}+\frac{5m}{3x}$
7. $\frac{4x}{3y}-\frac{2x}{5y}$
8. $1+\frac{(x+2)}{3}$
9. $\frac{x}{3}+\frac{(2x-1)}{4}$
10. $5m-\frac{(m-2)}{2}$
11. $\frac{x-3}{4}-\frac{x}{3}$
12. $\frac{4}{x}-\frac{5}{2x}+\frac{3}{4x}$
13. $\frac{x}{2}-\frac{(x-3)}{3}$
14. $\frac{2a+3b}{3}-\frac{a-2b}{2}$
15. $\frac{5-x}{5}-\frac{x-3}{2}$

Exercise 11.3 (All of the type found in CSE examination papers)

1. Express as a single fraction in its simplest form $\frac{3x}{4}-\frac{x}{8}$.
2. Simplify $\frac{3x^2y}{9xy^3}$.
3. Express $\frac{x}{4}-\frac{x-3}{5}$ as a single fraction, simplifying your answer as far as possible.
4. Express as a single fraction $\frac{3}{a}+\frac{2}{y}$.
5. Simplify $\frac{2x+10}{2}$.
6. Simplify $\frac{8x^6y^3}{12x^3y}$. (EA)
7. Express as a single fraction $\frac{x}{2}-\frac{2x}{3}+\frac{11x}{4}$.
8. Simplify $\frac{3}{p}+\frac{2}{3p}$.

12 OPERATIONS ON NUMBERS

- Given any two numbers there are various ways of OPERATING on them apart from adding, subtracting, multiplying and dividing.

Example

(a) If $a * b$ means $2a+b$, find the value of $4*3$.

We have $a=4$ and $b=3$, hence

$$4*3 = 2\times 4+3 = 8+3 = 11$$

(b) If $p \circ q$ means $\frac{1}{2}(p^3+q^2)$, find $2\circ 3$.

We have $p=2$ and $q=3$, hence

$$2\circ 3 = \tfrac{1}{2}(2^3+3^2) = \tfrac{1}{2}(8+9) = \tfrac{1}{2}\times 17 = 8\tfrac{1}{2}$$

Exercise 12.1

1. If $a\circ b$ means $3a+2b$, find $3\circ 1$.
2. If $m\circ n$ means $4m-n$, find $2\circ 1$.
3. If $p*q$ means $\frac{3}{4}(p+2q)$, find $2*3$.
4. If $a\circ b$ means $(a-b)^2$, find $7\circ 5$.
5. If $x*y$ means $\frac{1}{2}(x^3+y^2)$, find $2*4$.

Exercise 12.2 (All of the type found in CSE examination papers)

1. If $a\circ b$ means $2a+b$, work out the value of $3\circ 1$.
2. Suppose $x\circ y$ means $3x-2y$, work out the value of $3\circ 4$.
3. If $a*b$ means $\frac{1}{4}(a-b)$, calculate the value of $6*4$.
4. If $p\circ q = pq+1$, where both p and q are integers, calculate $6\circ 4$.
5. If $a\circ b$ means $(a+b)^2$, find the value of $3\circ 2$.
6. If $m*n$ means $\frac{1}{2}(m+n)$, calculate the value of $9*3$.

13 INDICES

- When MULTIPLYING powers of the same number together *add* their indices. Thus

$$a^2 \times a^5 = a^{2+5} = a^7$$

$$y^3 \times y^4 \times y^2 \times y^6 = y^{3+4+2+6} = y^{15}$$

- When DIVIDING powers of the same quantity *subtract* the index of the denominator (bottom part) from the index of the numerator (top part). Thus

$$\frac{a^5}{a^3} = a^{5-3} = a^2$$

$$\frac{x^3 \times x^5 \times x^4}{x^2 \times x^7} = \frac{x^{3+5+4}}{x^{2+7}} = \frac{x^{12}}{x^9} = x^{12-9} = x^3$$

- When RAISING THE POWER OF A QUANTITY to a power *multiply* the indices. Thus

$$(x^2y^3)^4 = x^{2\times4}y^{3\times4} = x^8y^{12}$$

$$\left(\frac{2p^3}{3q^2}\right)^2 = \frac{2^{1\times2}p^{3\times2}}{3^{1\times2}q^{2\times2}} = \frac{2^2p^6}{3^2q^4} = \frac{4p^6}{9q^4}$$

- A NEGATIVE INDEX indicates the reciprocal of the quantity. Thus

$$2^{-1} = \frac{1}{2}$$

$$7p^{-4} = \frac{7}{p^4}$$

- A FRACTIONAL INDEX indicates the root of a quantity. The numerator of the fractional index shows the power to which the quantity must be raised; the denominator shows the root which is to be taken. Thus

$$a^{2/3} = \sqrt[3]{a^2}$$

$$a^{1/2} = \sqrt{a}$$

(Note that for square roots the number indicating the root is usually omitted.)

- Any quantity RAISED TO THE POWER OF ZERO is equal to 1. Thus

$$a^0 = 1, \quad 198^0 = 1 \quad \text{and} \quad 5^0 = 1$$

Example

Find the values of: (a) 5^3 (b) 2^{-3} (c) $81^{1/4}$.

(a) $5^3 = 5 \times 5 \times 5 = 125.$

(b) $2^{-3} = \frac{1}{2^3} = \frac{1}{8}.$

(c) $81^{1/4} = (3^4)^{1/4} = 3^{4 \times 1/4} = 3^1 = 3.$

Exercise 13.1

Simplify each of the following:

1. $a^3 \times a^4$
2. $p^2 \times p^3 \times p^5$
3. $y^3 \times y^5 \times y^7 \times y^8$
4. $3^4 \times 3^5$
5. $2 \times 2^3 \times 2^4$
6. $3a \times 2a^2 \times 4a^3$
7. $p^7 \div p^5$
8. $2^5 \div 2^3$
9. $(q^5 \times q^3) \div q^4$
10. $\frac{m^3}{m^5} \times \frac{m^7}{m^2}$
11. $\frac{t^4 \times t^9}{t^3 \times t^5}$
12. $\frac{am^3}{am^2}$
13. $(3b^2)^3$
14. $(2x^2y)^3$
15. $(5ab^2c^3)^4$
16. $\left(\frac{3p^2}{2q^3}\right)^4$.
17. Find the values of the following stating the answers as common fractions:
    (a) 10^{-1} (b) 3^{-2} (c) 2^{-4} (d) 5^{-3}.
18. Find the numerical values of the following:
    (a) $5^{1/2} \times 5^{1\frac{1}{2}}$ (b) $8^{1/3}$
    (c) $16^{1/4}$ (d) $27^{1/3}$.
19. Express as powers of 2:
    (a) 8^2 (b) 32^4 (c) 64^3.
20. Express as powers of a:
    (a) $\sqrt{a}$ (b) $\sqrt[3]{a^2}$ (c) $\sqrt[5]{a^4}$.

- Any number can be expressed as a value between 1 and 10 multiplied by a power of 10. A number expressed in this way is said to be in STANDARD FORM.

Note that $100 = 10^2$, $1000 = 10^3$ and $1\,000\,000 = 10^6$.

The index is found by counting the number of zeros to the left of the decimal point.

Also, $0.1 = 10^{-1}$, $0.01 = 10^{-2}$ and $0.001 = 10^{-3}$.

The negative index is found by adding 1 to the number of zeros following the decimal point.

$$563 = 5.63 \times 100 = 5.63 \times 10^2$$

$$75\,352 = 7.5352 \times 10\,000 = 7.5352 \times 10^4$$

$$0.0036 = 3.6 \div 1000 = 3.6 \div 10^3 = 3.6 \times 10^{-3}$$

Exercise 13.2

Write the following in standard form:

1. 827
2. 730
3. 17 632
4. 8 036 000
5. 0.03
6. 0.0056
7. 0.0006
8. 0.000 007.

Write the following as ordinary numbers:

9. 3.2×10^2
10. 5×10^3
11. 1.87×10^6
12. 2×10^{-3}
13. 5.67×10^{-1}
14. 3.2×10^{-2}.

Exercise 13.3 (All of the type found in CSE examination papers)

1. Rewrite the number 3 500 000 in standard form.
2. Find the value of $4.3 \times 5 \times 10$, giving the answer in standard form.
3. The number 27 900 can be written in the standard form $A \times 10^4$. Find the value of A. (EA)
4. Write 5.01×10^{-3} as a decimal number, not in standard form. (EA)
5. Simplify $(8 \times 10^{20}) \div 4$. (EA)
6. Write 0.000 943 in standard form.
7. Give the answer to the following as a decimal number: $\dfrac{2^2 \times 2^5 \times 2^6}{10 \times 10 \times 10}$.
8. Find the value of $64^{1/3}$.
9. If $p = 3 \times 10^3$ and $q = 2 \times 10^2$, work out the values of the following giving the answer in standard form:
   (a) pq (b) $p + q$. (SW)
10. Find the values of:
    (a) $(-3)^2$ (b) 3^0 (c) 10^{-1}. (EA)
11. Write the number 8 500 000 in standard form.
12. State which is the greater, 7.59×10^3 or 3.2×10^4, and by how much.
13. Calculate the value of $10^{15} \div 10^{12}$, giving the answer as a power of 10.
14. Evaluate $16^{3/4}$. (EA)

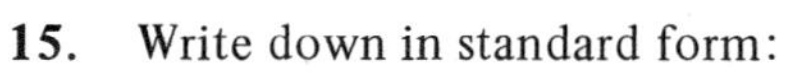

15. Write down in standard form:

(a) 87 000 (b) 0.073.

16. If $2^x = 32$, find the value of x.

17. Find the value of $16^{1/2}$. (EA)

18. Write 9×10^{-3} as a decimal number.

19. Find the product of 20.5 and 4000, giving the result in standard form.

20. Find the value of n in each of the following:

(a) $15^8 \times 15^n = 15^{11}$ (b) $3^7 \div 3^5 = 3^n$

(c) $53\,600 = 5.36 \times 10^n$.

Multi-choice questions 13

1. 3×10^3 equals

A 10 **B** 90 **C** 300 **D** 3000

E 9000 (EM)

2. $10^2 + 10^5$ equals

A 70 **B** 1100 **C** 100 100 **D** 10^7

E 10^{10} (EM)

3. In standard form, 43 000 is

A 0.43×10^5 **B** 4.3×10^4 **C** 43×1000

D 430×100 **E** any of these (EM)

4. $8^{2/3}$ is equal to

A 2 **B** 4 **C** $5\frac{1}{3}$ **D** 16

5. Written in standard form, 0.0063 is

A 63×10^{-4} **B** 6.3×10^3

C 6.3×10^{-3} **D** 0.63×10^{-2}

6. 4.05×10^6 is equivalent to

A 4 050 000 **B** 405 000 **C** 400 005

D 405 000 **E** none of these (WY)

7. Find the value of $2^3 \times 2^2$.

A 1024 **B** 256 **C** 64 **D** 32

E 24 (EA)

8. Find the value of 10^{-2}.

A -20 **B** 0.01 **C** $\frac{1}{8}$ **D** 8

9. 2.68×10^{-3} is equal to

A 0.002 68 **B** 0.0268 **C** 0.804

D 26.8 **E** 80.4

10. Find the value of $2p^3$ when $p = -2$.

A -64 **B** -16 **C** -12 **D** 16

14 LOGARITHMS

- A LOGARITHM consists of two parts
  (i) A whole number part called the *characteristic* which depends upon the size of the number
  (ii) A decimal part called the *mantissa* which is obtained directly from logarithm tables.

- For a number greater than 1, the CHARACTERISTIC is found by subtracting 1 from the number of figures to the left of the decimal point. Thus

$$\log 1.678 = 0.2248$$
$$\log 16.78 = 1.2248$$
$$\log 167.8 = 2.2248$$
$$\log 1678 = 3.2248$$

For a number less than 1, the characteristic is found by adding 1 to the number of zeros following the decimal point. Thus

$$\log 0.5237 = \bar{1}.7191$$
$$\log 0.052\,37 = \bar{2}.7191$$
$$\log 0.005\,237 = \bar{3}.7191$$

(Note that $\bar{2}.7191$ means $-2+0.7191$ and $\bar{3}.7191$ means $-3+0.7191$.)

- The table of ANTILOGARITHMS contains the numbers which correspond to a given logarithm.
  (i) Only the mantissa (decimal part) of the logarithm is used in the table.
  (ii) For a logarithm with a positive characteristic, the number of figures to the left of the decimal point is found by adding 1 to the characteristic. Thus to find the number whose logarithm is 2.3948, we find from the antilog tables that the number corresponding to .3948 is 2482. Since the characteristic is 2, the number must have three figures to the left of the decimal point. The number is therefore 248.2. (Note that $\log 248.2 = 2.3948$.)
  (iii) For a logarithm with a negative characteristic, the number of zeros following the decimal point is one less than the numerical value of the negative characteristic. Thus to find the number whose logarithm is $\bar{2}.3759$, we find from the antilog tables that the number corresponding to .3759 is 2376. Since the characteristic is $\bar{2}$, the number must have one zero following the decimal point. The number is therefore 0.023 76. (Note that $0.023\,76 = \bar{2}.3759$.)

Exercise 14.1

Write down the logarithms of the following numbers:

1. 8, 80, 800, 8000, 80 000
2. 6.5, 65, 650, 6500, 65 000
3. 5.23, 52.3, 523, 5230, 52 300
4. 3.126, 31.26, 312.6, 3126, 31 260
5. 0.3135, 0.003 135, 0.000 313 5
6. 0.002 689, 0.000 026 89, 0.000 002 689.

Find the numbers corresponding to the following logarithms:

7. 0.58, 1.58, 2.58, 3.58
8. 5.2679, 3.2679, 1.2679, 0.2679
9. $\bar{1}.3841, \bar{2}.3841, \bar{3}.3841$
10. $\bar{5}.0789, \bar{3}.0789, \bar{2}.0789, \bar{1}.0789$.

- The RULE FOR MULTIPLICATION using logarithms is: Find the logs of the numbers to be multiplied and *add* them together. The required product is found by taking the antilog of this sum.

Example

Find the value of $19.37 \times 0.0562 \times 136.3$.

$$\begin{aligned} \text{If} \quad N &= 19.37 \times 0.0562 \times 136.3 \\ \log N &= \log 19.37 + \log 0.0562 + \log 136.3 \\ &= 1.2871 + \bar{2}.7497 + 2.1345 \\ &= 2.1713 \\ N &= 148.4 \end{aligned}$$

(by taking the antilog of 2.1713).

Number	*Log*	
19.37	1.2871	
0.0562	$\bar{2}.7497$	ADD
136.3	2.1345	
	2.1713	

- The RULE FOR DIVISION using logarithms is: *Subtract* the log of the denominator from the log of the numerator. The answer is found by taking the antilog of this difference.

Example

Find the value of $27.35 \div 4.62$.

$$\begin{aligned} \text{If} \quad N &= 27.35 \div 4.62 \\ \log N &= \log 27.35 - \log 4.62 \\ &= 1.4370 - 0.6646 \\ &= 0.7724 \\ N &= 5.920 \end{aligned}$$

(by taking the antilog of 0.7724).

Number	*Log*
27.35	1.4370
4.62	0.6646
	0.7724

- The RULE FOR POWERS using logarithms is: Find the log of the number and multiply it by the index denoting the power to which the number is to be raised. The answer is found by taking the antilog of this product.

Example

Find $(12.35)^{1.3}$.

If

$$\begin{aligned} N &= (12.35)^{1.3} \\ \log N &= 1.3 \times \log 12.35 \\ &= 1.3 \times 1.0917 \\ &= 1.4192 \\ N &= 26.25 \end{aligned}$$

(by taking the antilog of 1.4192)

- The RULE FOR FINDING THE ROOT of a number using logs is: Find the log of the number and *divide* it by the number denoting the root. The answer is found by taking the antilog of the result of the division.

Example

Find the value of $\sqrt[5]{16.25}$.

If

$$\begin{aligned} N &= \sqrt[5]{16.25} \\ \log N &= \log 16.25 \div 5 \\ &= 1.2109 \div 5 \\ &= 0.2422 \\ N &= 1.746 \end{aligned}$$

(by taking the antilog of 0.2422)

Exercise 14.2

Use logarithms to find values for the following:

1. 1.435×6.271
2. 18.26×24.75
3. $8.192 \times 11.24 \times 182.3$
4. 0.635×17.63
5. 0.0317×0.235
6. $15.26 \div 3.271$
7. $0.325 \div 1.327$
8. $11.21 \div 0.523$
9. $(2.165)^3$
10. $(2.473)^{1.4}$
11. $\sqrt[3]{18.24}$
12. $\sqrt[4]{3.275}$.

Exercise 14.3 (All of the type found in CSE examination papers)

1. Use logarithms to calculate $178 \div 90.5$.
2. Find the value of x if $\log x = 0.4671$.
3. Use tables to find the logarithms of:
   (a) 8.963 (b) 780 (c) 0.59.

4. Use log tables to calculate $\frac{35.82 \times 63.78}{682.9}$

5. Use log tables to calculate $(23.54)^3$.

6. For what number is 0.8793 the logarithm?

7. Use logarithms to calculate $\frac{16.8 \times 8.02}{102}$. (EA)

8. If $\log x = 2.3670$, find x. (EA)

9. Use logarithms to calculate:
   (a) 18.9×23.4 (b) $(18.3)^3$
   (c) $\sqrt[3]{54.75}$ (d) $698 \div 3.14$.

10. If $\log x = 2.1576$, $\log y = 1.8387$ and $\log z = 0.6218$, use tables to find:
    (a) x (b) xyz (c) $\log z^3$ (d) y.

Multi-choice questions 14

1. The logarithm of a number x is 0.314. What is the logarithm of the number $100x$?
   A 0.314 00 **B** 0.628 **C** 2.314 **D** 31.4

2. When using logarithms, what is the characteristic of the number 0.5063?
   A $\bar{2}$ **B** $\bar{1}$ **C** 0 **D** 1

3. The logarithm of 5.63 is 0.750 51. Hence the logarithm of 563 is
   A $\bar{2}.750\,51$ **B** $\bar{1}.750\,51$ **C** 1.750 51 **D** 2.750 51

4. $3.5 \times 2.43 = 8.505$. The logarithm of 8.505 is 0.929 67. What is the logarithm of the answer to 35×24.3?
   A $\bar{2}.929\,67$ **B** $\bar{1}.929\,67$ **C** 1.929 67 **D** 2.929 67
   E none of these (WY)

5. The logarithm of 25 is 1.397 94, hence the logarithm of 5 is
   A $\bar{1}.397\,94$ **B** 0.198 97 **C** 0.397 94 **D** 0.698 97
   E 2.795 88 (WY)

6. The logarithm of 21.7×10^{-3} is
   A $\bar{3}.336$ **B** $\bar{2}.336$ **C** $\bar{1}.336$ **D** 0.336

7. If $\log x = 3.4875$, then the value of x is
   A greater than 1000 **B** between 100 and 1000
   C less than 100 **D** between 10 and 100

8. If $\log 8 = 0.903$, then $\log 0.08$ equals
   A $\bar{2}.903$ **B** $\bar{1}.903$ **C** 0.903 **D** 1.903
   E 2.903

15 SQUARES, SQUARE ROOTS AND RECIPROCALS

- When a number is multiplied by itself, the result is called the SQUARE OF THE NUMBER. Thus

$$\text{The square of } 8 = 8 \times 8 = 8^2 = 64$$

A table of squares gives the squares of numbers from 1 to 10 but it can be used to find the squares of numbers outside that range.

Example

(a) Find $(78.3)^2$.

$$(78.3)^2 = 78.3 \times 78.3 = 7.83 \times 10 \times 7.83 \times 10$$
$$= (7.83)^2 \times 100$$

From the table of squares $(7.83)^2 = 61.31$.

$$(78.3)^2 = 61.31 \times 100 = 6131$$

(b) Find $(0.0324)^2$.

$$(0.0324)^2 = 3.24 \times \frac{1}{100} \times 3.24 \times \frac{1}{100}$$
$$= (3.24)^2 \div 10\,000$$

From the table of squares $(3.24)^2 = 10.50$.

$$(0.0324)^2 = 10.50 \div 10\,000 = 0.001\,05$$

- The SQUARE ROOT OF A NUMBER is the number whose square equals the given number. Thus since $5^2 = 25$, $\sqrt{25} = 5$.

There are two tables of square roots: one gives square roots of numbers between 1 and 10 and the other gives square roots of numbers between 10 and 100. Thus

$$\sqrt{8.362} = 2.891$$
$$\sqrt{78.35} = 8.852$$

The square roots of numbers outside these ranges can also be found from the tables.

Example

(a) Find $\sqrt{2637}$.

Mark off the figures in pairs to the left of the decimal point. Each pair of figures is called a *period*. Thus 2637 becomes 26′37. The first period is 26 so look up $\sqrt{26.37} = 5.135$. To position the decimal point in the answer remember that for each period there will be one figure to the left of the decimal point in the answer. Thus

$$\sqrt{2637} = 51.35$$

(b) Find $\sqrt{0.0632}$.

In the case of numbers less than 1, mark off in pairs to the right of the decimal point. Thus 0.0632 becomes 0.06′32. The first period following the decimal point is 06 so look up $\sqrt{6.32} = 2.514$. For each zero pair in the original number there will be one zero in the answer. Hence

$$\sqrt{0.0632} = 0.2514$$

(c) Find $\sqrt{0.000\,032\,16}$.

Marking off the zeros in pairs gives 0.00′00′32′16.

$$\sqrt{32.16} = 5.671$$

$$\sqrt{0.000\,032\,16} = 0.005\,671 \quad \text{(since there are two zero pairs)}$$

- The SQUARE ROOT OF A PRODUCT is the product of the square roots. Thus

$$\sqrt{9 \times 16} = \sqrt{9} \times \sqrt{16} = 3 \times 4 = 12$$

- The SQUARE ROOT OF A FRACTION is the square roots of the numerator and denominator taken separately. Thus

$$\sqrt{\frac{25}{16}} = \frac{\sqrt{25}}{\sqrt{16}} = \frac{5}{4}$$

- The RECIPROCAL OF A NUMBER is $\frac{1}{\text{number}}$. Thus the reciprocal of 2 is $\frac{1}{2}$ and the reciprocal of 19.83 is $\frac{1}{19.83}$.

A table of the reciprocals of numbers gives the reciprocals of numbers from 1 to 10 but it can be used to find the reciprocals of numbers outside of this range.

Example

(a) Find the reciprocal of 129.

$$\text{Reciprocal of } 129 = \frac{1}{129} = \frac{1}{1.29} \times \frac{1}{100}$$

From the table

$$\frac{1}{1.29} = 0.7752$$

$$\frac{1}{129} = 0.7752 \div 100 = 0.007\,752$$

(b) Find the reciprocal of 0.6375.

$$\text{Reciprocal of } 0.6375 = \frac{1}{0.6375} = \frac{10}{6.375}$$

From the table

$$\frac{1}{6.375} = 0.1568$$

$$\frac{1}{0.6375} = 10 \times 0.1568 = 1.568$$

Exercise 15.1

Find: (a) the square, (b) the square root, (c) the reciprocal of each of the following numbers:

1.	4.2	**8.**	89.93	**15.**	0.062
2.	7.38	**9.**	400	**16.**	0.003 178
3.	6.329	**10.**	736	**17.**	0.3172
4.	8.939	**11.**	928.3	**18.**	0.006 231
5.	39	**12.**	716.3	**19.**	0.000 281
6.	43.2	**13.**	0.26	**20.**	0.000 006.
7.	57.68	**14.**	0.352		

Find the square root of each of the following:

21.	9×25	**23.**	16×49	**25.**	$\frac{9}{25}$
22.	11×11	**24.**	$\frac{1}{16}$	**26.**	$\frac{49}{36}$

Exercise 15.2 (All of the type found in CSE examination papers)

1. Find the values of $(14.53)^2$ and $\sqrt{196}$. (EM)
2. Find the reciprocal of 8.76.
3. Find $\sqrt{784}$. (EA)
4. (a) Write down the whole number which is nearest to $\sqrt{14}$.
   (b) Write down the whole number which is nearest to $\sqrt{220}$.

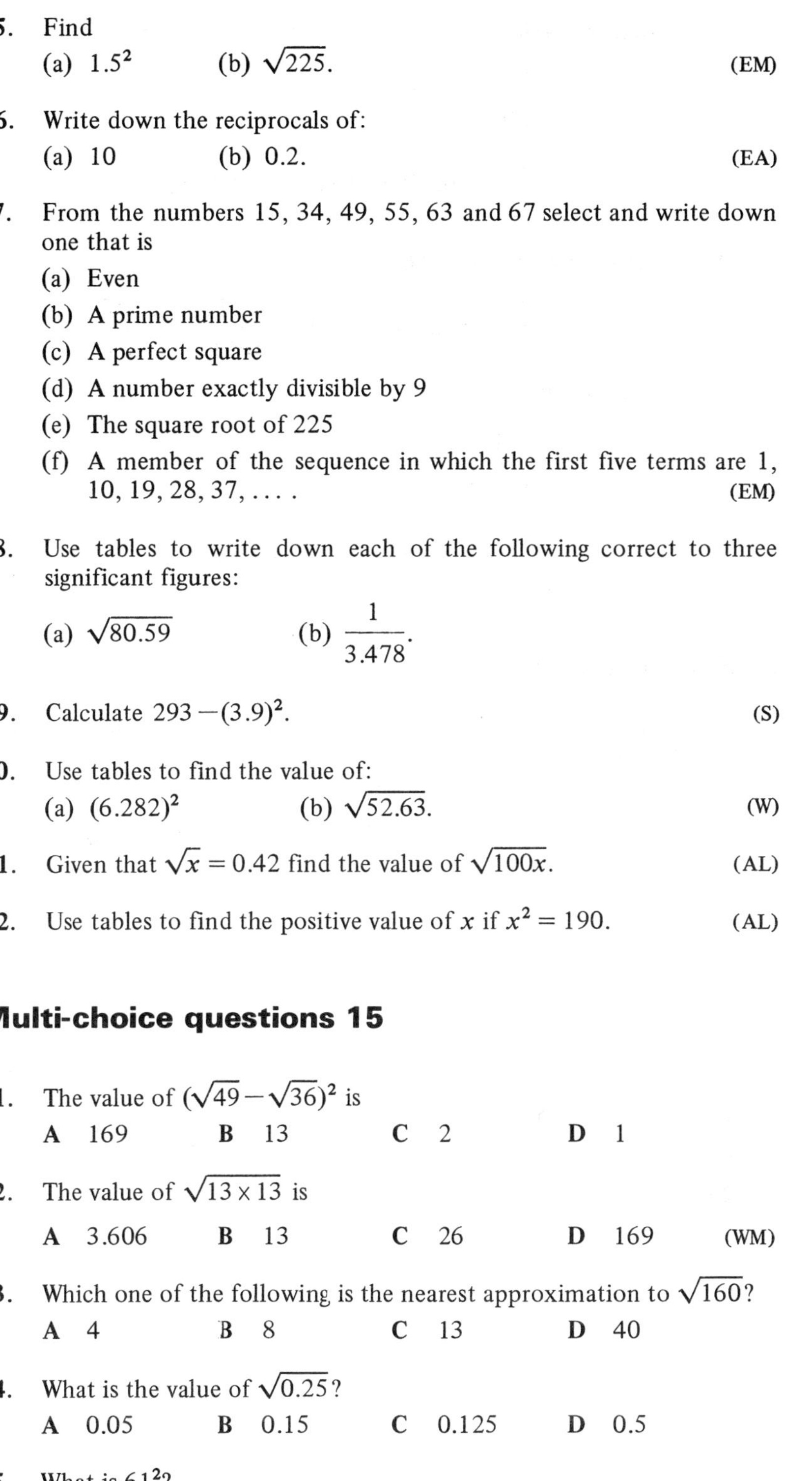

5. Find
(a) 1.5^2 (b) $\sqrt{225}$. (EM)

6. Write down the reciprocals of:
(a) 10 (b) 0.2. (EA)

7. From the numbers 15, 34, 49, 55, 63 and 67 select and write down one that is
(a) Even
(b) A prime number
(c) A perfect square
(d) A number exactly divisible by 9
(e) The square root of 225
(f) A member of the sequence in which the first five terms are 1, 10, 19, 28, 37, ... (EM)

8. Use tables to write down each of the following correct to three significant figures:
(a) $\sqrt{80.59}$ (b) $\dfrac{1}{3.478}$.

9. Calculate $293-(3.9)^2$. (S)

10. Use tables to find the value of:
(a) $(6.282)^2$ (b) $\sqrt{52.63}$. (W)

11. Given that $\sqrt{x}=0.42$ find the value of $\sqrt{100x}$. (AL)

12. Use tables to find the positive value of x if $x^2=190$. (AL)

Multi-choice questions 15

1. The value of $(\sqrt{49}-\sqrt{36})^2$ is
A 169 **B** 13 **C** 2 **D** 1

2. The value of $\sqrt{13\times 13}$ is
A 3.606 **B** 13 **C** 26 **D** 169 (WM)

3. Which one of the following is the nearest approximation to $\sqrt{160}$?
A 4 **B** 8 **C** 13 **D** 40

4. What is the value of $\sqrt{0.25}$?
A 0.05 **B** 0.15 **C** 0.125 **D** 0.5

5. What is 61^2?
A 122 **B** 361 **C** 3601 **D** 3721 (EA)

6. Find the value of $\frac{1}{0.2}+\frac{1}{0.25}$.

A 2 **B** 2.5 **C** 4.5 **D** 9
E 45

7. $(1\frac{2}{3})^2$ is equal to

A $2\frac{7}{9}$ **B** $2\frac{4}{9}$ **C** $1\frac{4}{9}$ **D** $\frac{4}{6}$

8. $\frac{10^2-5^2}{10^2+5^2}$ is equal to

A $\frac{3}{5}$ **B** $\frac{2}{5}$ **C** $\frac{1}{3}$ **D** $\frac{1}{5}$

16 LINEAR EQUATIONS

- LINEAR EQUATIONS contain only the first power of the unknown quantity. Thus

$$3x-5 = 7 \quad \text{and} \quad \frac{x}{7} = 8$$

are both examples of linear equations.

- The SOLUTION of an equation is that value of the unknown which, when substituted into the equation, makes the left hand side equal to the right hand side.

- In SOLVING EQUATIONS:
  (i) The same quantity may be added to or subtracted from both sides
  (ii) Each side may be multiplied or divided by the same quantity.

Equations requiring multiplication

Example

Solve $\dfrac{x}{2} = 3$.

Multiply each side by 2. Then

$$x = 2\times 3$$

$$x = 6$$

(*Check*: substituting for x in the LHS, $\frac{6}{2} = 3$. Hence LHS = RHS and therefore the solution is correct.)

Equations requiring division

Example

Solve $5x = 10$.

Divide both sides by 5. Then

$$x = \frac{10}{5}$$

$$x = 2$$

Equations requiring multiplication and division

Example

Solve $\dfrac{3x}{2} = 12$.

Multiply both sides by 2. Then

$$3x = 2 \times 12$$
$$3x = 24$$

Divide both sides by 3. Then

$$x = \frac{24}{3}$$
$$x = 8$$

Equations requiring addition and/or subtraction

Example

Solve $4x - 5 = 7$.

Add 5 to both sides. Then

$$4x - 5 + 5 = 7 + 5$$
$$4x = 12$$
$$x = 3$$

The operation of adding 5 to both sides can be accomplished by taking the 5 to the RHS and changing its sign. Thus

$$4x - 5 = 7$$
$$4x = 7 + 5$$
$$4x = 12$$
$$x = 3$$

Equations containing the unknown on both sides

When an equation contains the unknown on both sides, group all the terms containing the unknown on one side of the equation and all the other terms on the other side.

Example

Solve $5x + 8 = 3x - 2$.

$$5x - 3x = -2 - 8$$
$$2x = -10$$
$$x = -5$$

Equations containing brackets

In equations of this type, remove the brackets first and then solve the resulting equation using the methods given previously.

Example

Solve $3(4x-5)+3 = 2(x+3)-7$.

$$\begin{aligned} 12x-15+3 &= 2x+6-7 \\ 12x-12 &= 2x-1 \\ 10x &= 11 \\ x &= \frac{11}{10} \end{aligned}$$

Equations containing fractions

The first step is to get rid of the denominators by multiplying each term of the equation by the LCM of the denominators. The resulting equation is then solved by using the methods shown previously.

Example

Solve $\frac{x}{2}-\frac{3x}{5} = \frac{7}{4}$.

The LCM of the denominators 2, 4 and 5 is 20. Multiplying each term of the equation by 20 gives

$$\begin{aligned} 10x-12x &= 35 \\ -2x &= 35 \\ x &= -\frac{35}{2} \end{aligned}$$

Exercise 16.1

Solve each of the following equations:

1. $\frac{x}{5} = 2$
2. $\frac{x}{3} = 7$
3. $2x = 8$
4. $3x = 9$
5. $x-3 = 5$
6. $2x-1 = 7$
7. $5x-8 = 2$
8. $3x-4 = x+6$
9. $5x-20 = 3x-8$
10. $5p+11 = 25-p$
11. $3(x+1) = 8$
12. $2(x-3)-(x+2) = 5$

13. $5(m+2)-3(m-5)=29$

14. $\frac{x}{3}=\frac{2}{5}$

15. $\frac{m}{3}+\frac{m}{5}=2$

16. $\frac{4}{x}=\frac{2}{3}$

17. $3x+\frac{3}{8}=4+\frac{2x}{3}$

18. $\frac{x}{2}+\frac{x}{3}+\frac{x}{4}=\frac{2}{5}$.

Exercise 16.2 (All of the type found in CSE examination papers)

1. Solve the equation $3(x-1)-4(2x+3)=15$.
2. Solve the equations:
   (a) $7x-9=5$ (b) $\frac{1}{2}(x-8)=3$.
3. Solve the equation $5(m-2)=15$.
4. Solve the equations:
   (a) $3x+7=5$ (b) $\frac{x}{3}-1=5$.
5. Solve the equations:
   (a) $5x-3=12$ (b) $3x=5(9-x)$ (c) $\frac{3x+12}{5}=6$.
6. Solve the equations:
   (a) $7n=-35$ (b) $\frac{-8}{x}=2$.
7. Solve the equations:
   (a) $3x+7=22$ (b) $2x-5=-1$.
8. If $3x-7=14$, find the value of x.
9. If $\frac{0.15}{p}=0.05$, find the value of p.
10. Solve the following equations:
    (a) $\frac{a}{3}-2=0$ (b) $3b-2=0$ (c) $3-c=2$
    (d) $\frac{3}{d}-2=0$ (e) $\frac{3}{e}=\frac{2}{3}$. (EA)
11. If $4x-3=5$, find the value of $7x$.
12. Solve the equations:
    (a) $12(x-4)=4(x+12)$ (b) $\frac{x}{4}+\frac{x}{5}=9$. (EA)

Multi-choice questions 16

1. If $3(2x-5)-2(x-3)=3$ then x is equal to

   A $1\frac{1}{4}$ B 3 C $2\frac{3}{4}$ D 6

2. If $\dfrac{x-5}{3}=\dfrac{x+2}{2}$ the value of x is

   A -16 B -7 C 7 D 16

3. If $\dfrac{3-2y}{4}=\dfrac{y}{3}$ then y is equal to

   A -3 B $\frac{9}{10}$ C $\frac{10}{9}$ D 3

4. If $\dfrac{5x}{6}-\dfrac{3x}{4}=\dfrac{1}{12}$, then the value of x is

   A -8 B -2 C -1 D 1

5. What is the value of p when $3p+8=5p-16$?

   A 1 B 3 C 4 D 12

6. If $a=x-1$ and $b=2x-3$, when $a=b$ the value of x is

   A -4 B -2 C $-\frac{2}{3}$ D 2

7. If $9x-6=24-3x$, then x equals

   A $1\frac{1}{2}$ B 2 C $2\frac{1}{2}$ D 3

8. If $2(2q-1)=30$, find the value of q.

   A $\frac{1}{2}$ B 7 C 8 D 15

9. Consider the equation $y=3x-2$. When $y=7$, the value of x is

   A $1\frac{2}{3}$ B 3 C 6 D 27

10. If $2(x+6)-3(x-3)=3$ then x is equal to

    A $-8\frac{1}{2}$ B -7 C 0 D 18

17 SIMULTANEOUS EQUATIONS

- Consider the equations

$$3x+2y = 12$$
$$2x-y = 1$$

Each equation contains the unknown quantities x and y. The solutions are the values of x and y which satisfy both equations simultaneously. Equations such as these are called SIMULTANEOUS EQUATIONS.

- The METHOD OF ELIMINATION is frequently used to solve simultaneous equations.

Example

Solve the simultaneous equations

$$3x+5y = 26 \quad (1)$$
$$2x+3y = 16 \quad (2)$$

If we multiply equation (1) by 2 and equation (2) by 3, we will obtain the same coefficient of x in both equations thereby allowing us to eliminate x from the equations. Thus

$$6x+10y = 52 \quad (3)$$
$$6x+9y = 48 \quad (4)$$

Subtracting equation (4) from equation (3) gives

$$(6x-6x)+(10y-9y) = (52-48)$$
$$y = 4$$

To find x we substitute for y in either equation (1) or equation (2). Substituting for y in equation (1) gives

$$3x+(5\times 4) = 26$$
$$3x+20 = 26$$
$$3x = 6$$
$$x = 2$$

To check the solutions substitute the values of x and y in equation (2) and see if the LHS = RHS. Thus

$$\text{LHS} = (2\times2)+(3\times4) = 4+12 = 16 = \text{RHS}$$

Hence the solutions, $x = 2$ and $y = 4$, are correct.

(Note that there would be no point in checking in equation (1) because this was used to find the value of x.)

Example

Solve the simultaneous equations

$$2x-3y = 4 \qquad (1)$$

$$5x+y = 27 \qquad (2)$$

With these equations we can obtain the same coefficient of y in both equations by multiplying equation (2) by 3 and leaving equation (1) as it is. Thus

$$2x-3y = 4 \qquad (1)$$

$$15x+3y = 81 \qquad (3)$$

To eliminate y from the equations we add equations (1) and (3).

$$(2x+15x)+(-3y+3y) = (4+81)$$

$$17x = 85$$

$$x = 5$$

To find y we substitute for x in equation (1). Thus

$$(2\times5)-3y = 4$$

$$10-3y = 4$$

$$-3y = -6$$

$$y = 2$$

Check in equation (2)

$$\text{LHS} = (5\times5)+2 = 25+2 = 27 = \text{RHS}$$

Hence the solutions, $x = 5$ and $y = 2$, are correct.

Exercise 17.1

Solve the following pairs of simultaneous equations:

1. $x+2y = 11$
   $x-2y = 3$
2. $3x+2y = 18$
   $2x-y = 6$
3. $2x+5y = 27$
   $3x+2y = 13$
4. $3x+2y = 13$
   $2x+3y = 12$
5. $2x+3y = 20$
   $5x-3y = 22.$

Exercise 17.2 (All of the type found in CSE examination papers)

1. Solve the simultaneous equations:

$$3x-y = 8$$
$$x+y = 4.$$

2. Solve the simultaneous equations:

$$5x-y = 18$$
$$3x+y = 14.$$

3. Solve the simultaneous equations:

$$3x-2y = 0$$
$$x-2y = -4.$$

4. Solve the simultaneous equations:

$$2x-y = 7$$
$$5x-3y = 16.$$

5. Solve the simultaneous equations:

$$4x-3y = 16$$
$$2x+3y = 26.$$

6. If $p-q = 8$ and $2p-q = 20$, calculate the value of p. (EA)

7. Given the two equations:

$$3x+2y = 23$$
$$2x+3y = 22$$

(a) By adding the two equations find the value of $x+y$.
(b) By subtracting the two equations find the value of $x-y$.
(c) Hence find the values of x and y.

8. Solve the following equations for x and y:

$$2x-3y = -1$$
$$2x+3y = 17.$$

Multi-choice questions 17

1. The solutions of the simultaneous equations $2x-y = 2$ and $x+y = -5$ are

A $x = 1, y = -4$ **B** $x = -1, y = -4$
C $x = -1, y = 4$ **D** $x = 1, y = 4$

2.

$$2x-y = 3 \quad (1)$$
$$2x-2y = 2 \quad (2)$$

The result of subtracting equation (2) from equation (1) is

A $-3y = 1$ **B** $4x-3y = 1$
C $y = 1$ **D** $4x+y = 1$

3. If $x+y=3$ and at the same time $2x-y=3$, find the value of x.

A 1 **B** 2 **C** 3 **D** 4

4. By eliminating x from the equations

$$2x-5y = 8 \quad (1)$$
$$2x-3y = -7 \quad (2)$$

one of the following equations is obtained

A $-8y=1$ **B** $-2y=15$

C $-8y=15$ **D** $-2y=1$

5. What is the value of x in the simultaneous equations $x+y=6$; $x-y=2$?

A 3 **B** 4 **C** 5 **D** 6

6. By eliminating y from the simultaneous equations

$$3x-4y = -10 \quad (1)$$
$$x+4y = 8 \quad (2)$$

one of the following equations is obtained

A $2x=-18$ **B** $4x=-18$

C $2x=-2$ **D** $4x=-2$

7. $x+2y=7$; $2x-5y=-4$. The solution of these simultaneous equations is

A $x=-3, y=-2$ **B** $x=1, y=3$

C $x=3, y=2$ **D** $x=-1, y=4$

8. $3x-2y=5$; $4x-y=10$. The solutions of these simultaneous equations are

A $x=-3, y=22$ **B** $x=-3, y=-22$

C $x=3, y=2$ **D** $x=3, y=-2$

18 QUADRATIC EQUATIONS

- An equation of the type $ax^2+bx+c=0$ is called a QUADRATIC EQUATION. The constants a, b and c can take any numerical value. The following are all examples of quadratic equations

$$x^2-4 = 0 \quad \text{in which} \quad a = 1, \quad b = 0 \quad \text{and} \quad c = -4$$

$$3x^2-5x+7 = 0 \quad \text{in which} \quad a = 3, \quad b = -5 \quad \text{and} \quad c = 7$$

$$2x^2-8x = 0 \quad \text{in which} \quad a = 2, \quad b = -8 \quad \text{and} \quad c = 0$$

- A quadratic equation always has two SOLUTIONS. It is possible for the two solutions to be the same and it is also possible for one of the solutions to be zero. The solutions of a quadratic equation are often called the *roots* of the equation.

- A quadratic equation can be SOLVED BY FACTORISATION. We make use of the fact that if the product of two factors is zero, then one of the factors must be zero. Thus if $ab=0$, then either $a=0$ or $b=0$. To solve the quadratic equation, the expression ax^2+bx+c is written as the product of two factors.

Example

(a) If $x^2=4$

$$x = \pm\sqrt{4}$$

$$x = \pm 2$$

(Note that the square root of a number has two possible values, one positive and the other negative. In the example we see that $(+2)^2=4$ and $(-2)^2=4$.)

(b) If $x^2-5x=0$

$$x(x-5) = 0$$

Either $\quad x = 0$

or $\quad x-5 = 0$, giving $x = 5$

Hence the roots are $x=0$ and $x=5$.

(It is incorrect to say that the solution is $x=5$. The solution $x=0$ must also be stated.)

(c) If $2x^2+5x-12=0$

$$(2x-3)(x+4) = 0$$

Either $2x-3 = 0$, giving $x = 1\frac{1}{2}$

or $x+4 = 0$, giving $x = -4$

Hence the roots of the equation are $x = 1\frac{1}{2}$ and $x = -4$.

(d) If $x^2-10x+25=0$

$$(x-5)^2 = 0$$

With this equation the two roots are the same, i.e. $x = 5$ and we say that the equation has equal roots.

Exercise 18.1

Solve the following quadratic equations:

1. $x^2-9=0$
2. $2x^2-50=0$
3. $x(x-5)=0$
4. $x^2+3x=0$
5. $2x^2-6x=0$
6. $(x+4)(x-3)=0$
7. $(2x-3)(3x+4)=0$
8. $x^2+5x+6=0$
9. $x^2-5x+6=0$
10. $x^2+x-20=0$
11. $2x^2+7x+3=0$
12. $3x^2-13x-10=0$.

Exercise 18.2 (All of the type found in CSE examination papers)

1. Solve the equation $x^2-10x+25=0$. (EA)
2. (a) Find the factors of $x^2-7x+12$.
   (b) Hence solve the equation $x^2-7x+12=0$.
3. If $x^2-25=0$, find the two solutions for x.
4. (a) Factorise x^2+3x-4, given that one factor is $(x-1)$.
   (b) Find two values for x such that $x^2+3x-4=0$. (S)
5. Solve for x the equation $2x^2-7x+5=0$.
6. Solve the equation $(x+4)(x-3)=0$.
7. Find the values of x which satisfy the equation $x^2-36=0$.
8. $(x-4)(x+3)=0$. Write down both solutions to this equation. (EA)
9. Solve the equations:
   (a) $(x-2)(x+5)=0$ (b) $x^2-2x-35=0$. (EA)
10. $x^2-16=0$. Write down both solutions to this equation.

Multi-choice questions 18

1. Which of the following is a solution of the equation $x^2-3x-10=0$?

   A -10 B -5 C -2 D 2
   E 10

2. Find the solutions of the equation $(2x-5)(x+3)=0$.

   A $2\frac{1}{2}, 3$ B $2\frac{1}{2}, -3$ C $-2\frac{1}{2}, -3$ D $-2\frac{1}{2}, 3$

3. If $2x^2+4x=16$, which one of the following could be a value of x?

   A -4 B -2 C 4 D 8 (EA)

4. Consider the equation $2x^2-3x+1=0$. Which one of the following could be a value for x which satisfies the equation?

   A 2 B 1 C 0 D -1

5. What are the solutions of the equation $x^2-3x-4=0$?

   A $-1, 4$ B $1, -4$ C $-1, -4$ D $-1, 3$

6. $x=-3$ is one root of the equation $x^2+10x+21=0$. What is the value of the other root?

   A -7 B 3 C 7 D 21

7. $p=x^2+2x-3$. A value of x which makes $p=0$ is

   A 2 B 1 C 0 D -1

19 FORMULAE

- A FORMULA is an equation which shows the relationship between two or more quantities. The statement that $E = IR$ is a formula for E in terms of I and R. The value of E is found by substituting the given values of I and R.

Example

If $v = u + at$, find the value of v when $u = 20$, $a = 3$ and $t = 5$.

Substituting the given values we have

$$v = 20 + (3 \times 5) = 20 + 15 = 35$$

Exercise 19.1

1. If $V = Ah$, find the value of V when $A = 8$ and $h = 4$.
2. The formula $K = Wa + b$ is used in engineering. Find K when $W = 25$, $a = 3$ and $b = 5$.
3. $S = 90(n-4)$ is a formula used in geometry. Find S when $n = 8$.
4. If $P = \dfrac{RT}{V}$, find the value of P when $R = 56$, $T = 18$ and $V = 7$.
5. $E = \dfrac{mv^2}{2g}$ is a formula used in physics. Find E when $m = 220$, $v = 8$ and $g = 10$.

- The formula $n = p + cr$ has the symbol n as its subject. By rearranging the formula we could make r the subject and we are said to have TRANSPOSED THE FORMULA for r.

The rules for transforming a formula are as follows

(i) Clear fractions
(ii) Clear brackets
(iii) Isolate the required subject.

Example

Transpose the formula $T = \dfrac{12(D-d)}{L}$ to make D the subject.

(i) Clear fractions by multiplying throughout by L

$$TL = 12(D-d)$$

(ii) Remove the bracket by dividing both sides by 12

$$\frac{TL}{12} = D-d$$

(iii) Isolate D by taking d to the LHS

$$\frac{TL}{12}+d = D$$

It is usual to place the subject on the LHS

$$D = \frac{TL}{12}+d$$

Exercise 19.2

Transpose the following formulae:

1. $C = \pi d$ for d
2. $PV = c$ for P
3. $I = PRT$ for T
4. $I = \frac{E}{R}$ for E
5. $S = \frac{ts}{T}$ for s
6. $x = \frac{p}{q}$ for q
7. $P = \frac{RT}{V}$ for V
8. $p = P + 15$ for P
9. $v = u + at$ for u
10. $n = p + cr$ for r
11. $a = \frac{7}{3+x}$ for x
12. $V = \frac{2R}{R-r}$ for r
13. $x = \frac{dh}{P-Q}$ for P
14. $p = \frac{2n+5}{x+3}$ for n.

Exercise 19.3 (All of the type found in CSE examination papers)

1. Consider the formula $S = 180(n-2)$.
   (a) Find the value of S when $n = 8$.
   (b) Find the value of n when $S = 720$.
   (c) Make n the subject of the formula. (EA)
2. $P = 2a + 2b$ is the formula for finding the perimeter of a rectangle.
   (a) Calculate P when $a = 3.5$ and $b = 2.4$.
   (b) Calculate a when $P = 30$ and $b = 4\frac{3}{4}$.
   (c) Make b the subject of the formula. (EA)
3. Given that $V = \pi r^2 h$, calculate V when $\pi = 3.14$, $r = 3$ and $h = 19$.
4. $3p = (q+r)s$.
   (a) Find s in terms of p, q and r.
   (b) Find q in terms of p, r and s.

5. $3k = kx + 7$.
   (a) Calculate x when $k = 4$.
   (b) Express x in terms of k. (S)

6. Rearrange the formula $y = 5x - 2$ to give x in terms of y.

7. Find the value of E if $E = \frac{1}{2}mv^2$ when $m = 20$ and $v = 5$.

8. The formula for simple interest is $I = \frac{PRT}{100}$. Express R in terms of I, P and T.

9. Make x the subject of the formula $y = mx + c$. (EA)

10. $a * b = ab - 2$:
    (a) Calculate $6 * 4$ (b) $3 * (6 * 4)$.

Multi-choice questions 19

1. The formula $v = u + at$ is used in physics. Express t in terms of u, v and a.

   A $\frac{v-u}{a}$ B $v - u = a$ C $\frac{uv}{a}$ D $\frac{u-v}{a}$

2. If $y = 5x - 4$, an expression for x is

   A $\frac{y-5}{-4}$ B $\frac{y+5}{4}$ C $\frac{y+4}{5}$ D $y + 4$

3. Given that $V = \frac{abh}{3}$, the expression for h in terms of a, b and V is

   A $\frac{ab}{3V}$ B $\frac{V-3}{ab}$ C $3V - ab$ D $\frac{3V}{ab}$

4. If $F = 2x^2 + 4x$, what is the value of F when $x = -2$?

   A -16 B 0 C 8 D 16 (EA)

5. If $M = 3(x + y)$, then

   A $y = \frac{M}{3} - x$ B $y = M - 2x$

   C $y = 3M - x$ D $y = M - \frac{x}{3}$

6. If $d = b^2 - 4ac$ then c expressed in terms of the other letters is

   A $\frac{b^2-d}{4a}$ B $\frac{d-b^2}{4a}$ C $d - b^2 + 4a$ D $\frac{d}{b^2-4a}$

7. If $s = \frac{u+v}{2}t$ and $s = 100$ when $u = 6$ and $v = 4$, calculate the value of t.

   A 0.05 B 5 C 20 D 190 (NW)

20 SIMPLE INTEREST

- To calculate the amount of SIMPLE INTEREST the following formula is used

$$I = \frac{PRT}{100}$$

where

P = the amount invested or borrowed (the *principal*)
R = the rate of interest charged per annum
T = the period of the investment or loan in years
I = the amount of simple interest

- The formula for simple interest may be TRANSPOSED to give

$$P = \frac{100I}{RT}$$

$$R = \frac{100I}{PT}$$

$$T = \frac{100I}{PR}$$

Example

(a) A person borrows £2000 for 4 years at 15% simple interest. Calculate the amount of interest that the person will pay.

We are given that $P = £2000$, $R = 15\%$ and $T = 4$ years and we have to find I.

$$I = \frac{PRT}{100} = \frac{2000 \times 15 \times 4}{100} = £1200$$

(b) The interest on £2400 invested for 5 years is £840. What is the rate of interest?

We are given $P = £2400$, $T = 5$ years and $I = £840$ and we have to find R.

$$R = \frac{100I}{PT} = \frac{100 \times 840}{2400 \times 5} = 7\%$$

Exercise 20.1

1. Find the simple interest on £1400 invested for 3 years at 11% per annum.
2. Find the simple interest on £1200 invested for 6 months at 12% per annum.
3. In what length of time will £1000 be the interest on £5000 invested at 5% per annum simple interest?
4. The simple interest on £1200 invested for 5 years is £420. What is the rate per cent per annum?
5. What principal is needed so that the interest will be £96 if it is invested for 5 years at 12% per annum simple interest?

Exercise 20.2 (All of the type found in CSE examination papers)

1. Find the simple interest on £700 invested for 4 years at 12% per annum.
2. (a) Calculate the simple interest on £750 for 5 years at 9%.
   (b) What sum of money will yield a simple interest of £150 at 10% for 3 years?
3. Calculate the simple interest on £850 invested at 8% per annum for 5 years.
4. Calculate how long £2400 must be invested at 8% simple interest per annum to give interest of £96.
5. Mark invested £500 in a Building Society giving an annual rate of 9% interest. He took out the interest at the end of 2 years and spent it. How much money did he spend?

21 AREAS AND PERIMETERS

- The AREA OF A PLANE FIGURE is measured by seeing how many square units it contains. 1 square metre is the area inside a square having a side of 1 metre; 1 square centimetre is the area contained inside a square having a side of 1 centimetre, etc. The standard abbreviations are

$$1 \text{ square metre} = 1 \text{ m}^2$$
$$1 \text{ square centimetre} = 1 \text{ cm}^2$$
$$1 \text{ square millimetre} = 1 \text{ mm}^2$$

- The PERIMETER is the distance round the plane figure. Table 21.1 gives the formulae for the areas and perimeters of some simple geometrical shapes.

Table 21.1

Figure	*Diagram*	*Formulae*
Rectangle	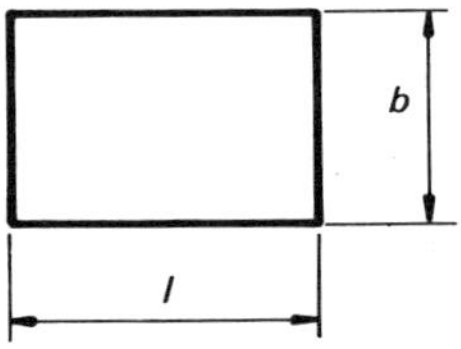	Area $= l \times b$ Perimeter $= 2l + 2b$
Parallelogram	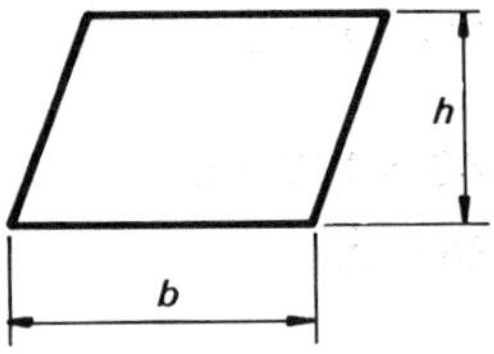	Area $= b \times h$
Triangle	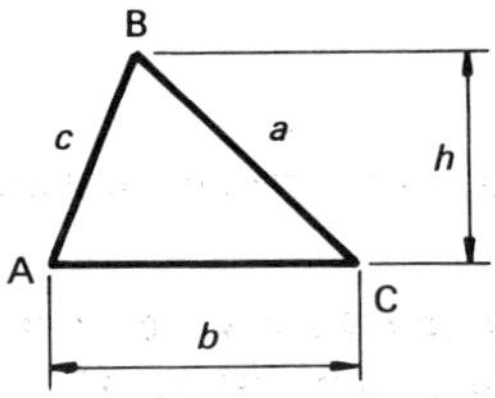	Area $= \frac{1}{2} \times b \times h$ Area $= \sqrt{s(s-a)(s-b)(s-c)}$ where $s = \dfrac{a+b+c}{2}$

Table 21.1 continued

Figure	*Diagram*	*Formulae*
Trapezium	a, h, b	Area $= \frac{1}{2} \times h \times (a+b)$
Circle	r	Area $= \pi r^2$ Circumference $= 2\pi r = \pi d$ $\left(\pi = 3.142 \quad \text{or} \quad \frac{22}{7}\right)$
Sector of a circle	$\theta°$	Area $= \pi r^2 \times \frac{\theta}{360}$ Length of arc $= 2\pi r \times \frac{\theta}{360}$

Example

(a) A trapezium has parallel sides which are 8 cm and 12 cm long respectively. If the distance between these parallel sides is 6 cm, calculate the area of the trapezium.

$$\begin{aligned} A &= \tfrac{1}{2} \times h \times (a+b) \\ &= \tfrac{1}{2} \times 6 \times (8+12) \\ &= \tfrac{1}{2} \times 6 \times 20 \\ &= 60\ \text{cm}^2 \end{aligned}$$

(b) A circle has a diameter of 42 cm. Taking $\pi = \frac{22}{7}$. calculate: (i) the circumference, (ii) the area.

(i) Circumference $= \pi d = \frac{22}{7} \times 42 = 132$ cm.

(ii) Area $= \pi r^2 = \frac{22}{7} \times 21^2 = 1386\ \text{cm}^2$.

Exercise 21.1

1. Find:

(a) the perimeter and (b) the area of a rectangle which is 8 cm long and 6 cm wide.

2. A rectangular lawn is 64 m long and 46 m wide. A path 3 m wide is made round the lawn. What is the area of the path?

3. The area of a rectangle is $144\ \text{m}^2$ and its length is 24 m. What is its width?

4. A carpet has an area of $49\ \text{m}^2$. If it is square, find the length of a side of the carpet.

5. How many square tiles each of side 15 cm are needed to cover a floor which is 9 m long and 24 m wide?

6. Find the area of a parallelogram which has a base 8 cm long and a vertical height of 5 cm.

7. A triangle has a base 8 cm long and a vertical height of 5 cm. Calculate its area.

8. The area of a triangle is $20\ \text{m}^2$. If its base is 8 m, find its vertical height.

9. A trapezium has parallel sides which are 5 cm and 7 cm long respectively. If the distance between these parallel sides is 12 cm, calculate the area of the trapezium.

10. A circle has a radius of 14 cm. Taking $\pi = \frac{22}{7}$, calculate:
    (a) its area (b) its circumference.

11. A circle has a diameter of 18 mm. Taking $\pi = 3.14$, calculate:
    (a) its circumference (b) its area.

12. A sector of a circle has a centre angle of 210°. If its radius is 21 cm, calculate:
    (a) the area of the sector (b) the arc subtended.

13. A pipe has an outside diameter of 25 mm and a bore of 18 mm. What is its cross-sectional area?

14. Find the length of arc and the area of a sector of a circle which subtends an angle of 122°, if its radius is 9 cm.

Exercise 21.2 (All of the type found in CSE examination papers)

1. Calculate the area of a circle with a diameter of 3 cm. Take $\pi = 3.14$.

2. A rectangular piece of plywood 20 cm by 12 cm has equal squares of side 4 cm cut out at two corners. The final shape is as shown in Fig. 21.1. Calculate:
    (a) the perimeter (b) the area of the shape.

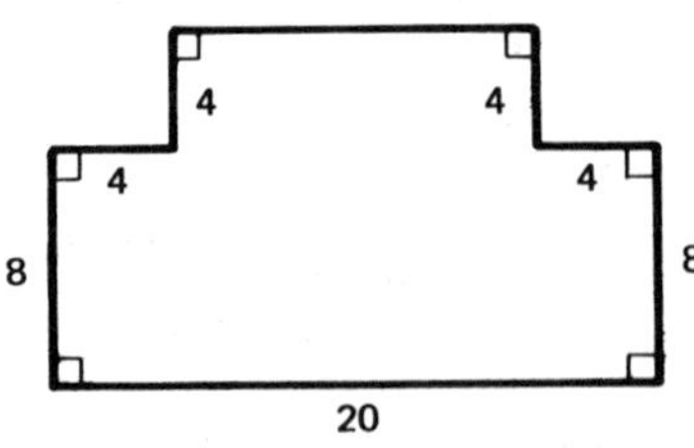

Fig. 21.1

3. A square has a perimeter of 40 cm. Calculate:
   (a) its length of side (b) its area.

4. Fig. 21.2 shows a trapezium. Calculate its area.

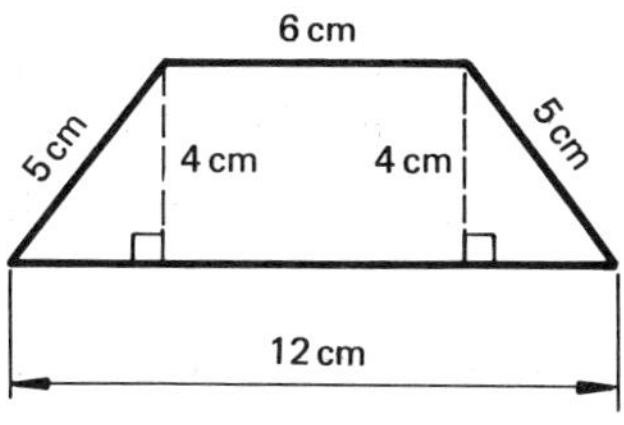

Fig. 21.2

5. In Fig. 21.3:
   (a) Calculate the length of AB
   (b) Determine the area of the triangle ADE. (EA)

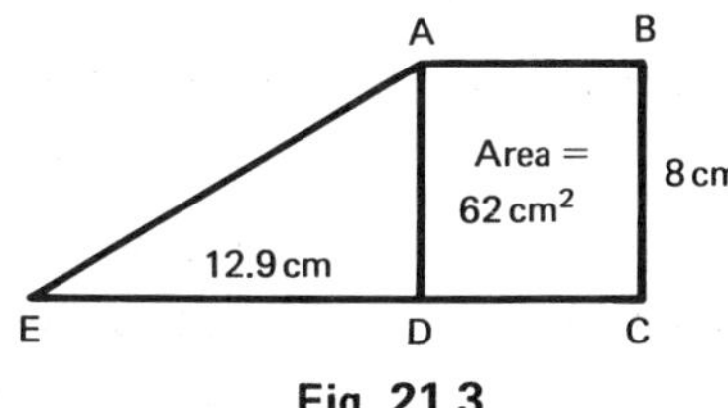

Fig. 21.3

6. The circumference of a circle is 44 cm. Taking $\pi = \frac{22}{7}$, calculate its radius.

7. Fig. 21.4 shows a rectangle. Calculate the area of the shaded triangle. (AL)

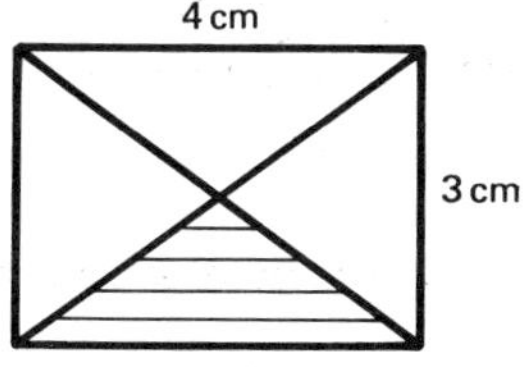

Fig. 21.4

8. Fig. 21.5 shows a parallelogram. Calculate:
   (a) its perimeter (b) its area.

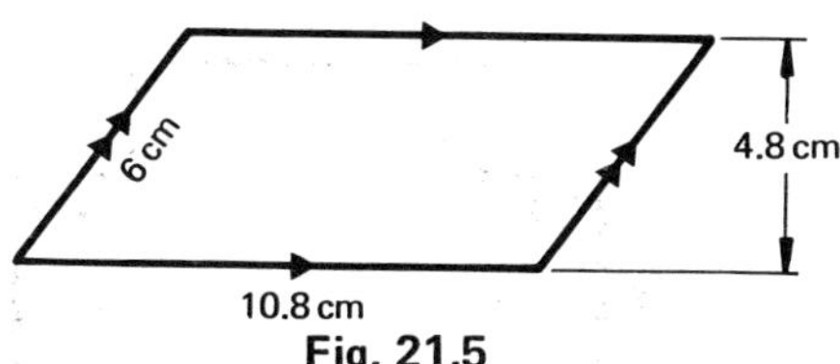

Fig. 21.5

9. A rectangular tablecloth is 150 cm wide and 200 cm long. Find the area of the tablecloth in square metres.

10. (a) Calculate the area of the sector shown in Fig. 21.6 giving the result correct to 2 significant figures.

(b) What is the length of the arc. (Take $\pi = 3\frac{1}{7}$.) (EA)

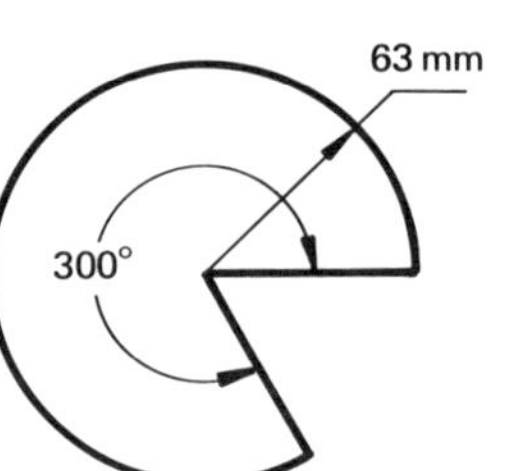

Fig. 21.6

11. In Fig. 21.7, ABCD is a rectangle 36 cm by 12 cm. The shaded triangles are cut off. Find the area of HEFGCD.

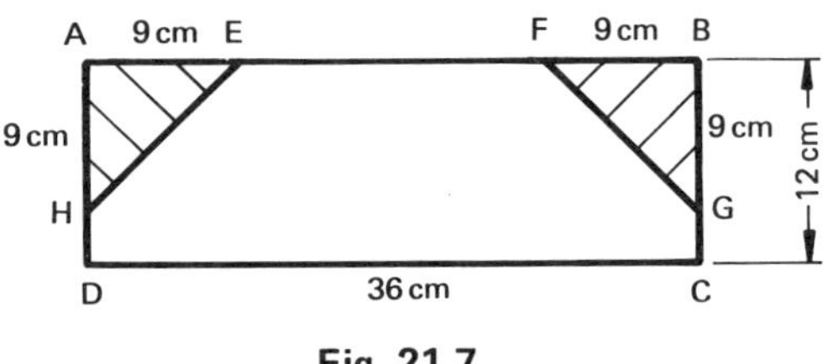

Fig. 21.7

12. The circumference of a circle is 72 cm. Find the length of an arc of this circle formed by an angle of 72° at the centre.

13. A square has an area of 81 cm^2. Calculate its perimeter. (S)

14. (a) Calculate the area of a circle with a diameter of 7 cm.

(b) Calculate the radius of a circle with a circumference of 88 cm. (Take $\pi = \frac{22}{7}$.)

Multi-choice questions 21

1. Find the perimeter of a square whose area is 36 cm^2.

   A 6 cm **B** 18 cm **C** 24 cm **D** 36 cm

2. Find the cost of a rectangular piece of wood 0.75 m by 2.7 m when wood costs £2.00 per square metre.

   A £40.50 **B** £6.75 **C** £4.05 **D** £0.40$\frac{1}{2}$

3. A rectangular lawn is 16 m long and 9 m wide. What is its area, in square metres?

   A 50 **B** 100 **C** 128 **D** 144

4. Taking $\pi = \frac{22}{7}$, the area of a circle of radius 14 cm is

A 176 cm² **B** 308 cm² **C** 352 cm² **D** 616 cm²

5. Taking $\pi = \frac{22}{7}$, the length of arc of the sector shown in Fig. 21.8 is

A $7\frac{1}{3}$ cm **B** $25\frac{2}{3}$ cm **C** 44 cm **D** 154 cm (AL)

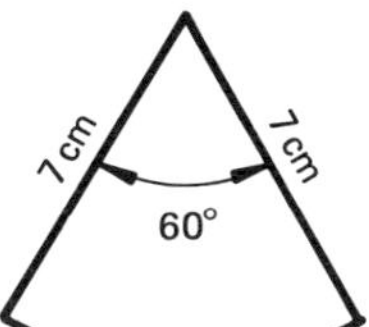

Fig. 21.8

6. How many square centimetres make 1 square metre?

A 10 **B** 100 **C** 1000 **D** 10 000 (AL)

7. In Fig. 21.9, ABCD is a rectangle. Calculate the area of ABED.

A 33 cm² **B** 40.5 cm² **C** 48 cm² **D** 52 cm²

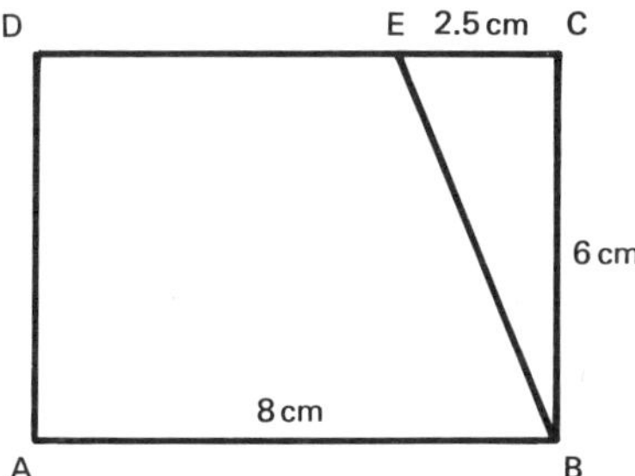

Fig. 21.9

8. A trapezium has parallel sides whose length are 18 cm and 22 cm. The distance between these parallel sides is 10 cm. What is the area of the trapezium?

A 200 cm² **B** 400 cm² **C** 495 cm² **D** 3960 cm²

9. A parallelogram has a base 10 cm long and a vertical height of 4 cm. What is its area in square centimetres?

A 14 **B** 20 **C** 28 **D** 40

10. An arc of a circle is 22 cm and the radius of the circle is 140 cm. Find the angle at the centre subtended by the arc. (Take $\pi = \frac{22}{7}$.)

A 9° **B** 18° **C** 90° **D** 180°

11. A ring has an outside diameter of 8 cm and an inside diameter of 4 cm. Its area, in square centimetres is

A $\pi(4-2)$ **B** $\pi(8-4)$

C $\pi(4^2-2^2)$ **D** $\pi(8^2-4^2)$

12. The triangle ABC (Fig. 21.10) is right-angled at B. What is its area?

A $8.5\ cm^2$ **B** $30\ cm^2$ **D** $32.5\ cm^2$ **D** $78\ cm^2$ (AL)

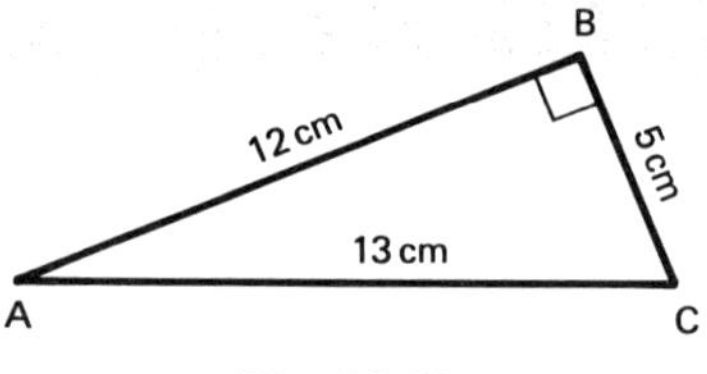

Fig. 21.10

22 VOLUMES AND SURFACE AREAS

- The VOLUME of a solid figure is found by seeing how many cubic units it contains.

  1 cubic metre is the volume contained inside a cube having a side of 1 metre; 1 cubic centimetre is the volume contained inside a cube having a side of 1 centimetre, etc. The standard abbreviations for units of volume are as follows

$$1 \text{ cubic metre} = 1 \text{ m}^3$$
$$1 \text{ cubic centimetre} = 1 \text{ cm}^3$$
$$1 \text{ cubic millimetre} = 1 \text{ mm}^3$$

- The SURFACE AREA of a solid figure is the total area of all its outside faces.

- Some FORMULAE FOR THE VOLUMES AND SURFACE AREAS OF SOLIDS are shown in Table 22.1.

Table 22.1

Solid	*Volume*	*Surface area*
Any solid having a uniform cross-section	Cross-sectional area × length of solid	Lateral surface + ends, i.e. (perimeter of cross-section × length of solid) + (total area of ends)
Cylinder	$\pi r^2 h$ 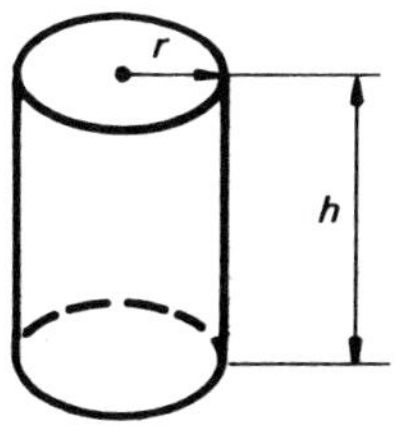	$2\pi r(h+r)$

Table 22.1 continued

Solid	*Volume*	*Surface area*
Cone	$\frac{1}{3}\pi r^2 h$ (h is the vertical height) 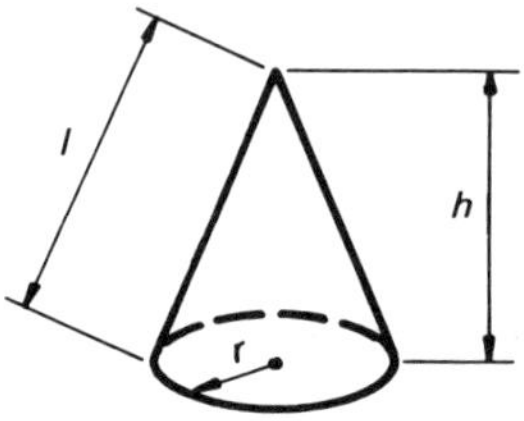	$\pi r l$ (l is the slant height)
Frustum of a cone	$\frac{1}{3}\pi h(R^2+Rr+r^2)$ (h is the vertical height) 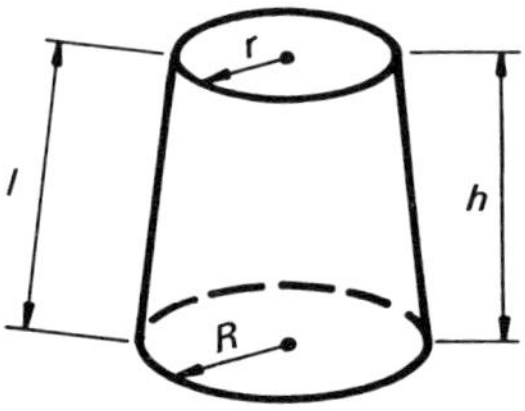	Curved surface area $=\pi l(R+r)$ Total surface area $=\pi l(R+r)+\pi R^2+\pi r^2$ (l is the slant height)
Sphere	$\frac{4}{3}\pi r^3$ 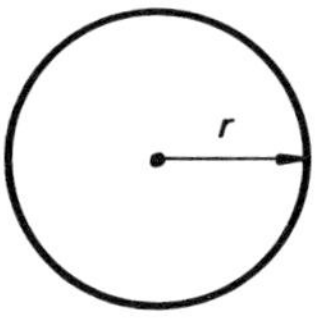	$4\pi r^2$
Pyramid	$\frac{1}{3}Ah$ 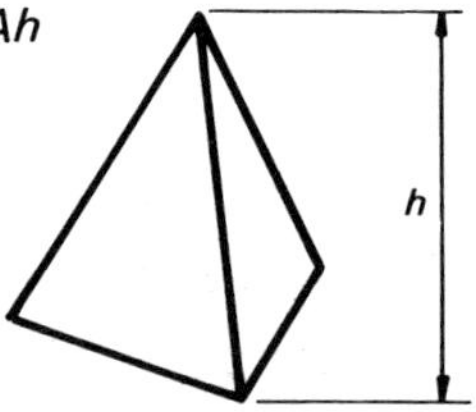	Sum of the areas of the triangles forming the sides plus the area of the base (A = area of base)

- The UNIT OF CAPACITY is the litre where

$$1 \text{ litre} = 1000 \text{ cm}^3$$

Example

(a) The solid shown in Fig. 22.1 has a constant cross-section which is a triangle having sides 15, 18 and 21 cm respectively. Calculate

(i) The lateral surface area of the solid

(ii) The total surface area of the solid

(iii) The volume of the solid.

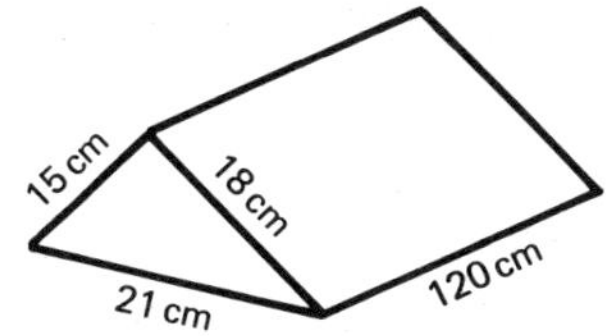

Fig. 22.1

(i) Perimeter of cross-section = $15 + 18 + 21 = 54$ cm.
Lateral surface area $= 54 \times 120 = 6480\ \text{cm}^2$.

(ii) To find the area of the ends use

$$A = \sqrt{s(s-a)(s-b)(s-c)}$$

$$s = \tfrac{1}{2}(15+18+21) = \tfrac{1}{2} \times 54 = 27\ \text{cm}$$

$$A = \sqrt{27 \times (27-15)(27-18)(27-21)}$$
$$= \sqrt{27 \times 12 \times 9 \times 6}$$
$$= 132.2\ \text{cm}^2$$

Total surface area = lateral surface area + area of ends
$= 6480 + (2 \times 132.2)$
$= 6480 + 264.4$
$= 6744\ \text{cm}^2$

(iii) Volume of solid = cross-sectional area × length
$= 132.2 \times 120 = 15\,864\ \text{cm}^3$

(b) A pyramid has a square base of side 8 cm and a vertical height of 12 cm. Calculate its volume.

To find the volume use the formula

Volume $= \tfrac{1}{3} \times$ area of base × vertical height
$= \tfrac{1}{3} \times (8 \times 8) \times 12$
$= \tfrac{1}{3} \times 64 \times 12$
$= 256\ \text{cm}^3$

(c) A rectangular tank is 4 m long, 2 m wide and 3 m high. How many litres of liquid can it hold?

In problems of this kind it is easiest to convert all the dimensions of the tank into centimetres, i.e. $400 \times 200 \times 300$ cm.

Volume = length × width × height
$= 400 \times 200 \times 300 = 24\,000\,000\ \text{cm}^3$

$$\text{Since 1 litre} = 1000\ \text{cm}^3$$

$$\text{Capacity of tank} = \frac{\text{volume}}{1000}$$

$$= \frac{24\,000\,000}{1000}$$

$$= 24\,000\ \text{litres}$$

Exercise 22.1

1. Find the volume and surface area of a block of wood which is 80 cm long, 20 cm wide and 15 cm high.

2. Find the volume of metal contained in a steel bar whose cross-section is shown in Fig. 22.2 if the bar is 200 mm long.

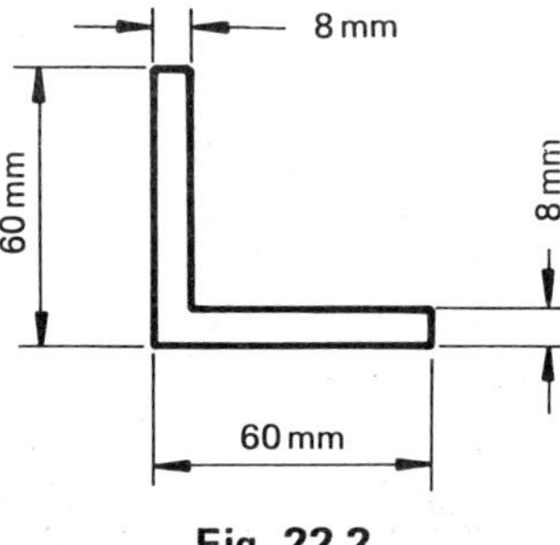

Fig. 22.2

3. Taking $\pi = \frac{22}{7}$, find the volume and total surface area of a cylinder whose radius is 14 cm and whose height is 20 cm.

4. Calculate the volume of metal contained in a pipe which has an outside diameter of 21 cm, a bore of 14 cm and a length of 200 cm. (Take $\pi = \frac{22}{7}$.)

5. Fig. 22.3 shows a triangular prism. Calculate its total surface area and its volume.

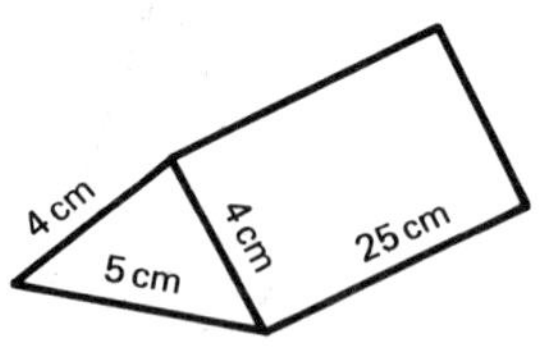

Fig. 22.3

6. Find the volume of a cone which has a base radius of 21 cm and a height of 12 cm. (Take $\pi = 3\frac{1}{7}$.)

7. A pyramid has a rectangular base 9 cm long and 6 cm wide. If its vertical height is 8 cm, calculate its volume.

8. A sphere has a diameter of 63 mm. Calculate its volume and surface area, taking $\pi = \frac{22}{7}$.

9. A rectangular tank is 3.2 m long, 1.8 m wide and 1.2 m high. How many litres of liquid will it hold?

10. A cylindrical water tank is 1.2 m high and it has a diameter of 0.9 m. How many litres of water does it hold when full? (Take $\pi = 3.14$.)

Exercise 22.2 (All of the type found in CSE examination papers)

1. Fig. 22.4 shows a prism ABCDEFGH with its cross-section FADE being a trapezium. Its dimensions are given in millimetres.
   (a) What is the area of the sloping face ABCD in square millimetres?
   (b) What is the area of the face FADE in square millimetres?
   (c) What is the volume of the prism in cubic millimetres?
   (d) What is the total surface area of the prism in square millimetres?
   (EA)

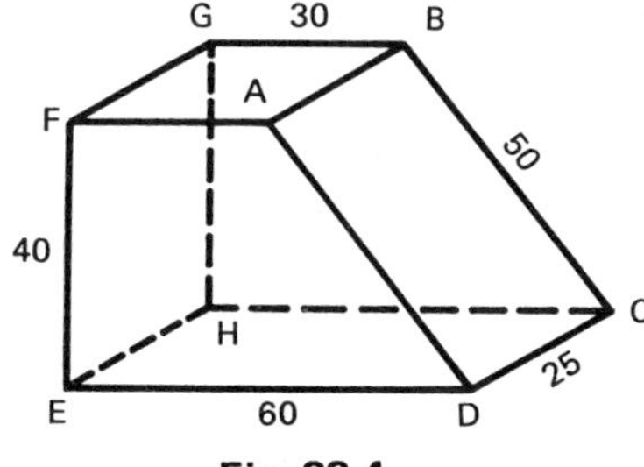

Fig. 22.4

2. Fig. 22.5 represents a solid cone whose base is a circle of radius 7 cm.
   (a) Using a value of $\pi = \frac{22}{7}$, calculate the area of the base.
   (b) If the volume of the cone is $308\,\text{cm}^3$, calculate the vertical height of the cone.

Fig. 22.5

3. What is the total surface area and the volume of a cube having edges 10 cm long? (EA)

4. The petrol tank of a car is a cuboid measuring 50 cm by 40 cm by 20 cm. How many litres will it hold when full. (EA)

5. A cube has a surface area of $216\ \text{cm}^2$. Find:
   (a) its volume (b) the total length of its edges. **(EA)**

6. A solid cylinder has a radius of 7 cm and a length of 10 cm. Calculate:
   (a) the area of the circular end
   (b) the volume of the cylinder. (Take $\pi = \frac{22}{7}$.)

7. A pyramid has a square base of side 3 cm and a vertical height of 5 cm. Calculate the volume of this pyramid.

8. A factory has its products packed in boxes 12 cm long, 4 cm wide and 6 cm high. These boxes are then packed in cartons 60 cm long, 20 cm wide and 30 cm high.
   (a) Calculate the volume of a carton.
   (b) How many boxes are needed to fill a carton completely?
   (c) If each box has a mass of 400 g and the full cartons are loaded on to a lorry which is only allowed to carry 10 t, how many boxes can be loaded? **(EA)**

9. Fig. 22.6 shows a prism of length 3 cm, width 1.7 cm and a vertical height 1.8 cm. Calculate:
   (a) the total surface area of the prism
   (b) the volume of the prism.

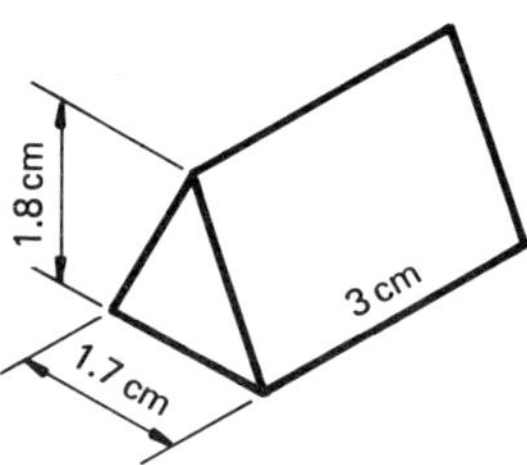

Fig. 22.6

10. A steel ingot whose volume is $2\ \text{m}^3$ is rolled out into a plate which is 15 mm thick and 1.75 m wide. Calculate the length of the plate.

Multi-choice questions 22

Fig. 22.7 represents a solid rectangular block. Use the diagram to answer questions 1, 2 and 3.

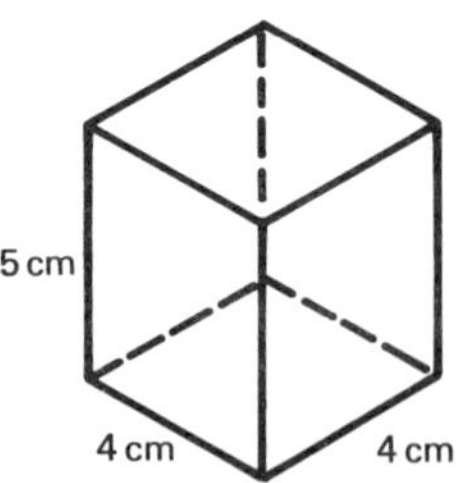

Fig. 22.7

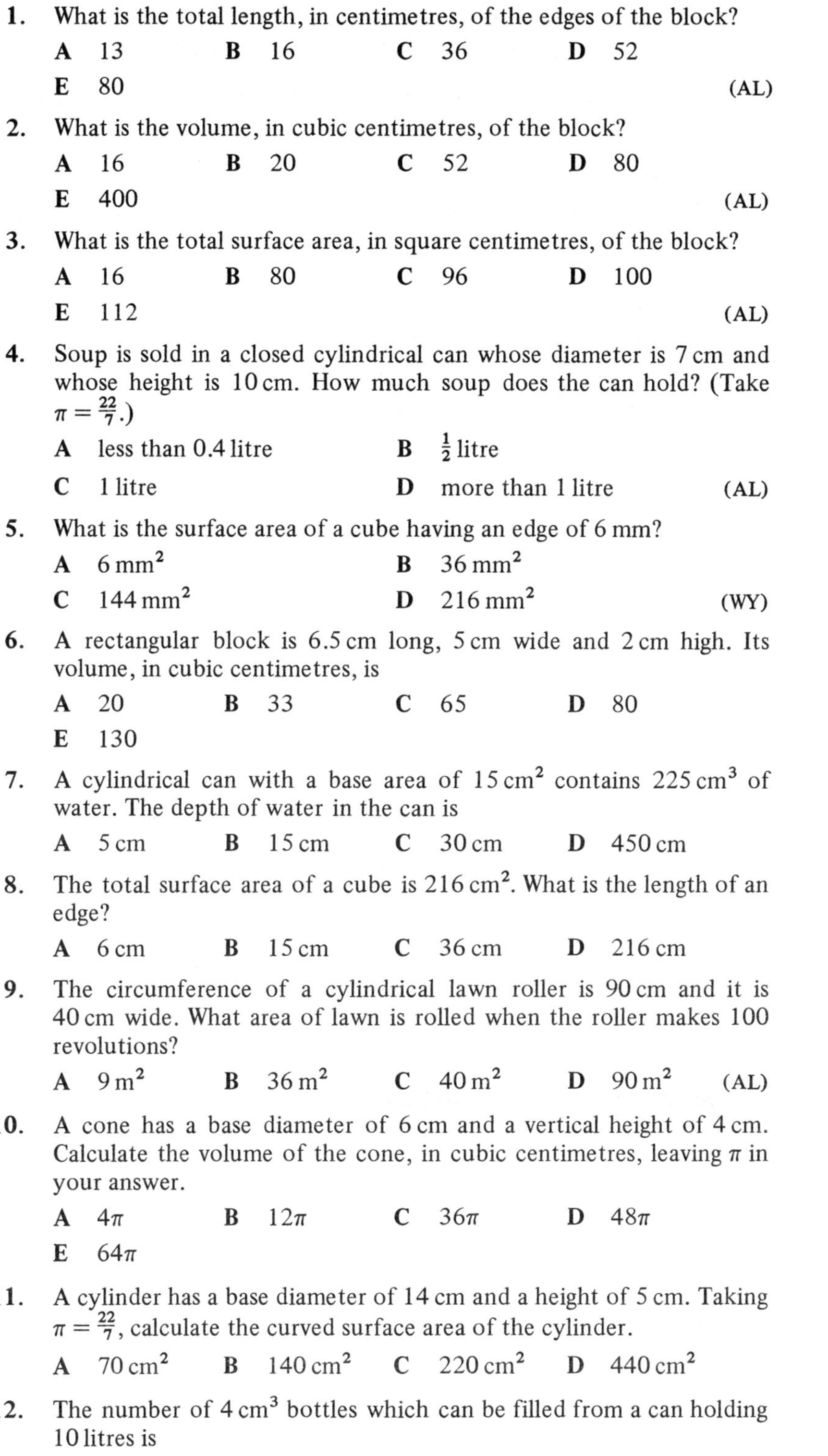

1. What is the total length, in centimetres, of the edges of the block?

A 13 **B** 16 **C** 36 **D** 52

E 80 (AL)

2. What is the volume, in cubic centimetres, of the block?

A 16 **B** 20 **C** 52 **D** 80

E 400 (AL)

3. What is the total surface area, in square centimetres, of the block?

A 16 **B** 80 **C** 96 **D** 100

E 112 (AL)

4. Soup is sold in a closed cylindrical can whose diameter is 7 cm and whose height is 10 cm. How much soup does the can hold? (Take $\pi = \frac{22}{7}$.)

A less than 0.4 litre **B** $\frac{1}{2}$ litre

C 1 litre **D** more than 1 litre (AL)

5. What is the surface area of a cube having an edge of 6 mm?

A $6\ \text{mm}^2$ **B** $36\ \text{mm}^2$

C $144\ \text{mm}^2$ **D** $216\ \text{mm}^2$ (WY)

6. A rectangular block is 6.5 cm long, 5 cm wide and 2 cm high. Its volume, in cubic centimetres, is

A 20 **B** 33 **C** 65 **D** 80

E 130

7. A cylindrical can with a base area of $15\ \text{cm}^2$ contains $225\ \text{cm}^3$ of water. The depth of water in the can is

A 5 cm **B** 15 cm **C** 30 cm **D** 450 cm

8. The total surface area of a cube is $216\ \text{cm}^2$. What is the length of an edge?

A 6 cm **B** 15 cm **C** 36 cm **D** 216 cm

9. The circumference of a cylindrical lawn roller is 90 cm and it is 40 cm wide. What area of lawn is rolled when the roller makes 100 revolutions?

A $9\ \text{m}^2$ **B** $36\ \text{m}^2$ **C** $40\ \text{m}^2$ **D** $90\ \text{m}^2$ (AL)

10. A cone has a base diameter of 6 cm and a vertical height of 4 cm. Calculate the volume of the cone, in cubic centimetres, leaving π in your answer.

A 4π **B** 12π **C** 36π **D** 48π

E 64π

11. A cylinder has a base diameter of 14 cm and a height of 5 cm. Taking $\pi = \frac{22}{7}$, calculate the curved surface area of the cylinder.

A $70\ \text{cm}^2$ **B** $140\ \text{cm}^2$ **C** $220\ \text{cm}^2$ **D** $440\ \text{cm}^2$

12. The number of $4\ \text{cm}^3$ bottles which can be filled from a can holding 10 litres is

A 25 **B** 250 **C** 2500 **D** 25 000

23 SCALES

- SCALES are used on maps and drawings of buildings, etc.

Scales may be expressed in one of two ways

(i) As a *ratio*, for example, 1 : 50 000. No units are involved when a scale is expressed in this way. Any distance measured off a map represents 50 000 times this distance on the ground.

Example

(a) The scale on a map is 1 : 25 000. Measured on the map, the length of a road is 20 cm. What is the true length of the road?

1 cm on the map represents 25 000 cm on the ground.

20 cm on the map represents 20 × 25 000 cm on the ground, i.e. $\frac{20 \times 25\,000}{100}$ m = 5000 m on the ground.

The length of the road is therefore 5000 m or 5 km.

(b) The scale on a map is 1 : 50 000. Measured on the ground, part of a motorway is 85 km. What distance will this be on the map?

50 000 cm = 500 m on the ground is represented by 1 cm on the map.

85 km = 85 000 m on the ground is represented by $\frac{85\,000}{500}$ = 170 cm.

The motorway is represented by 170 cm on the map.

(ii) As a *scale*, for example, 1 cm = 5 m. This means that 1 cm on a drawing or scale plan of a building represents an actual distance of 5 m.

Example

On the plan of a house, drawn to a scale of 1 cm = 2 m, the length of the lounge is 3 cm. What is the actual length of the lounge?

1 cm represents 2 m

3 cm represents 3 × 2 m = 6 m

Hence the actual length of the lounge is 6 m.

- When AREAS are to be found it is necessary to use the *square* of the scale.

Example

The plan of a church was made to a scale of 1 : 20. On the plan the area of the vestry is 750 cm². What is the actual area of the vestry?

$$750\text{ cm}^2 \text{ on the plan represents } 750 \times (20)^2\text{ cm}^2$$
$$= 300\,000\text{ cm}^2 = \frac{300\,000}{100 \times 100} = 30\text{ m}^2$$

- When VOLUMES are to be found it is necessary to use the *cube* of the scale.

Example

The model of a church is made to a scale of 1 : 25. The spire on the model has a volume of 0.5 m³. What is the actual volume of the spire?

$$0.5\text{ m}^3 \text{ on the model represents } 0.5 \times (25)^3\text{ m}^3$$
$$= 7810\text{ m}^3$$

Hence the spire has a volume of 7810 m³.

Exercise 23.1

1. The scale on a map is 1 : 20 000. What distance, in metres, does 4 cm on the map represent?

2. A street when measured on a plan made to 1 : 50 is 20 cm. What is its actual length in metres?

3. The plan of a house is made to a scale of 1 cm = 2 m. On the plan the kitchen measures 2 cm × 3 cm. What is the actual size of the kitchen?

4. The scale on a map is 2 cm = 1 km. The distance between two points on the map is 10.8 cm. How far apart, in kilometres, are the two points?

5. The plan of a house and garden is made to a scale of 1 : 50.
   (a) The length of the house is represented by 25 cm on the plan. Calculate its actual length.
   (b) The area of the kitchen floor on the plan is 90 cm². Calculate its actual area.
   (c) The height of the house is 7 m. What length will represent this height on the plan?

6. The model of a leisure centre is made to a scale of 1 : 20.
   (a) The model has a total length of 4 m. What is the actual length of the centre?
   (b) The actual floor area of a hall used for roller skating is $2000\,m^2$. What is this area on the model?
   (c) The volume of the building is $64\,000\,m^3$. What is the volume of the model?

7. The scale of a map is 5 cm to 1 km. On the map a field has an area of $250\,cm^2$. Calculate, in square kilometres, the actual area of the field.

8. A model train is made to a scale of 1 : 40. Its volume is $24\,cm^3$. What is the actual volume of the train?

Exercise 23.2 (All of the type found in CSE examination papers)

1. A map has a scale of 1 : 10 000. What is the length of a road, in metres, which is represented by 5 cm on the map. **(N)**

2. A photograph 5 cm high and 6 cm wide is enlarged to give a print 15 cm high. What is the area of the print? **(EA)**

3. A rectangular kitchen is $3\frac{1}{2}$ m wide, $2\frac{1}{2}$ m high and has a floor area of $14\,m^2$.
   (a) Calculate the length of the kitchen.
   (b) What is the volume of the kitchen?
   (c) The living room floor is a similar shape to the kitchen floor but it is 4 times its area. Find the dimensions of the living room floor. **(EA)**

4. The building plans of a house are drawn to a scale of 1 : 50.
   (a) Find the length of the house represented by a length of 20 cm on the plan.
   (b) Find the length, in centimetres, which represents the height of house on the plan, if the actual height of the house is 7 m.
   (c) The area of the kitchen floor on the plan is $80\,cm^2$. Find the area, in square metres, of the kitchen floor of the house. **(EA)**

5. The plan of a civic building is drawn to a scale of 1 : 20.
   (a) Calculate the actual length of the building, in metres, if the plan has a length of 2.3 m.
   (b) A window in the building is shown on the plan as having an area of $0.15\,m^2$. Find the actual window area.
   (c) The building has a volume of $4000\,m^3$. Find the volume of the building on the plan.

6. The plan of a house and garden is drawn to a scale of 1 : 50.
   (a) Determine the distance on the plan which represents an actual distance of 40 m.
   (b) On the plan, the width of the garden is shown as 75 cm. What is the actual width of the garden?
   (c) The dining room has an area of $30\,m^2$. Find the area in square centimetres of that part of the plan which represents the dining room.

24 GRAPHS

- To plot a graph the AXES OF REFERENCE are first drawn (Fig. 24.1). Their intersection, the point O, is called the *origin*. The vertical axis is frequently called the y-axis and the horizontal axis is then called the x-axis.

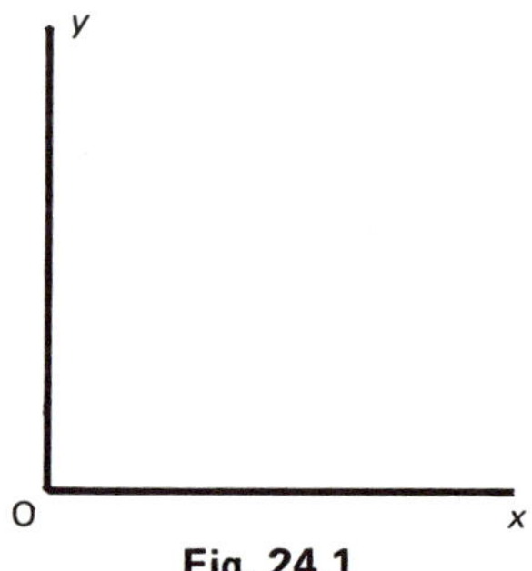

Fig. 24.1

- The number of units represented by a unit length along an axis is called the SCALE, for example, 1 cm = 5 units.

- COORDINATES are used to mark the points on a graph. In Fig. 24.2 the point P has been plotted so that $x = 4$ and $y = 6$. The values of 4 and 6 are called the ***rectangular coordinates*** of P. For brevity, the point P is said to be the point (4, 6). Note carefully that the value of x is always given first and because the order of the coordinates is important the values (4, 6) are called an *ordered pair*. Thus any pair of coordinates constitutes an ordered pair.

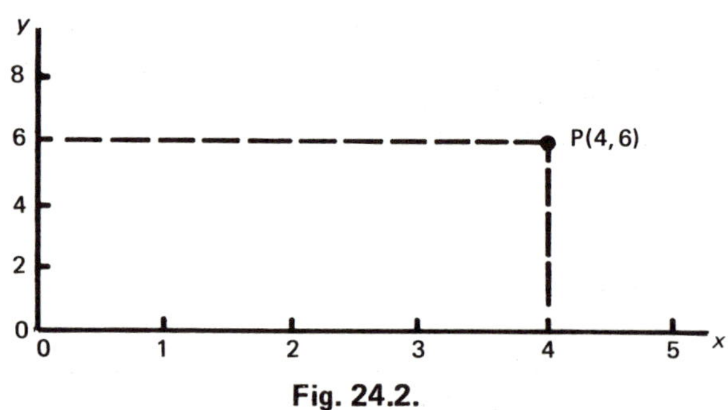

Fig. 24.2.

- Every GRAPH shows a relation between two sets of figures. The table below shows corresponding values of x and y.

x	0	2	4	6	8	10
y	4	10	16	22	28	34

To plot the graph we first draw the two axes of reference (Fig. 24.3). Suitable scales are then chosen to represent the values of x and y. Along the x-axis a scale of 1 large square to represent 2 units has been chosen while along the y-axis the scale is 1 large square to represent 5 units. The points are now plotted and we see that a straight line passes through all of them.

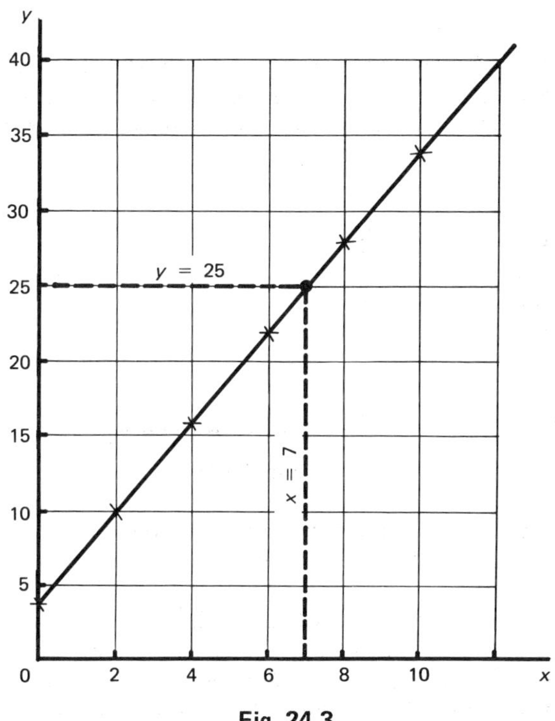

Fig. 24.3

When a graph is a straight line or a smooth curve it can be used to deduce corresponding values of x and y not given in the original table of values. Thus to find the value of y corresponding to $x = 7$, we draw the vertical and horizontal lines shown in the diagram and find that when $x = 7$, $y = 25$. Using a graph in this way to find values not given in the original table is called INTERPOLATION.

Exercise 24.1

1. Fig. 24.4 (p. 116) shows a graph of y plotted against x. Find:
   (a) the values of y when $x = 4$, $x = 7$ and $x = 11$
   (b) the values of x when $y = 7$, $y = 11$ and $y = 23$.
2. Fig. 24.5 (p. 116) shows a graph of distance plotted against time. From the graph, find:
   (a) the distances corresponding to times of 2, 3 and 5 s
   (b) the times corresponding to distances of 4, 28 and 54 m.

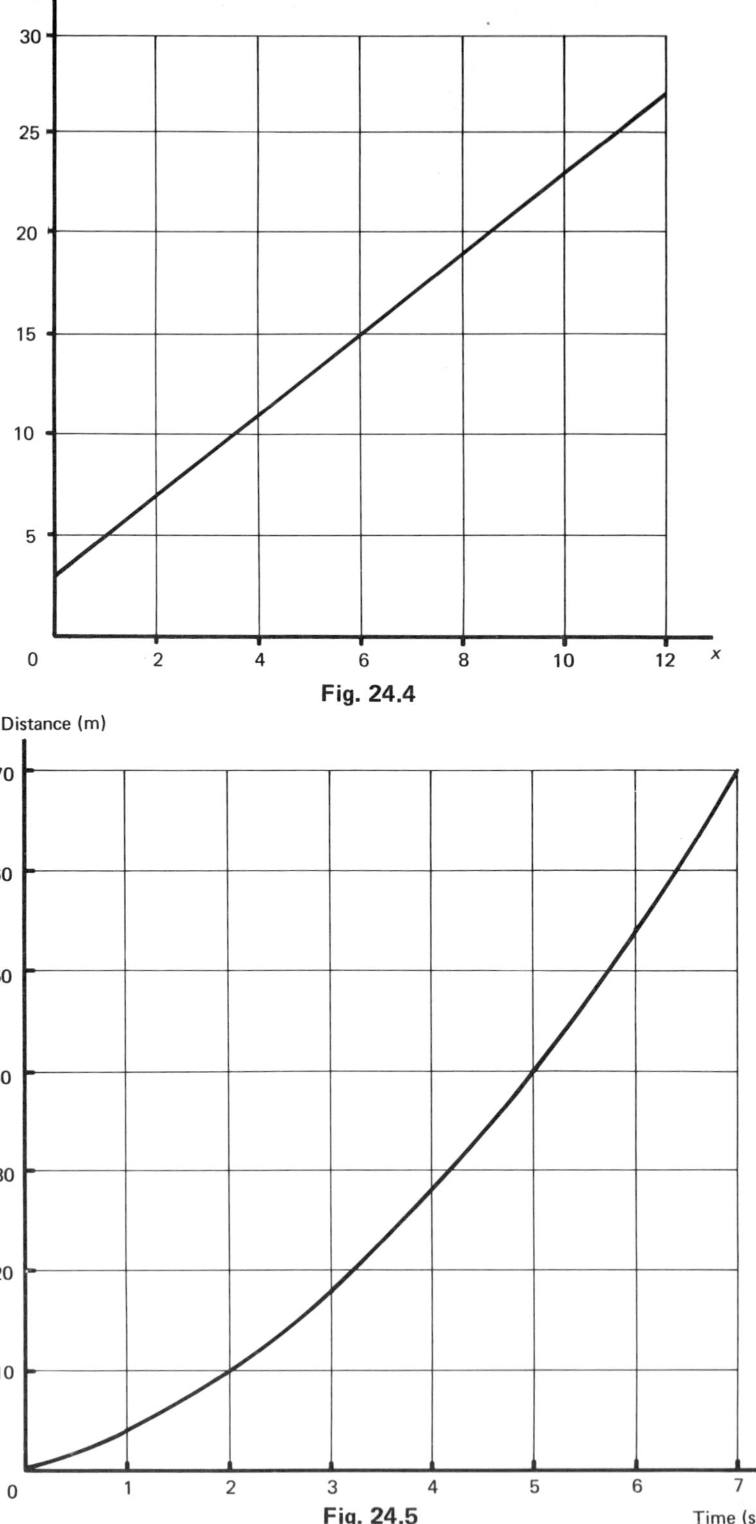

Fig. 24.4

Fig. 24.5

2	36
3	73
4	108
5	121
6	169
7	177
8	153
9	113
10	118
11	81
12	39

Score	(2 – 12)
No of poss. ways	
Prob	
decimal	
(theoretical) %	
actual frequency	
~~Predic~~ predicted %	

3. The values below show corresponding values of P and Q:

P	0	2	4	6	8	10
Q	2	12	22	32	42	52

Plot P horizontally and Q vertically using scales of 1 large square to 2 units horizontally and 1 large square to 10 units vertically. Draw the graph and use it to find the value of Q when $P = 5$ and the value of P when $Q = 37$.

4. The figures below show corresponding values for the units of electricity used and the cost in £. Draw a graph of this information with units used on the horizontal axis and cost on the vertical axis. Use a scale on the horizontal axis of 1 large square to represent 200 units and 1 large square to represent £5 on the vertical axis.

Units used	0	300	600	900	1200
Cost £	2	8	14	20	26

Use your graph to find:

(a) the cost of 800 units

(b) the number of units used when the cost is £12.

5. The speed of a body (v m/s) measured at various times (t s) was as follows:

t	2	4	6	8	10	12
v	6.4	7.7	9.0	10.3	11.7	13.0

Plot a graph of this information with t on the horizontal axis. Hence estimate the speed when $t = 7$ s and the time when $v = 9.5$ m/s. (Scales: horizontally use 1 large square to represent 2 s and vertically use 1 large square to represent 2 m/s.)

- A relation may also be shown in the form of a MAPPING DIAGRAM. The mapping diagram in Fig. 24.6 shows the relation between X and Y. The set of starting elements 0, 2, 4, 6, 8 is called the *domain* and the set of finishing elements 3, 7, 11, 15, 19 is called the *range*. The values connected by the arrows constitute ordered pairs and the relation may be shown in the form of a graph (Fig. 24.7, p. 118).

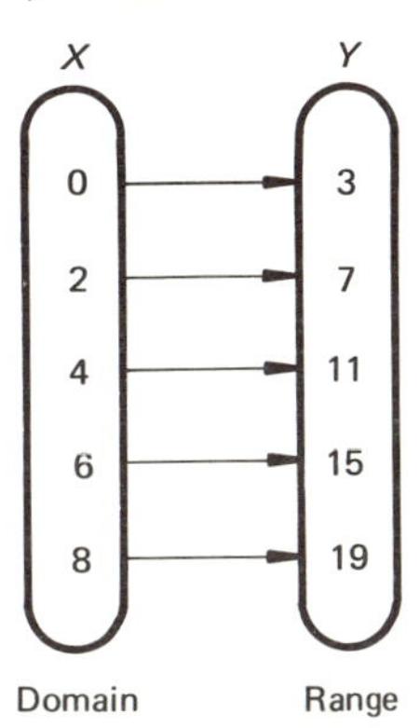

Fig. 24.6

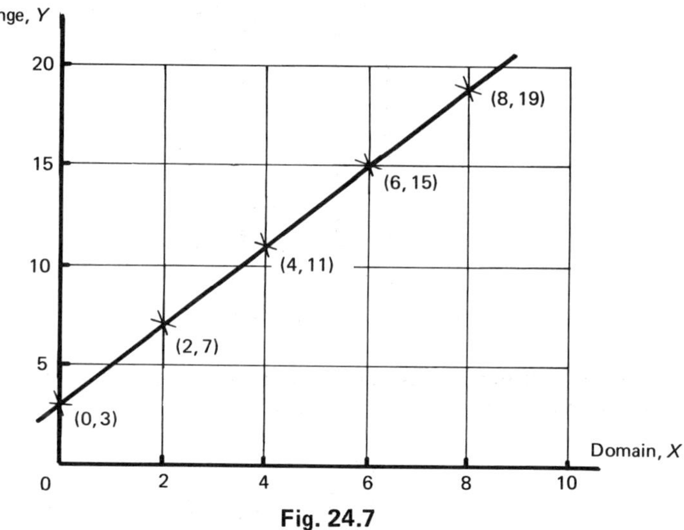

Fig. 24.7

Fig. 24.8 shows the domain and range which are related by $x \longmapsto 2x+5$ which is read as 'x is mapped on to $2x+5$'. The elements of the range are obtained by substituting the values of x in the domain into the expression $2x+5$. Thus when

$$x = 0: \quad 2x+5 = (2\times 0)+5 = 5$$
$$x = 1: \quad 2x+5 = (2\times 1)+5 = 7$$
$$x = 2: \quad 2x+5 = (2\times 2)+5 = 9$$
$$x = 3: \quad 2x+5 = (2\times 3)+5 = 11$$

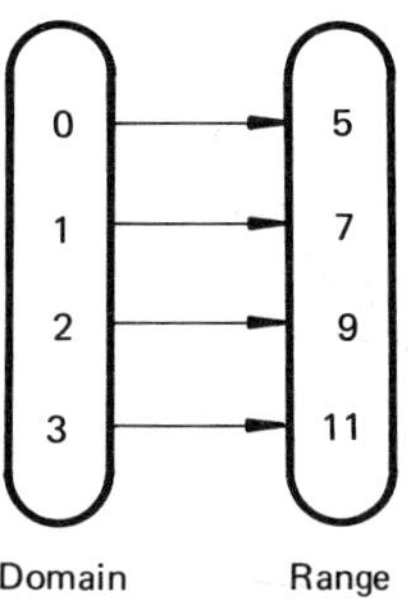

Fig. 24.8

- A FUNCTION is a relation in which one, and only one, arrowed line in a mapping diagram leaves each member of the domain. Thus the relations shown in Figs. 24.6 and 24.8 are functions.

In Fig. 24.9, only one arrowed line leaves each element of the domain and although two arrowed lines arrive at one of the elements of the range this relation is still a function. Such a diagram represents a *many-to-one mapping*.

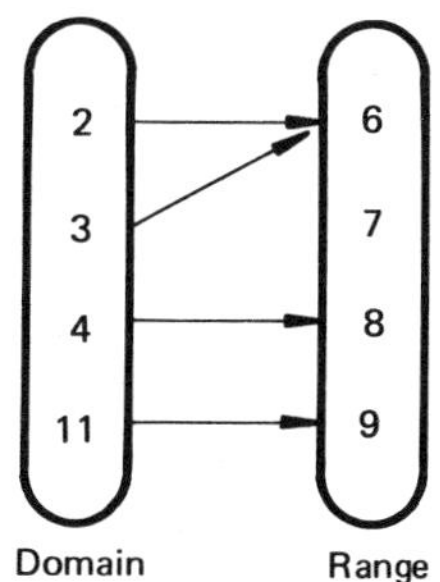

Fig. 24.9

The relation shown in Fig. 24.10 is *not* a function because two arrowed lines leave one of the elements in the domain.

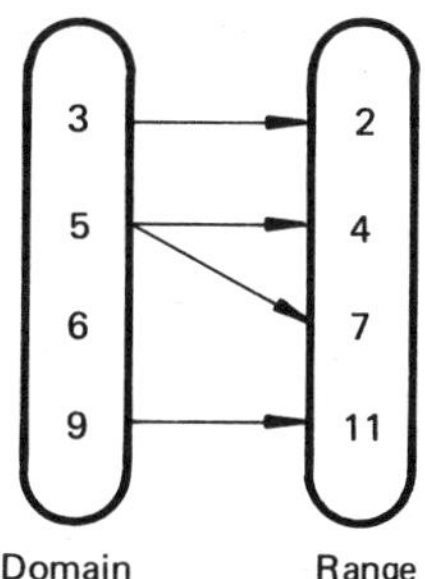

Fig. 24.10

If a relation is not a function then a vertical line will pass through two ordered pairs when the graph is drawn (Fig. 24.11).

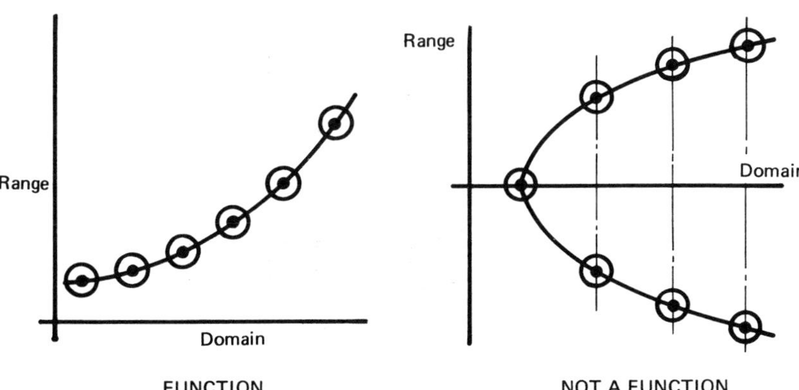

Fig. 24.11

- The FUNCTION NOTATION

$$f: x \longmapsto 3x+1$$

or

$$f(x) = 3x+1$$

is used to define the relation $x \longmapsto 3x+1$.

- Equations of the type $y = 2x + 7$, in which the highest power of x is the first, are called LINEAR FUNCTIONS because their graphs are straight lines.

Example

Plot the graph of $f(x) = 2x - 9$ for values of x between -2 and $+5$.

In order to draw the graph of a linear function only two points are needed but it is safer to plot three points, the third point acting as a check on the other two.

x	-2	0	$+5$
$f(x)$	-13	-9	$+1$

(Note that when $x = -2$, $f(-2) = 2 \times (-2) - 9 = -13$, etc.)

The graph is plotted in Fig. 24.12

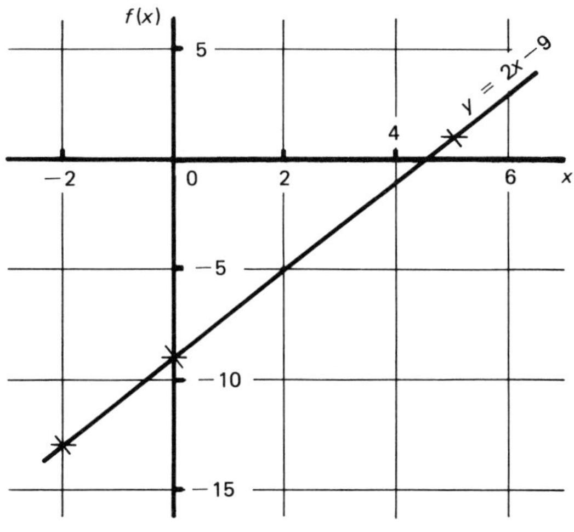

Fig. 24.12

- Every linear function can be written in the STANDARD FORM $y = mx + c$, where m is the gradient of the straight line and c is the intercept on the y-axis.

Example

Fig. 24.13 shows the graph of a linear function. Find its equation.

To find the gradient (i.e. the value of m) draw the right-angled triangle ABC.

$$\text{Gradient} = \frac{BC}{AC} = \frac{16}{4} = 4$$

Hence $m = 4$

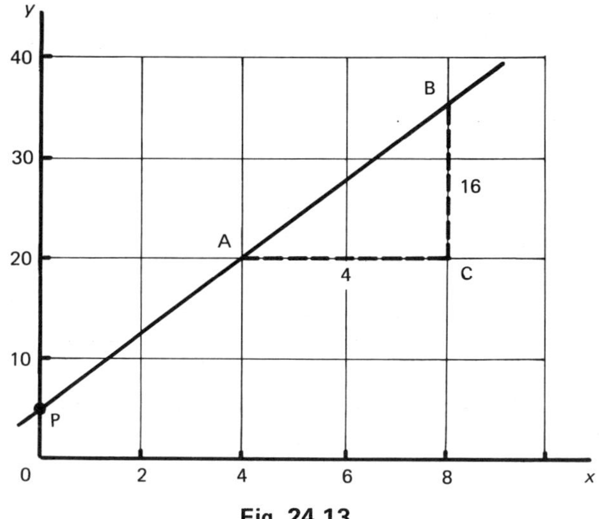

Fig. 24.13

The point P is where the straight line cuts the y-axis and the y-coordinate of P is called the intercept on the y-axis. The y-coordinate of P is 5 and hence $c = 5$. Therefore the equation of the straight line is

$$y = 4x + 5$$

Example

(a) The line $y = 3x + c$ passes through the point $(2, 8)$. Find the value of c.

When $x = 2$, $y = 8$, hence

$$8 = (3 \times 2) + c$$

$$8 = 6 + c$$

$$c = 2$$

(b) The straight line $y = mx + c$ passes through the points $(0, 2)$ and $(3, 17)$. What are the values of m and c?

Since the line passes through the point $(0, 2)$, the intercept on the y-axis is 2; hence $c = 2$.

The equation of the straight line is now

$$y = mx + 2$$

Since $y = 17$ when $x = 3$, hence

$$17 = 3m + 2$$

$$3m = 15$$

$$m = 5$$

Exercise 24.2

1. Copy and complete Fig. 24.14 if the relation is:
   (a) $x \mapsto 7x-4$ (b) $x \mapsto 3x$.

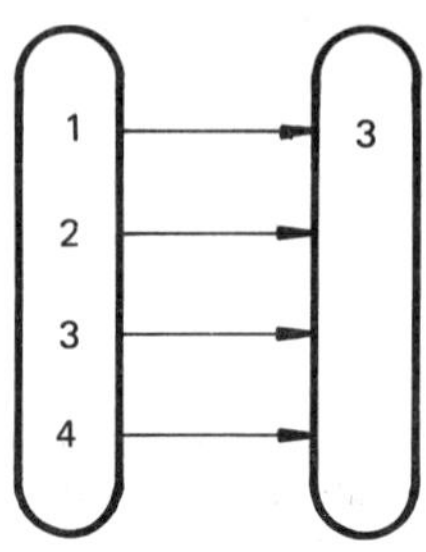

Fig. 24.14

2. For the relation $x \mapsto x^3$, what is 5 mapped on to?
3. With 1, 3, 4 and 7 as domain draw mapping diagrams for:
   (a) $x \mapsto 7-2x$ (b) $x \mapsto x^2-2x+3$.
4. If $f(x) = 5x-4$, find:
   (a) $f(-3)$ (b) $f(0)$ (c) $f(4)$.
5. If $f: x \mapsto 2x^2-3x$, find:
   (a) $f(-3)$ (b) $f(-1)$ (c) $f(0)$
   (d) $f(2)$ (e) $f(5)$.
6. Draw the graph of $f(x) = 7x-8$ for values of x between -3 and $+2$.
7. Draw the graph of $f(x) = 6-5x$ for values of x between -4 and $+3$.
8. Fig. 24.15 shows the graph of a straight line. Find its equation in the form $y = mx+c$.

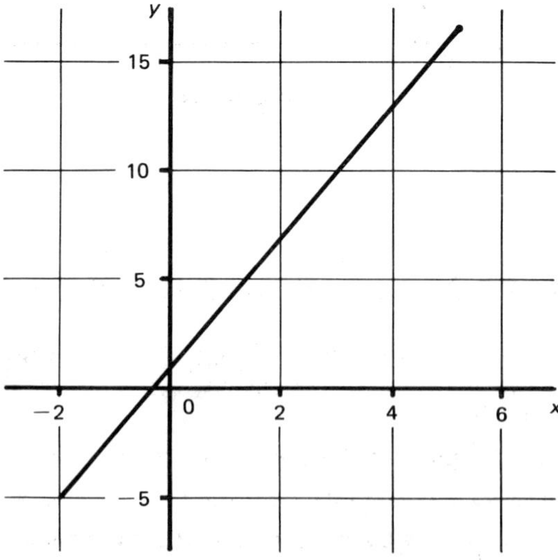

Fig. 24.15

9. The table below gives corresponding values of x and y:

x	5	7	8	12
y	7	11	13	21

Plot the graph and find the equation giving the relation between x and y. (Scales: horizontal 1 cm = 2; vertical, 1 cm = 5.)

10. The straight line $y = 3x + c$ passes through the point (5, 13). Find the value of c.

11. Using a scale of 1 cm to 1 unit on the x-axis and 1 cm to 2 units on the y-axis, draw a pair of axes on graph paper for values of x from -5 to $+5$ and for values of y from -10 to $+10$.

(a) Plot the points P(3, 10) and Q(-4, -4). Join the points P and Q.

(b) Use your graph to find the equation of the line PQ in the form $y = mx + c$.

12. The equation of the straight line AB (Fig. 24.16) may be expressed in the form $y = mx + c$. What is the value of m and what is the value of c?

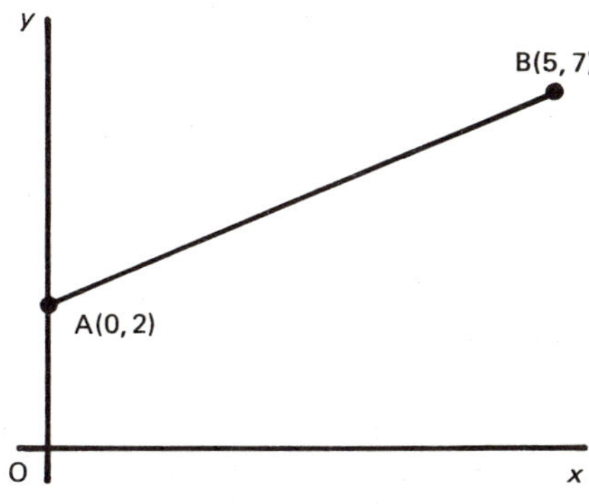

Fig. 24.16

Exercise 24.3 (All of the type found in CSE examination papers)

1. The graph shown in Fig. 24.17 (p. 124) shows the relation between the number of units used and the cost of a telephone bill. Use the graph to find:

(a) The telephone bill when 360 units are used.

(b) The number of units used when the bill is £14.75.

(c) If the bill is made up of a rental charge plus so much per unit used, find the rental charge. (EA)

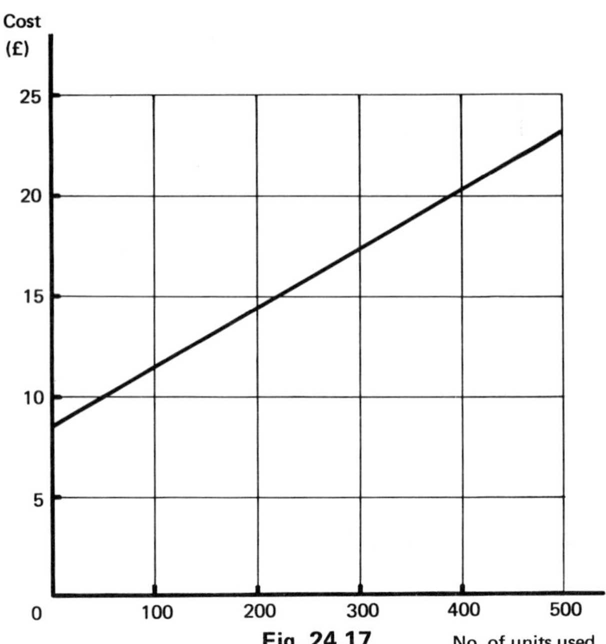

Fig. 24.17

2. Using a scale of 1 cm to 1 unit on the x-axis and 1 cm to 2 units on the y-axis, construct a pair of axes on graph paper for values of x from -6 to $+6$ and for values of y from -10 to $+10$.

   (a) On your axes plot the points A$(4, 8)$ and B$(-3, 1)$. Join A and B.

   (b) Use your graph to find:
   (i) The coordinates of the point where the line AB crosses the y-axis.
   (ii) The gradient of the line AB.

   (c) (i) On your axes plot the graph of $y = 7 - 2x$.
   (ii) Give the coordinates of the point where this line intersects the line AB.

3. The equation of the straight line AB (Fig. 24.18) is expressed in the form $y = mx + c$. Given A$(0, 3)$ and B$(6, 6)$:

   (a) What is the value of: (i) c (ii) m?

   (b) Calculate the area of the triangle ABC. (W)

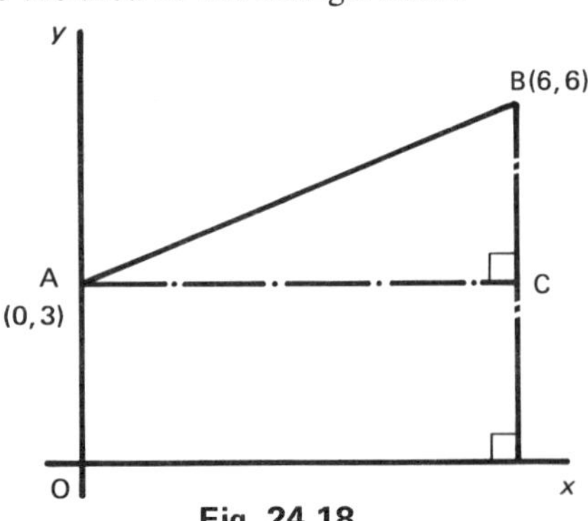

Fig. 24.18

4. In Fig. 24.19, RP is the line $y = 3x + 2$ and ON $= 4$ units. Calculate:
   (a) the length of PN (b) the length OR.

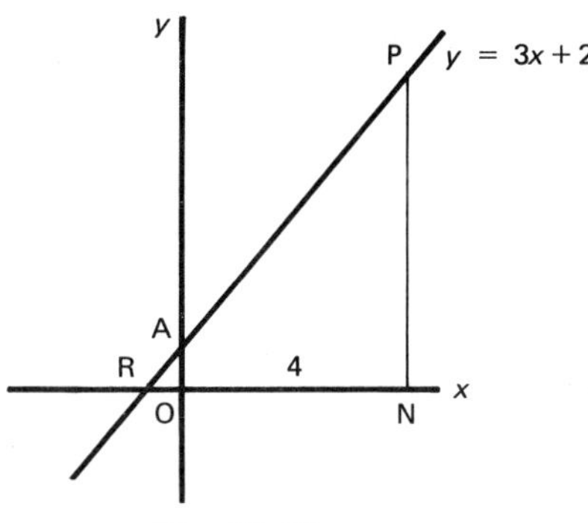

Fig. 24.19

5. Complete the mapping shown in Fig. 24.20 if $x \mapsto 3x - 1$. (S)

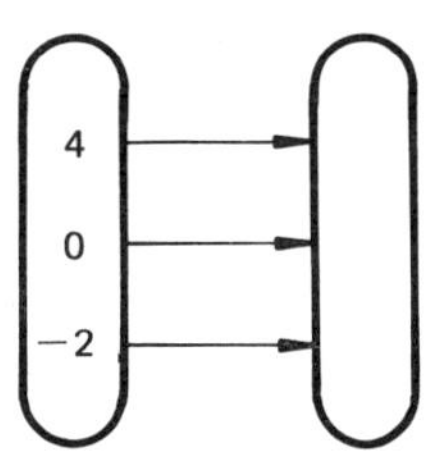

Fig. 24.20

6. The line $y = 5x + c$ passes through the point $(2, 12)$. Find the value of c. (WM)

7. Using the graph shown in Fig. 24.21:
   (a) Give the coordinates of point A.
   (b) What is the gradient of the line AB?
   (c) What is the equation of the line AB?
   (d) Plot on the graph the points D(3, 9) and E(−2, 4). (EA)

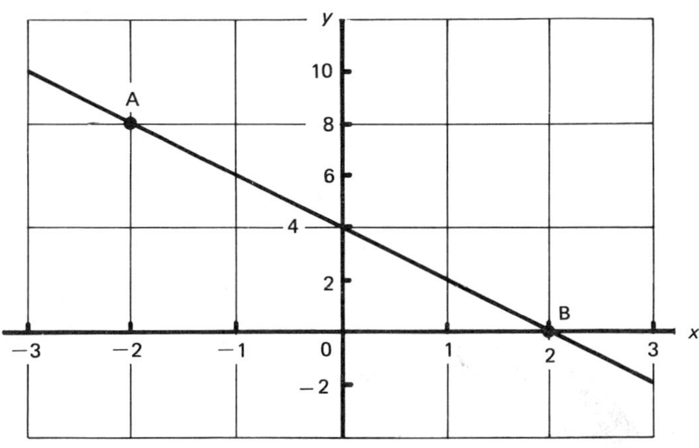

Fig. 24.21

8. Fig. 24.22 shows the straight line passing through the points A and B, whose equation is $y = \frac{1}{4}x + 5$.
   (a) Write down the coordinates of the point A.
   (b) Write down the coordinates of the point B.
   (c) Calculate the area of the triangle AOB.

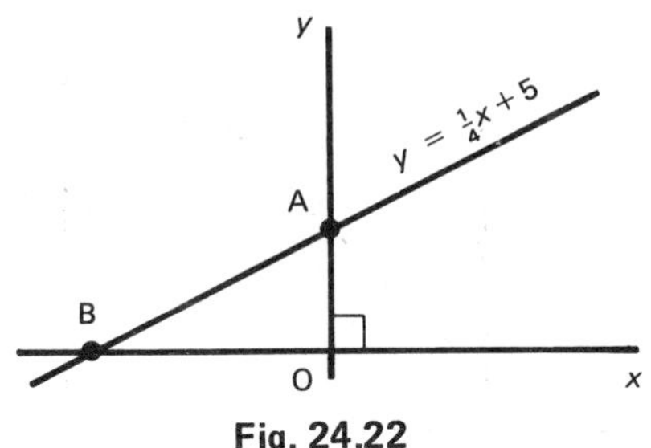

Fig. 24.22

Multi-choice questions 24

Fig. 24.23 shows the straight line $y = 2x + 1$. The points $P(2, a)$ and $Q(b, 9)$ both lie on this line. Use this information to answer questions 1, 2 and 3.

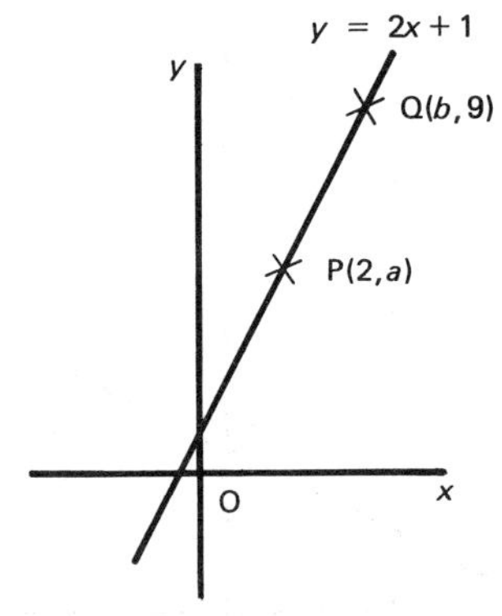

Fig. 24.23

1. What is the value of a?
   A 1   **B** 2   **C** 3   **D** 5   (AL)

2. What is the value of b?
   A 4   **B** 5   **C** 6   **D** 16   (AL)

3. What is the intercept of the straight line on the y-axis?
   A -1 **B** $\frac{1}{2}$ **C** 1   **D** 2   (AL)

4. In Fig. 24.24, if BM = 29, then the length of OM is

A 7 **B** 10 **C** 13 **D** 16

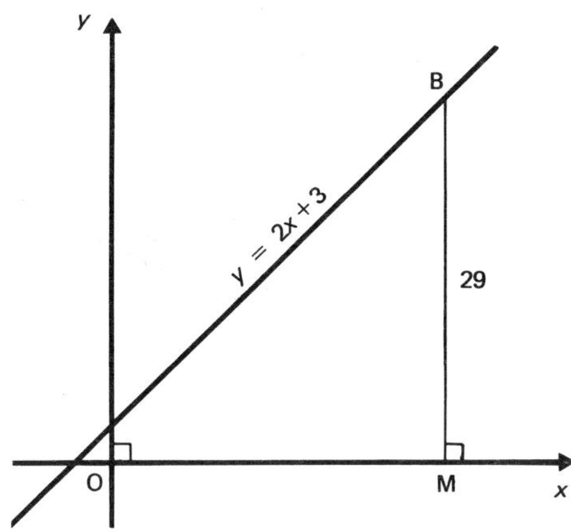

Fig. 24.24

5. The equation of the straight line shown in Fig. 24.25 is

A $y = x - 2$ **B** $y = x - 4$ **C** $y = x + 2$

D $y = 2x + 2$ **E** $y = 2x - 2$

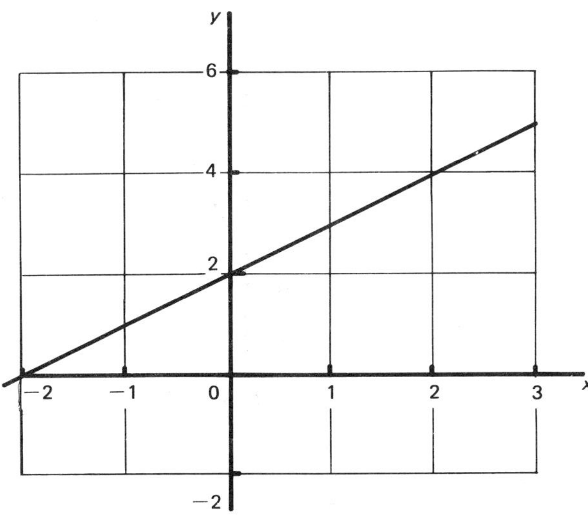

Fig. 24.25

6. Fig. 24.26 (p. 128) shows the height of water in a tank as it is being filled. By how many centimetres does the water rise in 3 s?

A $1\frac{1}{2}$ **B** 2 **C** $2\frac{1}{2}$ **D** 3

E $3\frac{1}{2}$

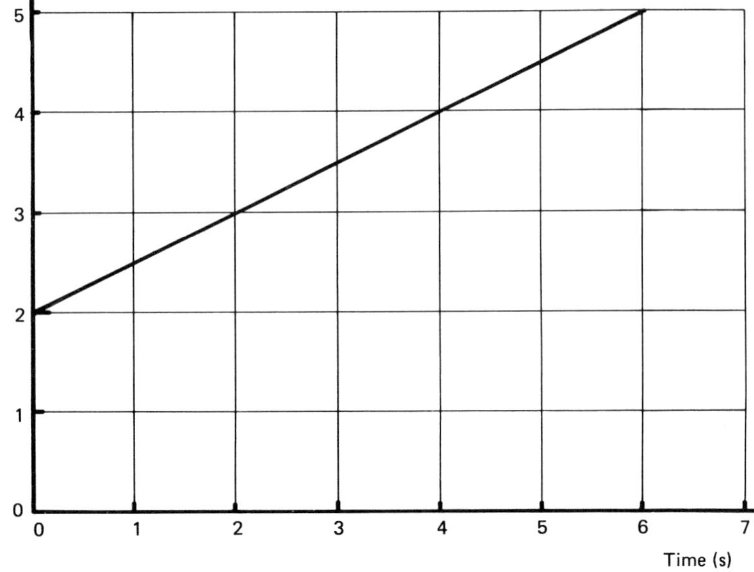

Fig. 24.26

7. $y = 4x - 5$ is the equation of a straight line. The gradient of the line is

   A -5 **B** -4 **C** 4 **D** 5

8. The equation of the straight line shown in Fig. 24.27 is

   A $y = 5 - 2\frac{1}{2}x$ **B** $y = 2\frac{1}{2}x - 5$

   C $y = 2\frac{1}{2}x - 2$ **D** $y = 2\frac{1}{2}x + 5$

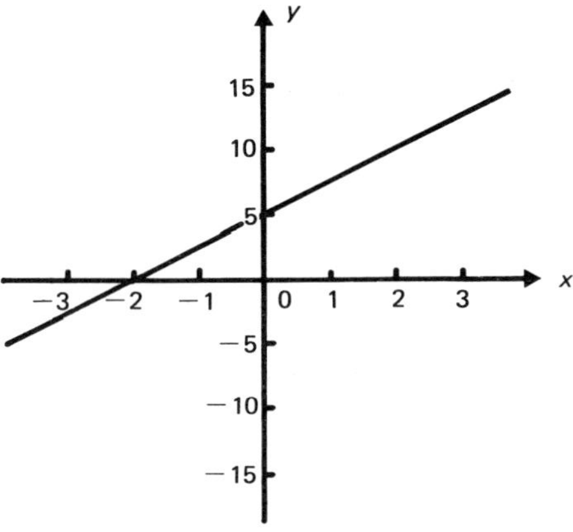

Fig. 24.27

25 INEQUALITIES

- The following SYMBOLS are used when dealing with inequalities

$>$ means 'greater than'

$\geqslant$ means 'equal to or greater than'

$<$ means 'less than'

$\leqslant$ means 'equal to or less than'

Note that the 'arrow' always points to the smaller quantity.

- The SOLUTIONS OF INEQUALITIES may be shown by means of a number line.

If $x<3$, all the possible values of x are shown in Fig. 25.1. The empty circle at the end of the line shows that $x=3$ is not included.

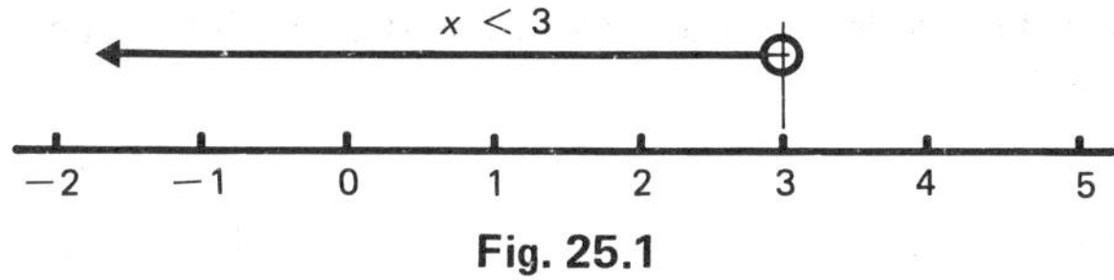

Fig. 25.1

Fig. 25.2 shows all the solutions for $x \geqslant -4$. Since $x=-4$ is included, the line ends in a solid circle.

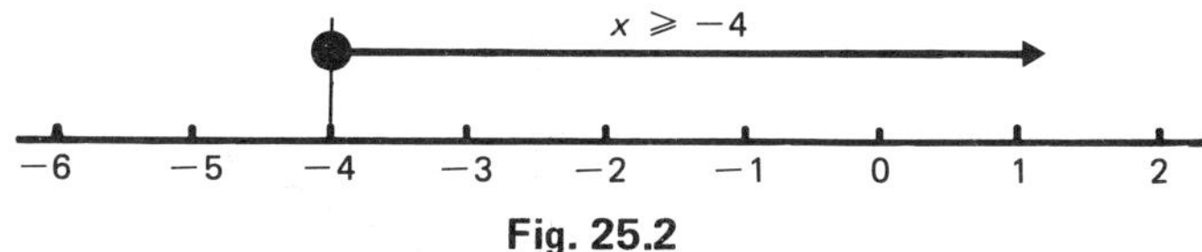

Fig. 25.2

Example

If x has to be a whole number, find the solution for $x \leqslant 3$ and $x>1$.

Representing both inequalities on a number line (Fig. 25.3) we see that the solution is that x must equal 2 or 3, because the arrowed lines representing the independent solutions for each inequality overlap. We say that the *solution set* is $\{2,3\}$.

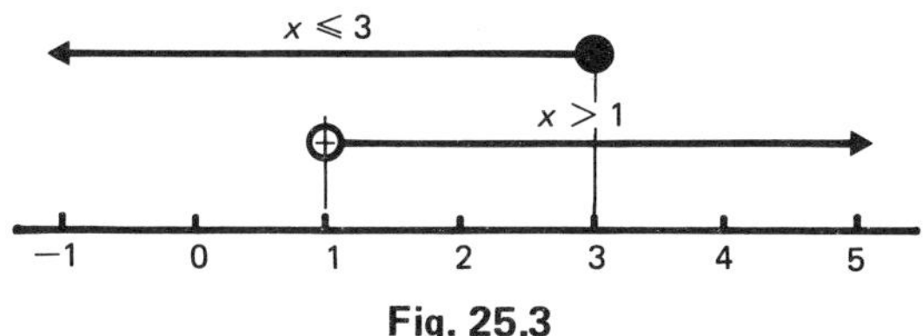

Fig. 25.3

- Inequalities may be COMBINED. For instance if we have $4 < x$ and $x < 7$, the inequalities can be combined to give $4 < x < 7$, which means that the value of x lies between 4 and 7.

Example

If x is an integer find the solution set for $-3 < x < 2$.

Drawing a number line (Fig. 25.4) we see that the solution set is $\{-2, -1, 0, 1\}$.

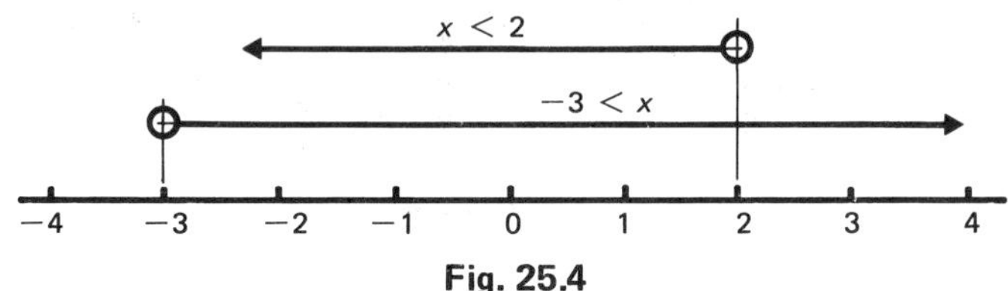

Fig. 25.4

- In SOLVING INEQUALITIES the following rules must be observed.
  (i) The same number may be added to or subtracted from both sides of the inequality.
  (ii) Multiplying or dividing both sides of the inequality by the same positive number leaves the inequality unaltered.

Example

Solve the inequality $5x + 17 > 2x + 29$.

Bringing the terms in x to the LHS and taking the numbers to the RHS gives

$$5x - 2x > 29 - 17$$

$$3x > 12$$

$$x > 4$$

- Linear inequalities may be illustrated by means of GRAPHS.

Example

Illustrate on a graph the inequalities:

(a) $x > -2$ (b) $y \leqslant 10$ (c) $y > 2x + 1$.

(a) In Fig. 25.5 draw the line $x = -2$. Since the solution does not include the points on the line $x = -2$, this is shown dashed. The solution of the inequality is shown by the shaded region.

(b) In Fig. 25.6 draw the line $y = 10$. Since the solution includes points on the line $y = 10$, this is drawn as a full line. The solution of the inequality is shown by the shaded region.

(c) In Fig. 25.7 draw the dashed line $y = 2x + 1$. The solution of the inequality is shown by the shaded region.

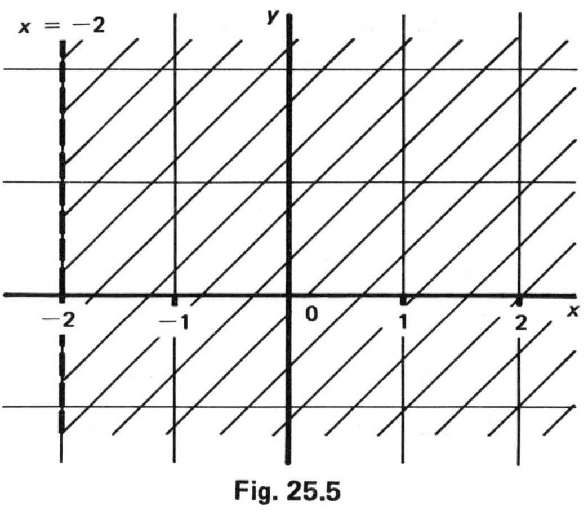

Fig. 25.5

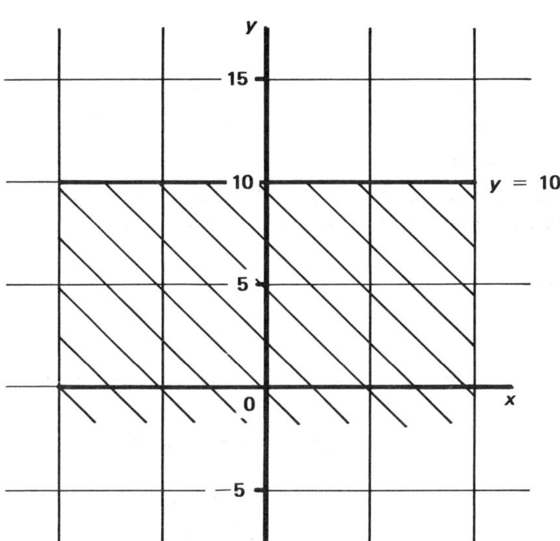

Fig. 25.6

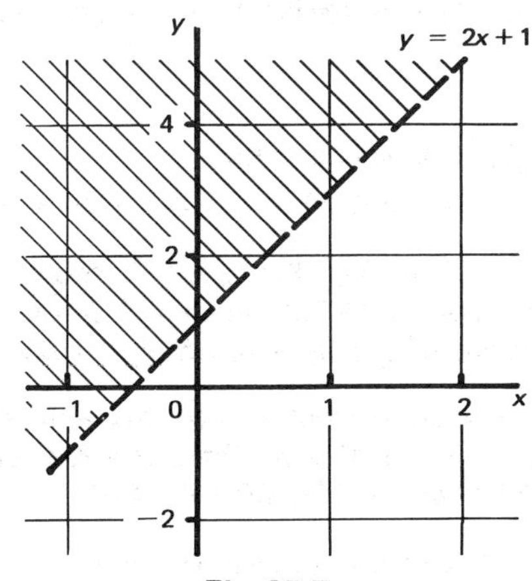

Fig. 25.7

Exercise 25.1

1. Insert one of the signs <, > or = between the following pairs of values so as to make true statements:

   (a) 2 5 (b) $\frac{3}{4}$ $\frac{1}{2}$ (c) $\frac{7}{8}$ $\frac{21}{24}$

   (d) $\frac{2}{3}$ $\frac{4}{9}$ (e) $\frac{3}{8}$ $\frac{11}{32}$

2. Use number lines to show the solutions for the following inequalities:

   (a) $x \leqslant 4$ (b) $x < 7$ (c) $x \geqslant 3$ (d) $x > -2$.

3. Find the solution sets for the following pairs of inequalities if x has to be an integer:

   (a) $4 < x < 9$ (b) $-2 \leqslant x < 3$

   (c) $5 \leqslant x \leqslant 8$ (d) $-4 < x < 5$.

4. Solve the following inequalities:

   (a) $3x \geqslant 12$ (b) $3x - 4 \leqslant 11$

   (c) $x - 5 > 7$ (d) $5x - 3 > 2x + 15$.

   (e) $5(x-2) < 15$ (f) $5(x+2) - 3(x-5) \geqslant 29$

   (g) $4(x-5) > 7 - 5(3-2x)$.

5. State the inequalities which are represented by the shaded regions shown in Fig. 25.8.

6. Represent the following inequalities on graph paper:

   (a) $y > 3x + 2$ (b) $y \leqslant 5 - 2x$ (c) $y < 2x - 3$.

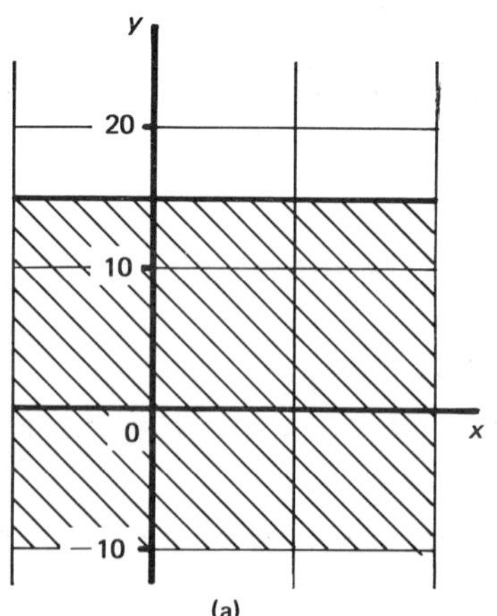

(a)

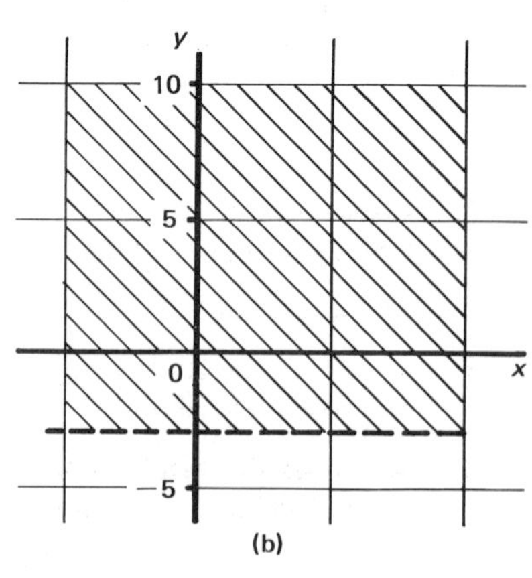

(b)

(Fig. 25.8 cont.)

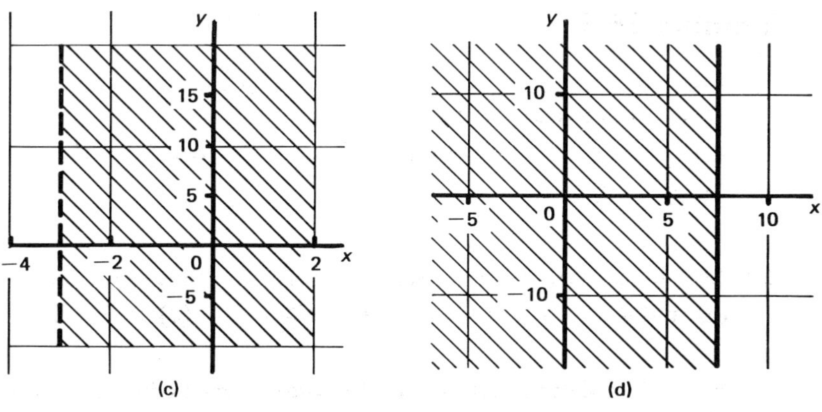

Fig. 25.8

Exercise 25.2 (All of the type found in CSE examination papers)

1. State the inequalities which represent the shaded regions shown in Fig. 25.9. (S)

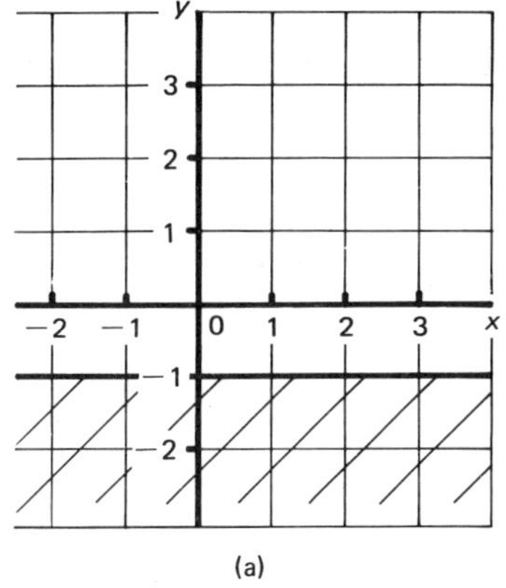

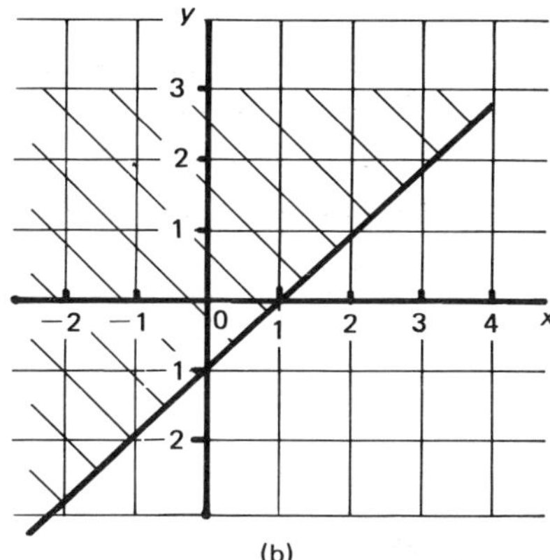

Fig. 25.9

2. Insert one of the signs <, > or = between the following pairs of fractions so as to make true statements:

   (a) $\frac{30}{40}$ $\quad$ $\frac{27}{36}$ $\qquad$ (b) $\frac{5}{16}$ $\quad$ $\frac{3}{8}$ (EA)

3. If x is a positive integer such that $10 < x < 100$, give the least possible value of x. (EA)

4. Which of the symbols <, = or > can be correctly inserted between the following pairs of statements?

   (a) $5^2 - 3^2$ $\quad$ 4^2 $\qquad$ (b) $9^2 \div 3^2$ $\quad$ 3^2 $\qquad$ (c) $\frac{5}{9} - \frac{1}{3}$ $\quad$ $\frac{4}{6}$

   (d) $\sqrt{14.4}$ $\quad$ 1.2 $\qquad$ (e) 0.5^2 $\quad$ 0.5^3 (W)

5. If x is a whole number and $2x-3>7$, find the smallest value of x.

6. Solve the inequality $1+5x>16$.

7. Solve the inequality $3x+8<5$.

8. Find all the whole number values of x for which $5 \leqslant 3x-1 \leqslant 11$.

9. If x has to be an integer greater than zero, find the solution sets for the following:
(a) $3x+2>11$ (b) $2x-1<6$. (EA)

10. If x is a positive integer (excluding zero), find the solution set for $2x+5 \leqslant 9$. (EA)

11. If x is a positive integer (excluding zero), find the solution set for $3x-5<7$. (EA)

12. If x is a positive integer (excluding zero), find the solution set for $7 \leqslant 2x+3 \leqslant 11$. (EA)

13. Given that $3(15-x)>(x+7)$, find the largest possible positive integer which satisfies this inequality.

14. Complete the following when x is a whole number:
(a) Given $3x>x+6$, $x=\dots$
(b) Given $4x<12-x$, $x=\dots$
(c) Given $4x>x+6$ and $2x<12-x$, $x=\dots$.

15. What is the largest possible integer solution for $13>2x+1>7$?

Multi-choice questions 25

1. If $2x=8$ and $y=2$, then
A $y=x$ **B** $y=2x$ **C** $y>x$ **D** $y<x$ (WM)

2. Which of the following inequalities is true for the point $(4, -2)$?
A $x+y>3$ **B** $x<3y$
C $x-y<10$ **D** $2x-3y<12$ (AL)

3. The solution of the inequality $9-6x>15$ is
A $x>4$ **B** $x>-1$ **C** $x<-1$ **D** $x>5$

Select the range of values which satisfies the inequalities in questions 4, 5 and 6.

4. If $3x>18$, then
A $x>4$ **B** $x>6$ **C** $x>15$ **D** $x>21$
E $x>54$ (EM)

5. If $x-2<6$, then
A $x<3$ **B** $x<4$ **C** $x<6$ **D** $x<8$
E $x<12$ (EM)

6. If $2x > x+3$, then

   A $x > 1$ B $x > 2$ C $x > 3$ D $x > 4$

   E $x > 0$ (EM)

7. Which set of inequalities describes the shaded region shown in Fig. 25.10?

   A $x \leqslant 0$ and $y \leqslant 10$ B $y \geqslant 0$ and $y \leqslant 10$

   C $y \leqslant 0$ and $y \geqslant 10$ D $y \leqslant 0$ and $y \geqslant 10$

   E $y \geqslant 0$ and $x \leqslant 10$ (Y)

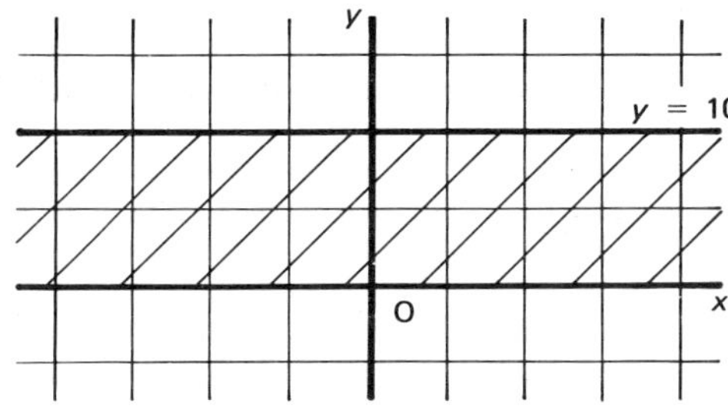

Fig. 25.10

8. If $p = 7$, $q = 5$ and $t = 6$, then

   A $q > t$ B $t > p$ C $q < p$ D $p+q < t$

26 TIME, DISTANCE AND SPEED

- The UNITS OF TIME are

$$60 \text{ seconds (s)} = 1 \text{ minute (min)}$$
$$60 \text{ minutes} = 1 \text{ hour (h)}$$
$$24 \text{ hours} = 1 \text{ day}$$

- The standard CLOCK face is marked off in hours from 1 to 12. Times between 12 midnight and 12 noon are called a.m. (e.g. 9.25 a.m.) and times between 12 noon and 12 midnight are called p.m. (e.g. 11.15 p.m.).

- The 24 HOUR CLOCK (Fig. 26.1) is also used. Times between 12 midnight and 12 noon are given the times 00 hours to 12 hours and times between 12 noon and 12 midnight are given the times 12 hours to 24 hours. Thus 3.50 a.m. is written 03 50 hours and 3.50 p.m. is written 15 50 hours.

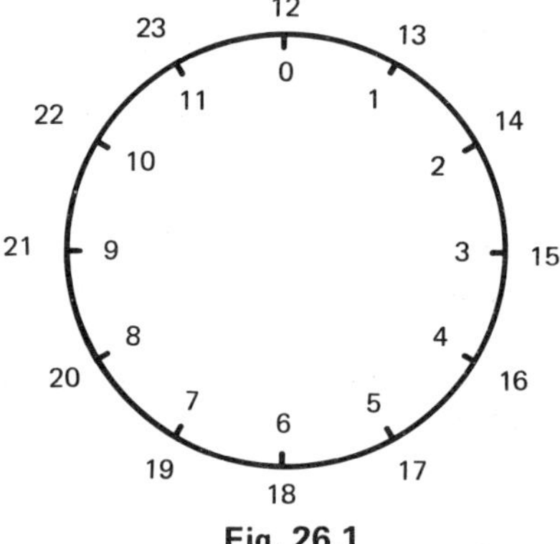

Fig. 26.1

Example

Find the length of time between
(a) 6.38 a.m. and 2.32 p.m.
(b) 02 45 hours and 17 26 hours.

(a) The best way is to use the 24 hour clock so that 2.32 p.m. becomes 14 32 hours. The problem then becomes

$$14\,32 - 06\,38 = 7 \text{ hours } 54 \text{ minutes}$$

(b) 17 26 − 2 45 = 14 hours 41 minutes.

Exercise 26.1

Find the length of time in hours and minutes between the following times:

1. 1.39 a.m. and 7.48 a.m.

2. 8.54 a.m. and 7.54 p.m.

3. 6.32 a.m. and 11.22 p.m.

4. 0.54 a.m. and 8.02 p.m.

5. 00 57 hours and 05 43 hours

6. 02 54 hours and 17 25 hours

7. 12 39 hours and 23 00 hours

8. 03 57 hours and 18 38 hours.

- The AVERAGE SPEED of a vehicle is defined as total distance travelled divided by total time taken. That is

$$\text{Average speed} = \frac{\text{distance travelled}}{\text{time taken}}$$

If the distance is in kilometres and the time in hours the speed will be measured in kilometres per hour (km/h). If the distance is in metres and the time in seconds the speed will be measured in metres per second (m/s).

Note that

$$\text{Time taken} = \frac{\text{distance travelled}}{\text{average speed}}$$

$$\text{Distance travelled} = \text{average speed} \times \text{time taken}$$

Example

(a) A train travels 320 km in 4 h. Calculate its average speed.

$$\text{Average speed} = \frac{320\text{ km}}{4\text{ h}} = 80\text{ km/h}$$

(b) A lorry travels at 45 km/h for 3 h. How far does it travel?

$$\text{Distance travelled} = 45\text{ km/h} \times 3\text{ h} = 135\text{ km}$$

(c) A girls walks for 2 km at a speed of 4 km/h. She then cycles for 12 km at a speed of 8 km/h. What was her average speed for the complete journey?

$$\text{Time taken for 2 km} = \frac{2\text{ km}}{4\text{ km/h}} = \tfrac{1}{2}\text{ h}$$

$$\text{Time taken for 12 km} = \frac{12\text{ km}}{8\text{ km/h}} = 1\tfrac{1}{2}\text{ h}$$

$$\text{Total time taken} = \tfrac{1}{2} + 1\tfrac{1}{2} = 2\text{ h}$$

$$\text{Total distance travelled} = 2 + 12 = 14\text{ km}$$

$$\text{Average speed} = \frac{\text{total distance travelled}}{\text{total time taken}}$$

$$= \frac{14\text{ km}}{2\text{ h}} = 7\text{ km/h}$$

Exercise 26.2

1. A train travels 400 km in 5 h. Calculate its average speed.
2. A car travels 350 km in 7 h. What is its average speed?
3. A car travels 200 km at an average speed of 40 km/h. How long does the journey take?
4. A cyclist travels 40 km at an average speed of 10 km/h. What is the time for the journey?
5. A train travels at 80 km/h for 3 h. How far has it travelled?
6. A car travels for 5 h at 60 km/h. What distance has it travelled?
7. A boy walks for 3 km at 6 km/h and then cycles for 6 km at 12 km/h. What is his average speed for the entire journey?
8. A car travels 136 km at an average speed of 32 km/h. On the return journey the speed is increased to 48 km/h. Calculate the average speed for the complete journey.

Exercise 26.3 (All of the type found in CSE examination papers)

1. (a) A clock at a station shows the departure time of an evening train to be 8.15. How would this time be written in a timetable which uses the 24 hour clock system?
   (b) The train arrives at its destination at 21 45 hours. For how long was it travelling?
   (c) The journey was 60 km. What was the average speed of the train? (EA)
2. A train left Sheffield at 16 54 h, taking 2 hours and 50 minutes for its journey to Halifax. At what time did the train arrive at Halifax?
3. Express buses leave Nottingham at 08 38 each day. One travels to London 140 miles away, another to Newcastle 157 miles away, a third to Birmingham 56 miles away.
   (a) The bus arrives in London at 12 20. How long does the journey take?
   (b) The journey to Newcastle takes $4\frac{1}{4}$ hours. At what time does the bus arrive in Newcastle?
   (c) The Birmingham bus travels at an average speed of 48 miles per hour. How long does the journey take? (EA)
4. The distance from Leicester to Edinburgh is 455 km.
   (a) How long will it take a motorist to make the journey if he travels at an average speed of 70 km/h?
   (b) At what average speed must he drive to complete the journey in 7 h? (EM)

5. A man travels for 2 h at a speed of 5 km/h and then for the remaining 12 km of his journey he travels at 4 km/h. Find:
   (a) How far he travels in the first 2 h
   (b) His average speed for the total journey. (SW)

6. A cross-channel ferry leaves Calais at 02 35 and arrives in Dover at 04 10. How long does the journey last? (EA)

7. A cyclist leaves town A at 9 a.m. to travel to town B which is 68 miles from A. If he travels at an average speed of 16 miles per hour, calculate:
   (a) The distance he has travelled from A at 12 noon
   (b) The time when he reaches B. (WM)

8. The following is part of a railway timetable for three trains A, B and C for the journey Birmingham–Coventry–London:

	Birmingham	*Coventry*	*London*
A	08 15	08 36	09 49
B	12 48	13 09	14 28
C	18 48	19 09	20 25

   (a) How long before midnight does train C arrive in London?
   (b) How long does train A take to go from Birmingham to Coventry?
   (c) How long does train B take to go from Birmingham to London? (Y)

9. A man kept a record of his visit by car to Mahos, a distance of 200 km from London.

Time	
07 00	Left London
08 30	Car broke down 45 km from London
09 00	Continued journey to Mahos
11 30	Arrived at Mahos
15 00	Left Mahos for return journey to London

   (a) Calculate his average speed in kilometres per hour before he broke down.
   (b) Find how far he was from Mahos when he broke down.
   (c) Find his average speed between 07 00 and 11 30. Give your answer correct to 1 decimal place.
   (d) Calculate the length of time he had in Mahos.
   (e) On his return journey from Mahos to London he averaged 50 km/h. Calculate the time of his arrival in London. (EA)

10. Calculate the number of minutes from 11.35 a.m. to 2.15 p.m. (EA)

Multi-choice questions 26

1. A hiker walks 33 km at an average speed of 6 km/h. How long does the walk take?

   A 5 hours 3 minutes **B** 5 hours 5 minutes
   C 5 hours 20 minutes **D** 5 hours 30 minutes
   E 5 hours 33 minutes

2. A journey lasts 45 minutes and the average speed is 24 km/h. The length of the journey, in kilometres, is

   A 8 **B** 16 **C** 18 **D** 32

3. A train travels 80 km at 100 km/h. The time, in minutes, taken for the journey is

   A 20 **B** 48 **C** 75 **D** 180 (AL)

4. A boy walks at a steady speed of x kilometres per hour. How many hours does it take him to walk $2x$ kilometres?

   A $2x^2$ **B** $2x$ **C** $\frac{1}{2}x$ **D** 2
   E $\frac{1}{2}$ (NW)

5. An aeroplane flies non-stop for $4\frac{1}{2}$ hours and travels 3240 km. Its average speed, in kilometres per hour, is

   A 3645 **B** 800 **C** 720 **D** 364.5

6. The number of hours between 08 15 and 14 45 is

   A $6\frac{1}{2}$ **B** $6\frac{3}{4}$ **C** $7\frac{1}{2}$ **D** $17\frac{1}{2}$
   E 25

27 ANGLES AND STRAIGHT LINES

- ANGLES are measured in degrees, seconds and minutes as follows

60 seconds (″)	=	1 minute (′)
60 minutes	=	1 degree (°)
90 degrees	=	1 right angle
360 degrees	=	1 revolution

Example

(a) Add together 43°17′38″ and 54°49′53″.

$$\begin{array}{r} 43°17'38'' \\ 54°49'53'' \\ \hline 98°\ 7'31'' \\ \hline \end{array}$$

(b) Subtract 28°39′ from 47°18′.

$$\begin{array}{r} 47°18' \\ 28°39' \\ \hline 18°39' \\ \hline \end{array}$$

Exercise 27.1

Add together the following angles:

1. 18°15′ and 27°29′
2. 37°42′ and 45°54′
3. 29°18′37″ and 42°39′48″
4. 27°5′49″ and 53°19′47″.

Subtract the following angles:

5. 18°19′ from 54°48′
6. 37°54′ from 46°8′
7. 17°19′27″ from 39°41′56″
8. 49°46′59″ from 78°18′27″.

- The various TYPES OF ANGLES are shown in Fig. 27.1.

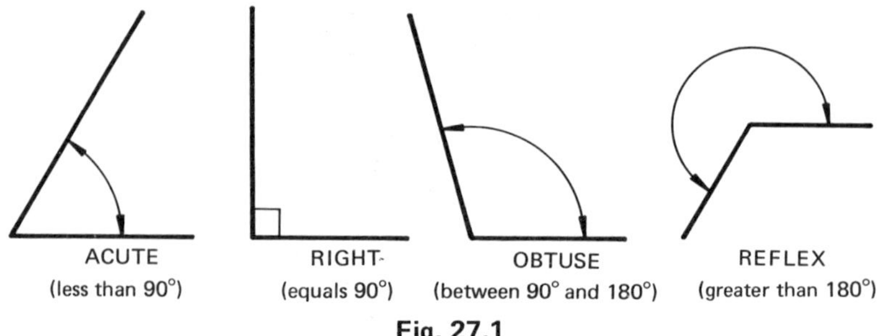

Fig. 27.1

- COMPLEMENTARY ANGLES are angles whose sum is $90°$. Thus $18°$ and $72°$ are complementary angles since $18° + 72° = 90°$.

- SUPPLEMENTARY ANGLES are angles whose sum is $180°$. Thus $103°$ and $77°$ are supplementary angles since $103° + 77° = 180°$.

- The TOTAL ANGLE ON A STRAIGHT LINE is $180°$ (Fig. 27.2). The angles A and B are called *adjacent* angles.

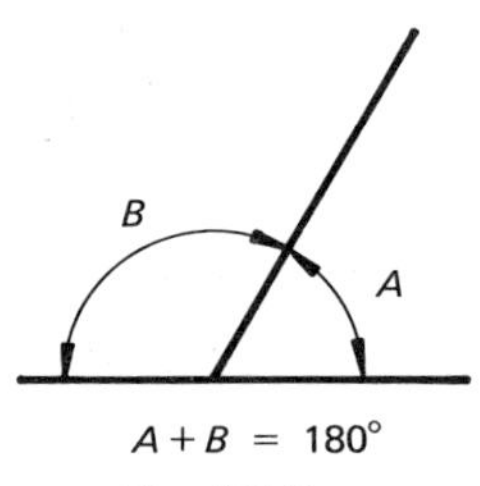

$A + B = 180°$

Fig. 27.2

- When two straight lines intersect, the VERTICALLY OPPOSITE angles are equal. (Fig. 27.3).

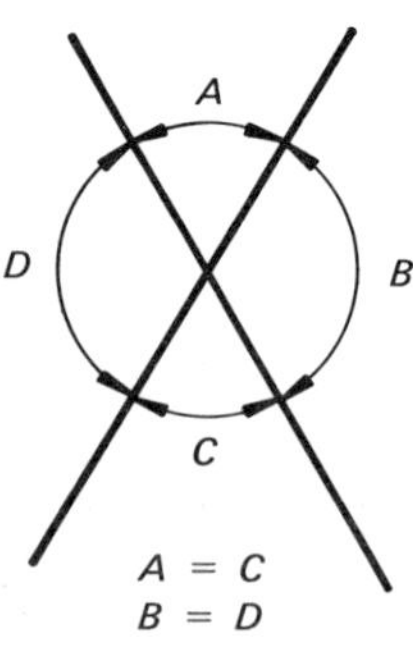

$A = C$
$B = D$

Fig. 27.3

- When two PARALLEL LINES are cut by a transversal (Fig. 27.4) the corresponding angles are equal: $a = l$, $b = m$, $c = p$ and $d = q$;

  the alternate angles are equal: $d = m$ and $c = l$;

  the interior angles are supplementary: $d + l = 180°$ and $c + m = 180°$.

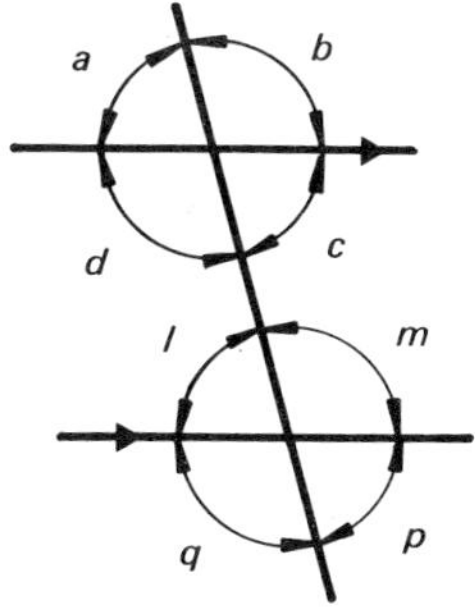

Fig. 27.4

Example

In Fig. 27.5, AB and CD are two parallel straight lines. Find the size of the angles marked a, b, c, d and e.

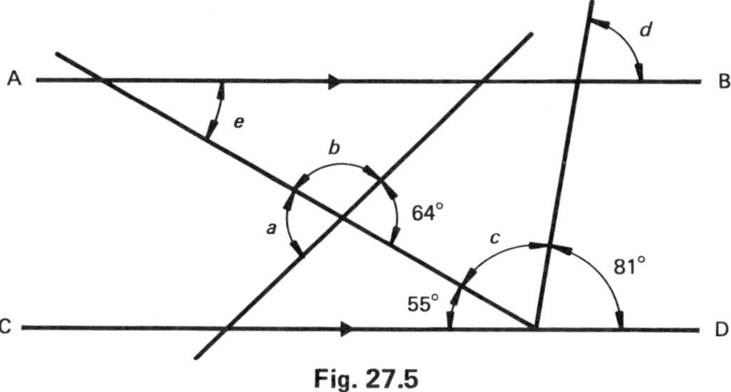

Fig. 27.5

$a = 64°$ (vertically opposite angles)

$b = 180° - 64° = 116°$ (sum of angles on straight line equals 180°)

$c = 180° - 55° - 81° = 44°$ (sum of angles on straight line equals 180°)

$d = 81°$ (AB ∥ CD)

$e = 55°$ (AB ∥ CD)

Exercise 27.2 (All of the type found in CSE examination papers)

1. Find the value of the angle a in Fig. 27.6. (S)

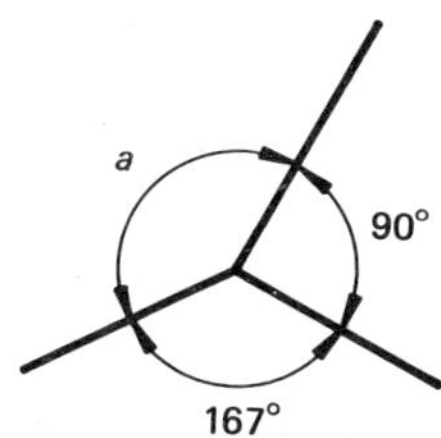

Fig. 27.6

2. In Fig. 27.7, calculate the angles marked x and y. (S)

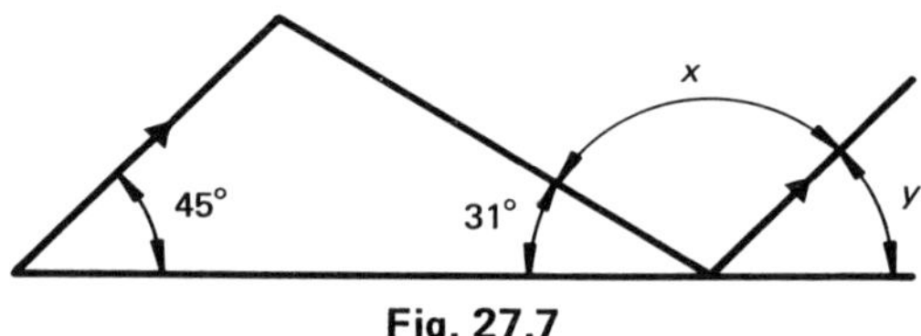

Fig. 27.7

3. In Fig. 27.8, AB and CD are parallel. Calculate the size of the angles marked m, n, p and q. (W)

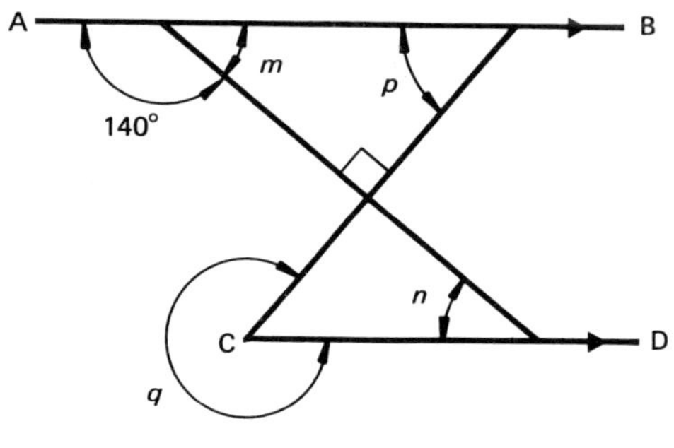

Fig. 27.8

4. In Fig. 27.9, AB and CD are parallel lines crossed by the line ST. Find the size of the angle marked y. (NW)

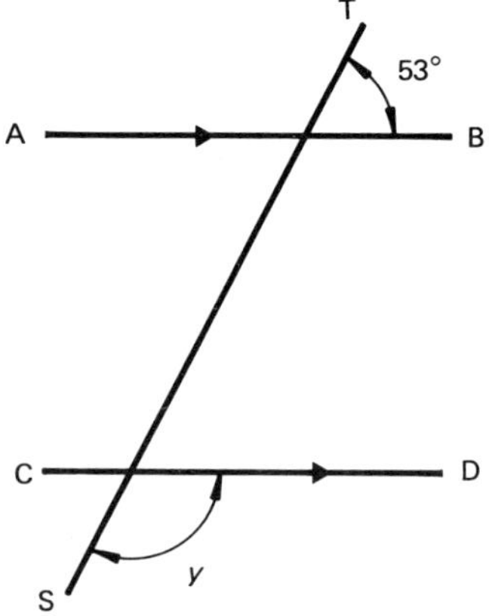

Fig. 27.9

5. ABC and DEF (Fig. 27.10) are two parallel straight lines with transversals BD, EC and BF. CE and BF intersect at 90°. Write down the size of the angles marked p, q, r and s. (W)

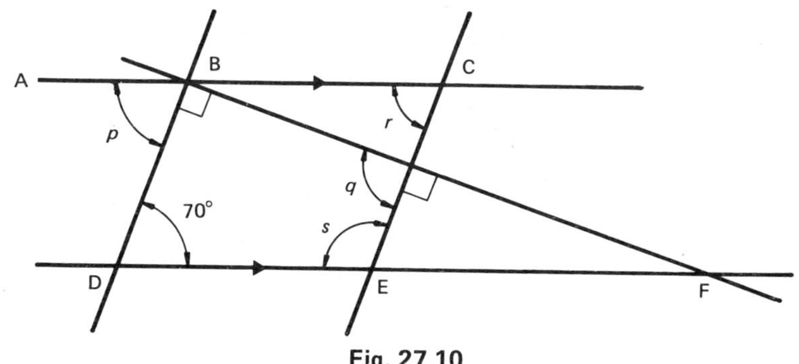

Fig. 27.10

6. In Fig. 27.11, AB is parallel to CD. State the value of the angles marked x and y.

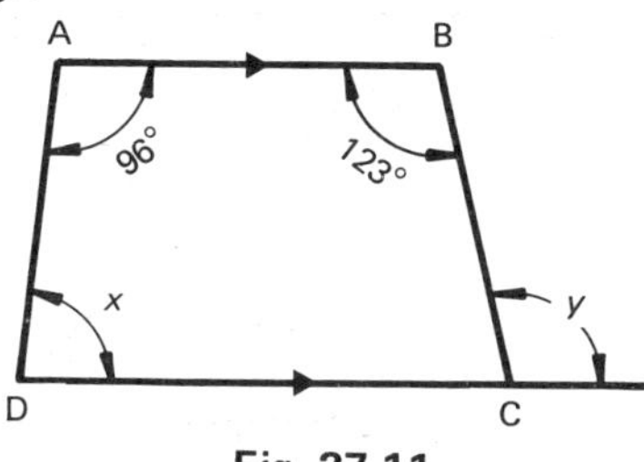

Fig. 27.11

7. In Fig. 27.12, calculate the angle marked a. (AL)

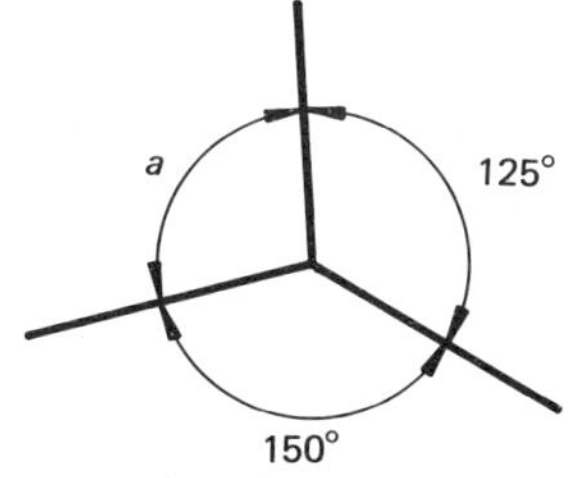

Fig. 27.12

8. In Fig. 27.13, ABCDE and FGH are two parallel straight lines with transversals BKG, FKCJ and GDJ. Find the size of the angles marked u, v, w, x, y and z. (W)

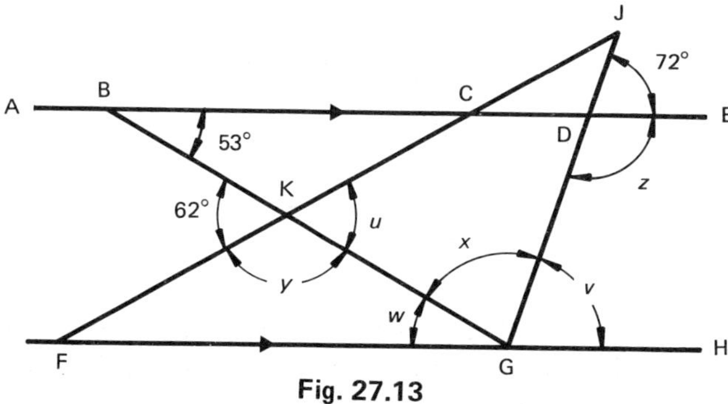

Fig. 27.13

Multi-choice questions 27

1. The value of x in Fig. 27.14 is

   A 30° **B** 60° **C** 120° **D** 130°
   E 300°

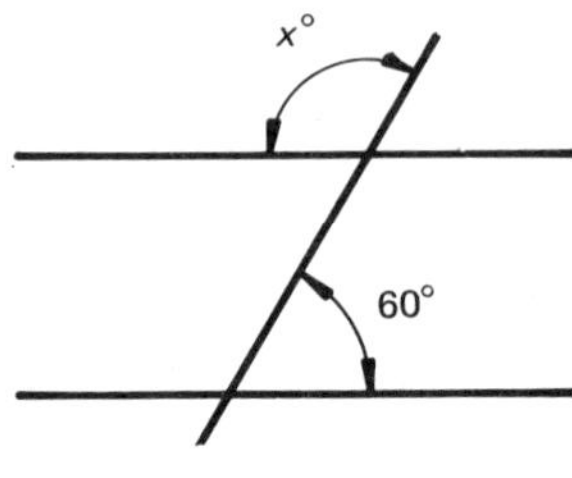

Fig. 27.14

2. What is the size of the angle a in Fig. 27.15?

   A 43° **B** 57° **C** 80° **D** 90°
   E 100° (AL)

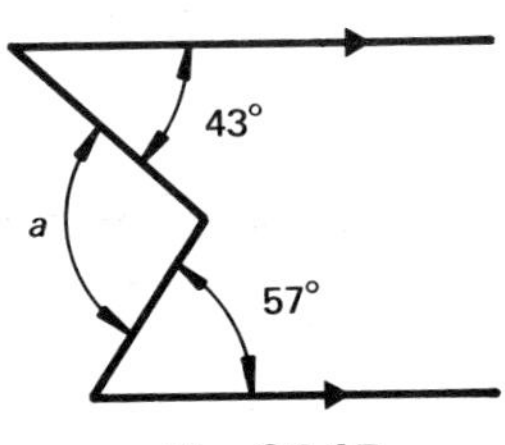

Fig. 27.15

3. In Fig. 27.16, PQ and RS are parallel and the point T lies in RS. What is the size of the angle X?

   A 20° **B** 40° **C** 60° **D** 70°
   E 80° (AL)

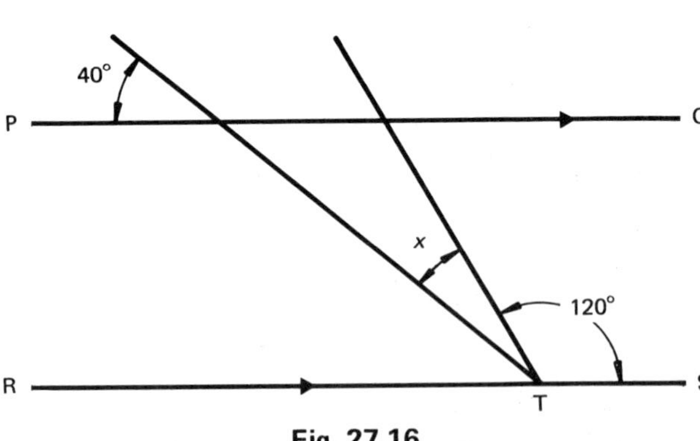

Fig. 27.16

4. The size of the angle marked a in Fig. 27.17 is

   A 73° **B** 107° **C** 117° **D** 146° (EA)

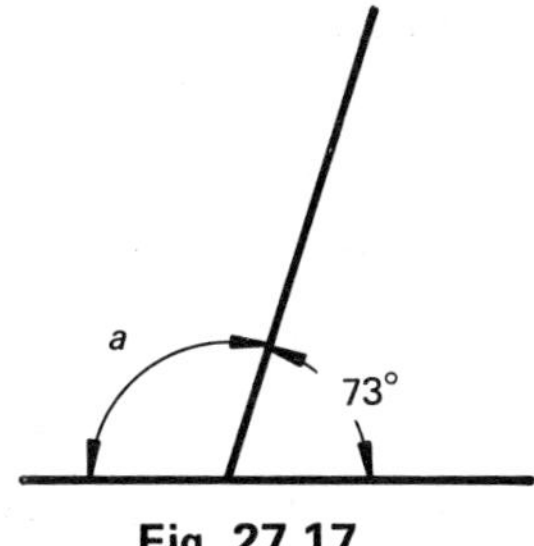

Fig. 27.17

5. The size of the angle marked b in Fig. 27.18 is

A 70° **B** 110° **C** 120° **D** 160° (EA)

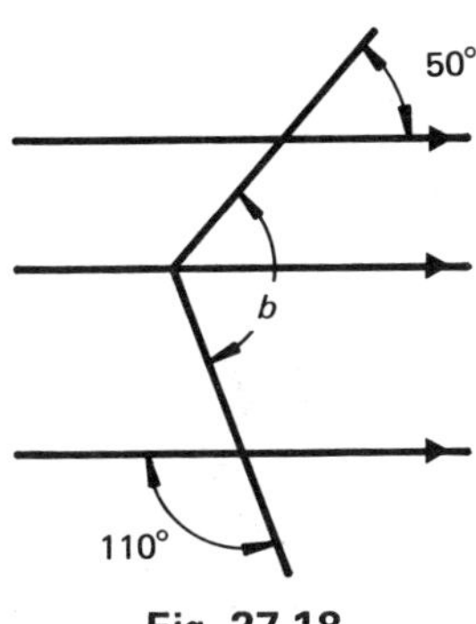

Fig. 27.18

In Fig. 27.19, the lines AB and CD are parallel and the point E lies on CD. Use this diagram to answer questions 6, 7 and 8.

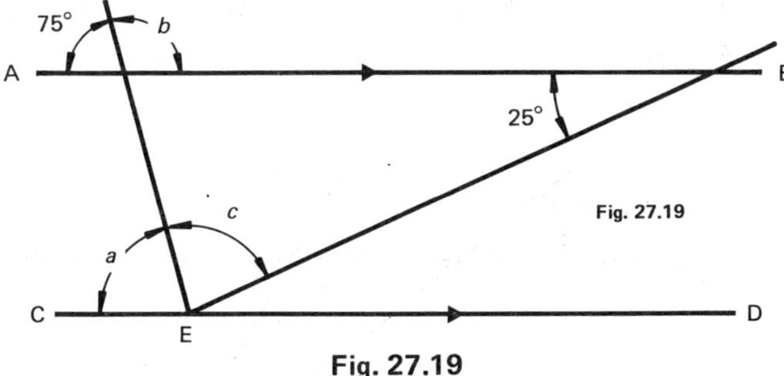

Fig. 27.19

6. The size of angle a is

A 25° **B** 50° **C** 75° **D** 80° (AL)

7. The size of the angle b is

A 80° **B** 100° **C** 105° **D** 150° (AL)

8. The size of angle c is

A 25° **B** 50° **C** 75° **D** 80° (AL)

9. In Fig. 27.20, calculate the value of x.

A 45° **B** $67\frac{1}{2}°$ **C** 90° **D** 135°

E 270° (NW)

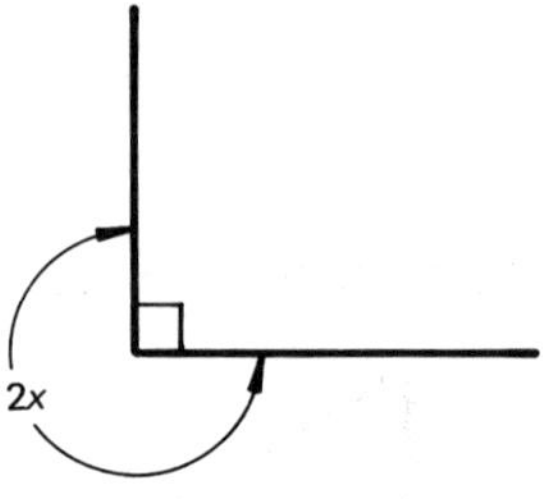

Fig. 27.20

10. The angle shown in Fig. 27.21 is

A acute **B** right **C** reflex **D** obtuse (WY)

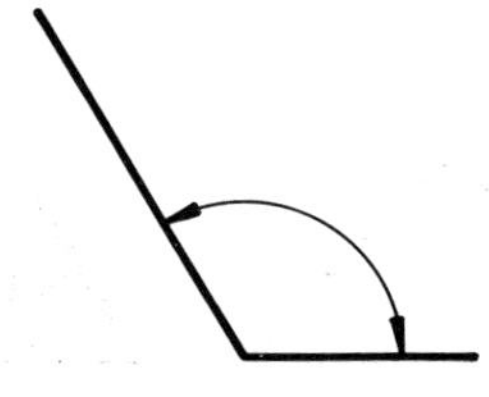

Fig. 27.21

28 TRIANGLES

- The various TYPES OF TRIANGLES are shown in Fig. 28.1.

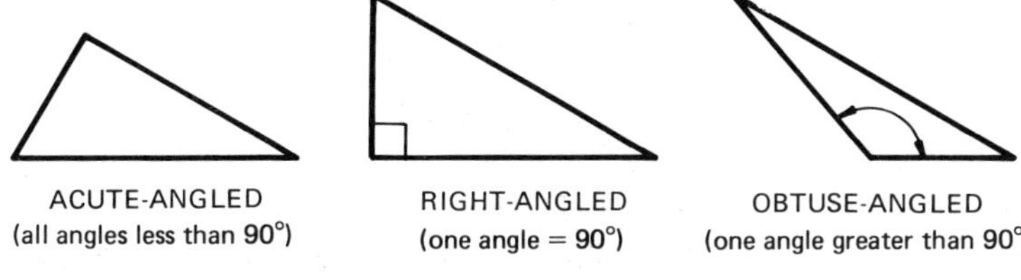

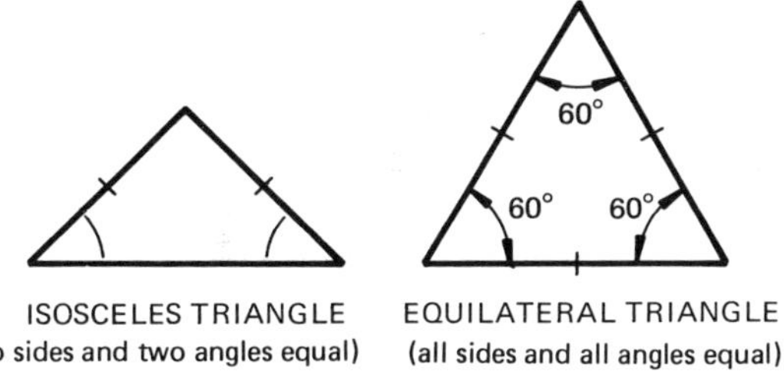

Fig. 28.1

- The SUM OF THE ANGLES OF A TRIANGLE is 180°. Thus in Fig. 28.2, $A+B+C=180°$.

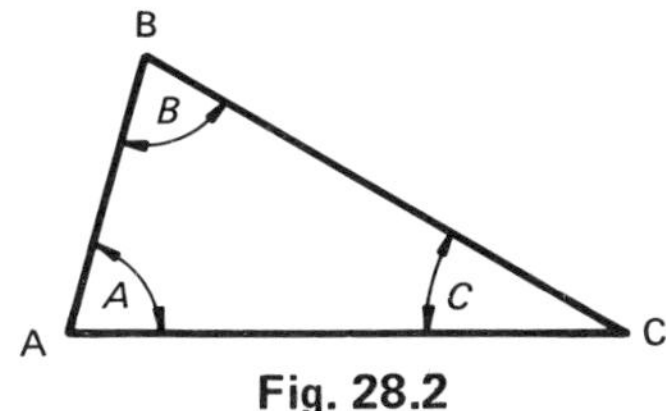

Fig. 28.2

- When the side of a triangle is produced, the EXTERIOR ANGLE so formed is equal to the sum of the opposite interior angles. Thus in Fig. 28.3, $\theta = A+B$.

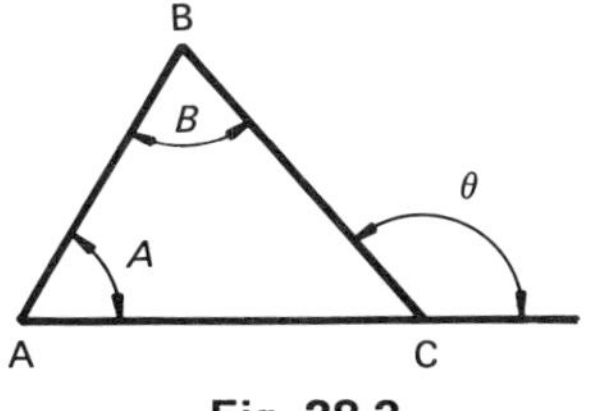

Fig. 28.3

Example

In Fig. 28.4, QP and RT are parallel lines. Find the size of the angles marked a and b.

$\angle SQR = 50°$ (vertically opposite angles)

$\angle QSR = 100°$ (alternate angles, OP ∥ RT)

$a = 180° - 50° - 100° = 30°$ (sum of the angles of a triangle equals 180°)

$b = 100° + 30° = 130°$ (exterior angle of triangle = sum of opposite interior angles)

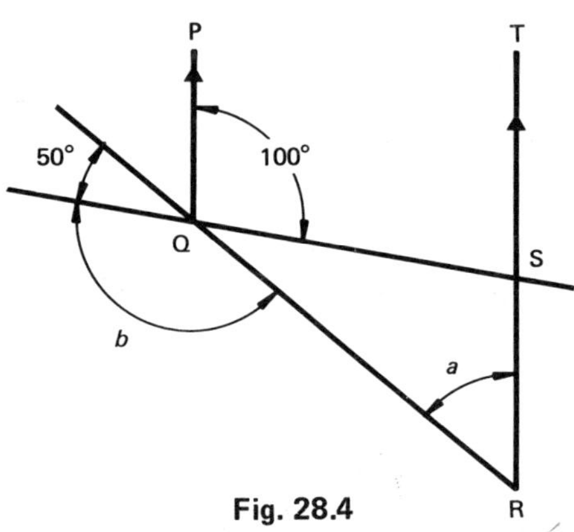

Fig. 28.4

Exercise 28.1

Find the angles marked x and y in Fig. 28.5.

1.
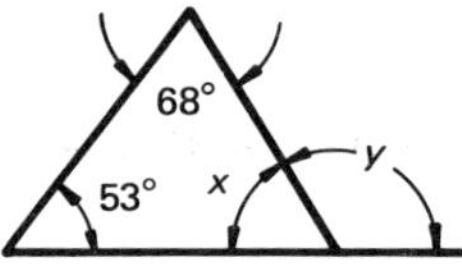

2.
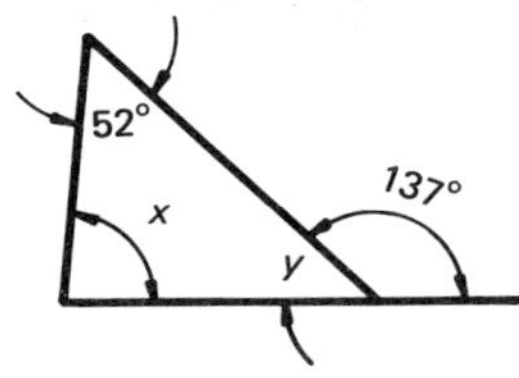

3.
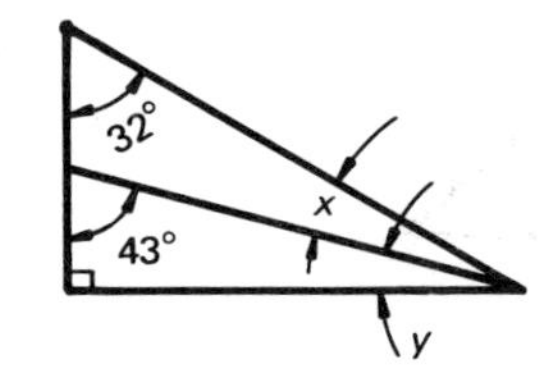

4.
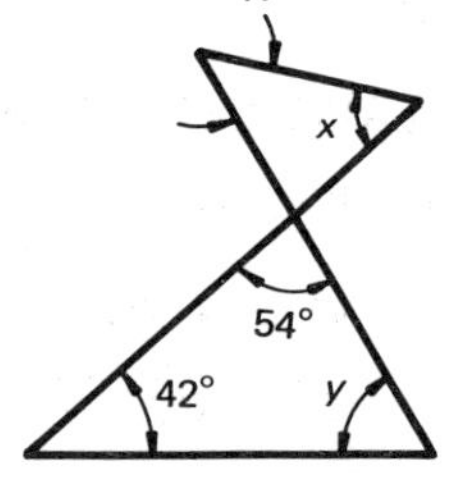

5.
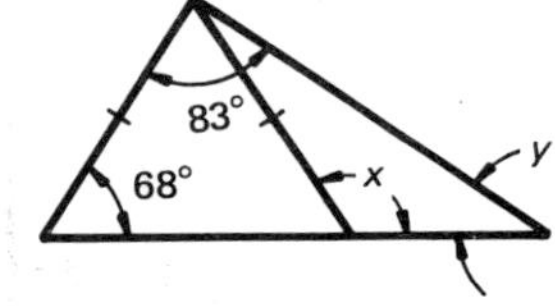

Fig. 28.5

- Fig. 28.6 shows the STANDARD NOTATION FOR A TRIANGLE. The three vertices are labelled A, B and C. The three angles are called by the same letters as the vertices. Then side $\boldsymbol{a}$ lies opposite angle A, side $\boldsymbol{b}$ lies opposite angle B and side c lies opposite angle C.

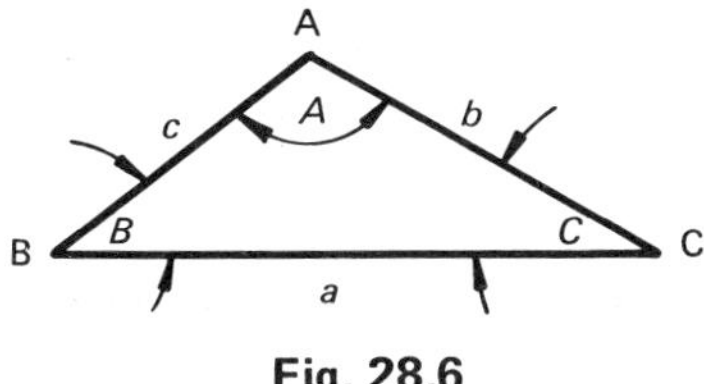

Fig. 28.6

- PYTHAGORAS' THEOREM states that in any right-angled triangle the square on the hypotenuse is equal to the sum of the squares on the other two sides. Thus in Fig. 28.7

$$a^2 = b^2+c^2$$

(Note that the *hypotenuse* is the longest side of the triangle and it lies opposite the right angle.)

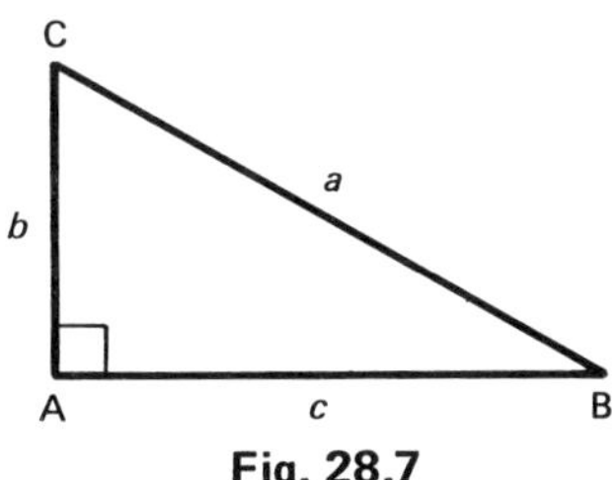

Fig. 28.7

Example

(a) In ΔABC, $\angle A = 90°$, $b = 3.5$ cm and $c = 4.2$ cm. Find $\boldsymbol{a}$.

From Fig. 28.8

$$a^2 = b^2+c^2 = 3.5^2+4.2^2 = 12.25+17.64 = 29.89$$

$$a = \sqrt{29.89} = 5.47 \text{ cm}$$

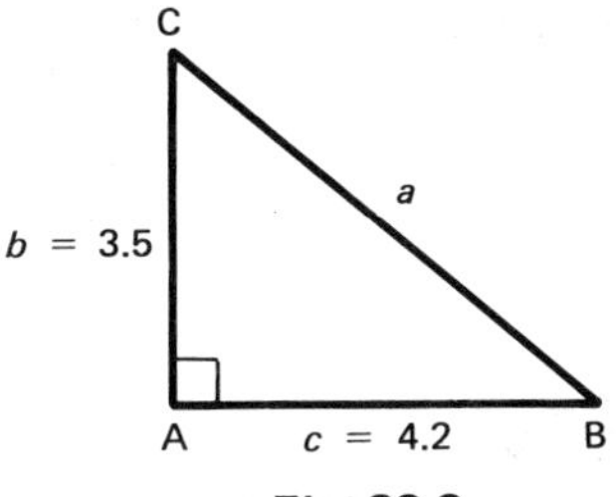

Fig. 28.8

(b) In ΔABC, $\angle A = 90°$, $a = 7.3$ cm and $b = 5.8$ cm. Find c.

From Fig. 28.9

$$c^2 = a^2 - b^2 = 7.3^2 - 5.8^2 = 53.29 - 33.64 = 19.65$$

$$c = \sqrt{19.65} = 4.43 \text{ cm}$$

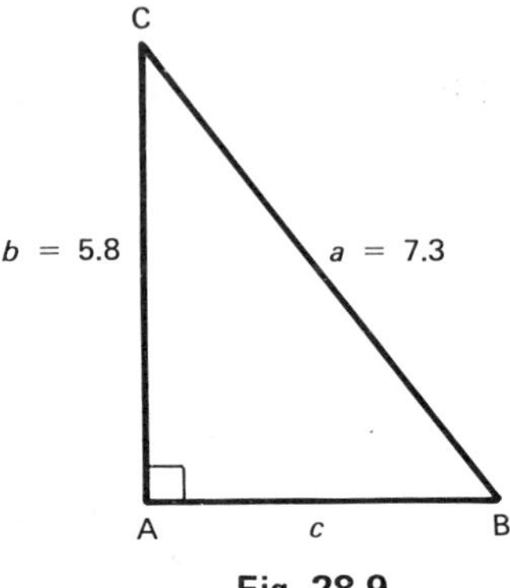

Fig. 28.9

- The PROPERTIES OF AN ISOSCELES TRIANGLE are as follows. The perpendicular dropped from the apex to the unequal side:
  (i) Bisects the unequal side
  (ii) Bisects the apex angle.

Thus in Fig. 28.10, BD = CD and $\angle$BAD = $\angle$DAC.

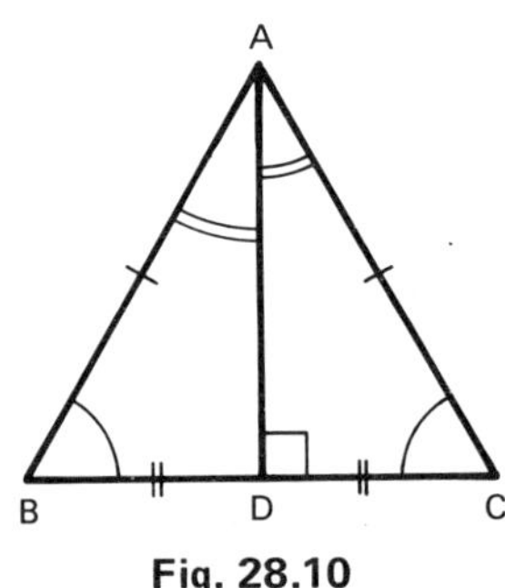

Fig. 28.10

Example

In Fig. 28.10, AB = AC = 11.2 cm and BC = 8.4 cm. Find the vertical height, AD, of the triangle ABC.

Since AD is a perpendicular dropped from A to BC

$$\text{BD} = \text{CD} = \tfrac{1}{2}\text{BC} = 4.2 \text{ cm}$$

In ΔABD, using Pythagoras' theorem

$$\begin{aligned} AD^2 &= AB^2 - BD^2 \\ &= 11.2^2 - 4.2^2 \\ &= 125.4 - 17.6 \\ &= 107.8 \\ &= \sqrt{107.8} = 10.4 \end{aligned}$$

Hence the vertical height of the triangle is 10.4 cm.

Exercise 28.2

Find the lengths of the sides marked x in Fig. 28.11.

1.

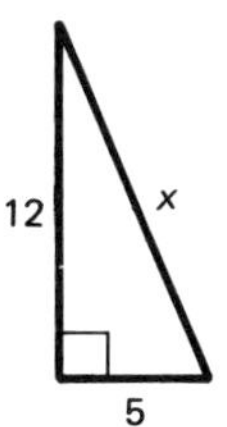

2.

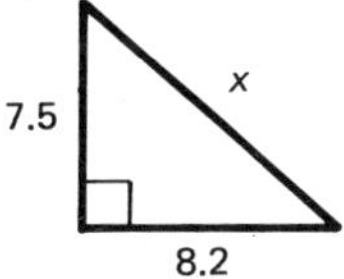

3.

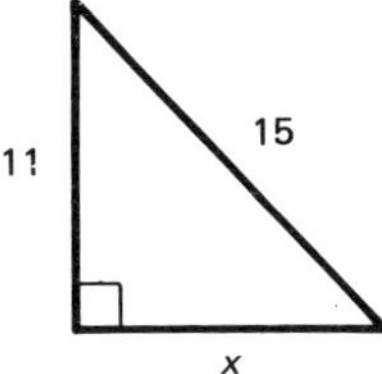

4.

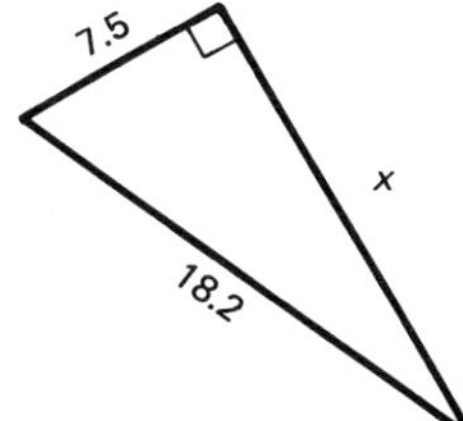

5.

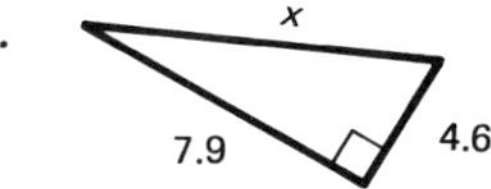

Fig. 28.11

6. Find the vertical height, AD, in Fig. 28.12.

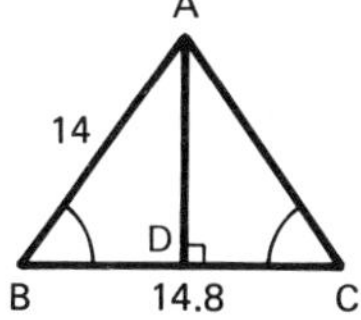

Fig. 28.12

7. Find the length of the side marked a in Fig. 28.13.

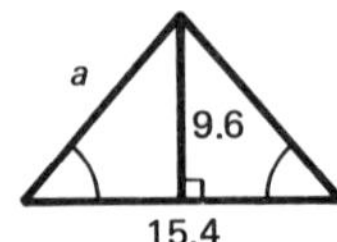

Fig. 28.13

8. Find the length of the base, BC, of the triangle ABC shown in Fig. 28.14.

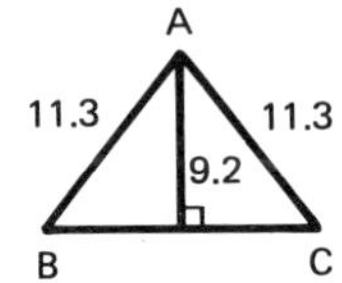

Fig. 28.14

9. Find the angles marked a in Fig. 28.15.

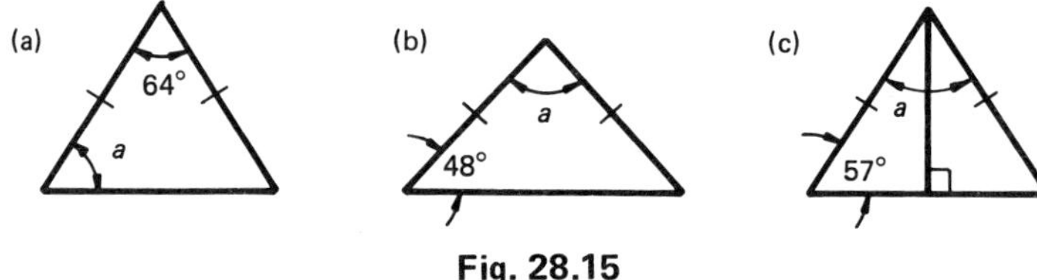

Fig. 28.15

10. Find the angles marked x and y in Fig. 28.16.

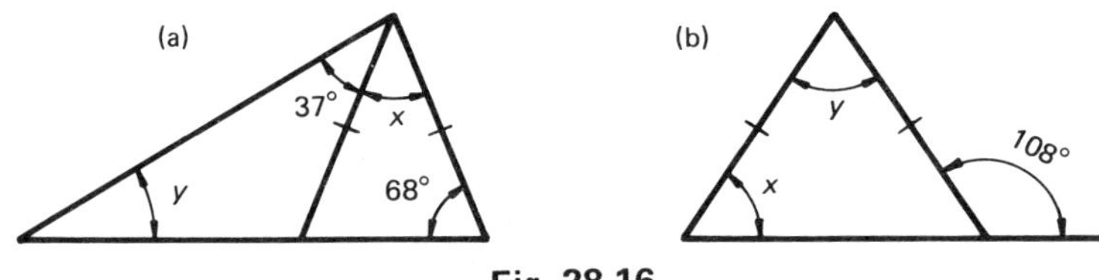

Fig. 28.16

Exercise 28.3 (All of the type found in CSE examination papers)

1. In Fig. 28.17, calculate the length of PQ. (EA)

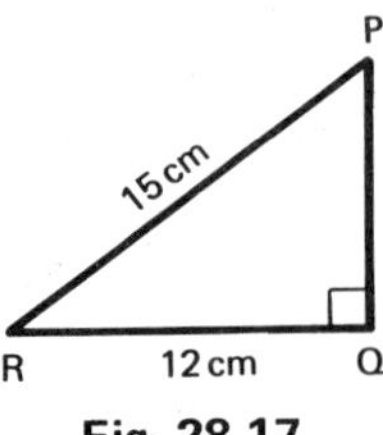

Fig. 28.17

2. In Fig. 28.18, calculate:

(a) the length of AB (b) the area of the triangle.

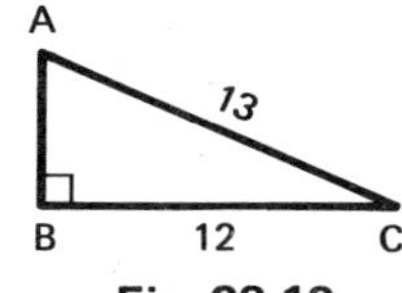

Fig. 28.18

3. In Fig. 28.19, calculate:

(a) the value of x (b) the value of y

(c) ∠ABC. (d) What type of triangle is ABD?

(EA)

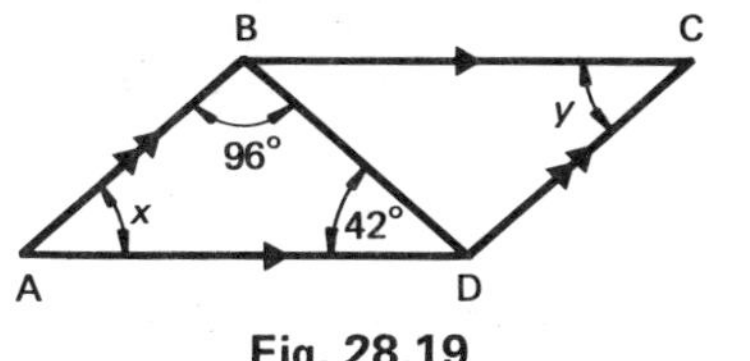

Fig. 28.19

4. In Fig. 28.20, calculate the side marked x. (Y)

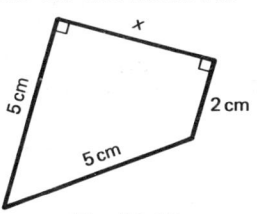

Fig. 28.20

5. In Fig. 28.21, calculate

(a) ∠C (b) ∠APQ. (EA)

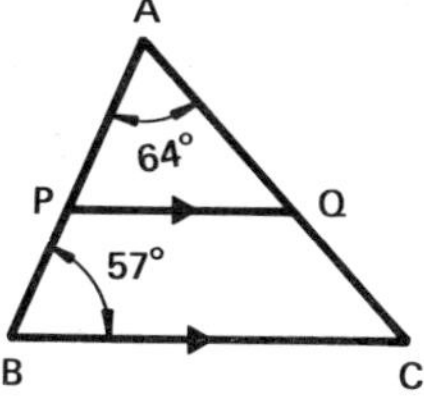

Fig. 28.21

6. Calculate the length of the third side of the right-angled triangle shown in Fig. 28.22.

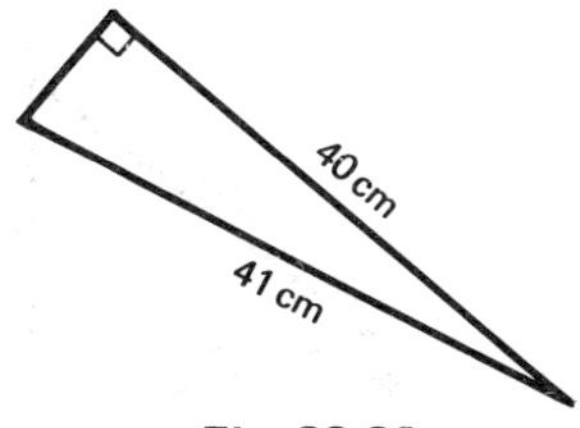

Fig. 28.22

7. A piece of wire 12 cm in length can be bent in different ways to form different triangles. The sides of the triangle must always be a whole number of centimetres.

Write down the lengths of the three sides if the triangle is:

(a) equilateral (b) isosceles (c) right-angled.

(SW)

8. In Fig. 28.23, calculate:

(a) the value of x (b) the angle BAK (c) the value of y.

(EA)

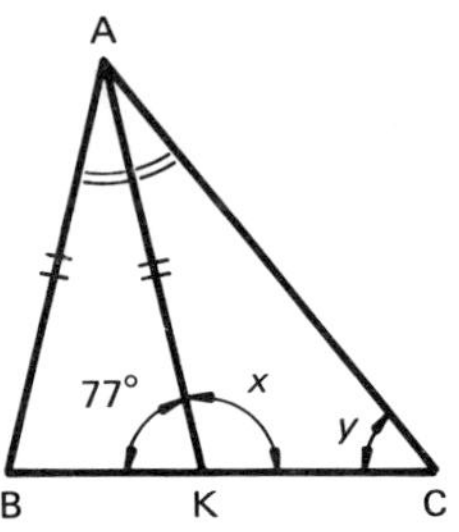

Fig. 28.23

9. Fig. 28.24 shows two parallel lines ABCD and EFG with two transversals BOF and EOC which intersect at 90° at O. If $\angle ABO = 140°$, calculate the sizes of the angles marked m, n, p and q. (W)

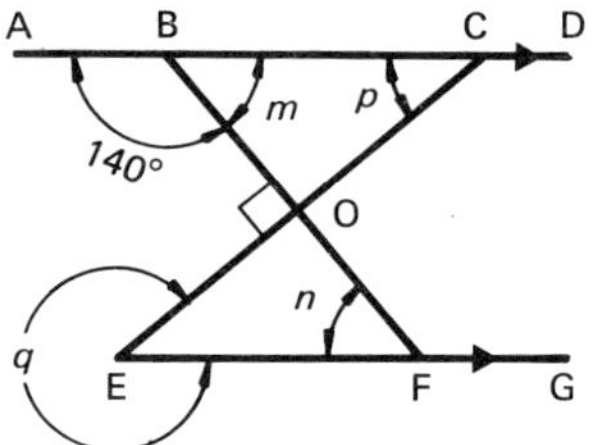

Fig. 28.24

10. In Fig. 28.25, write down the sizes of the angles marked p, q, r and s. (W)

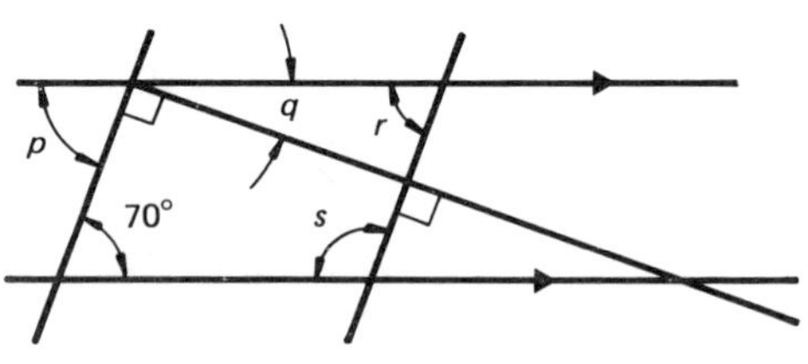

Fig. 28.25

Multi-choice questions 28

1. In Fig. 28.26, what is the length of LM?

A 3 cm **B** 9 cm **C** 27 cm **D** 57.28 cm (WY)

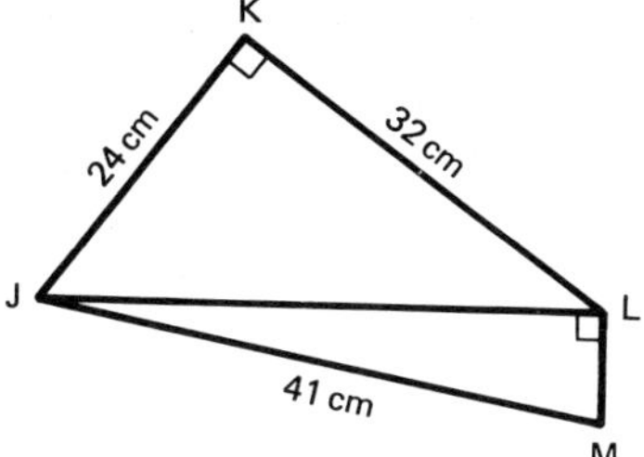

Fig. 28.26

2. In Fig. 28.27, the size of angle YXZ is

A 25° **B** 35° **C** 60° **D** 85°

E impossible to find (WM)

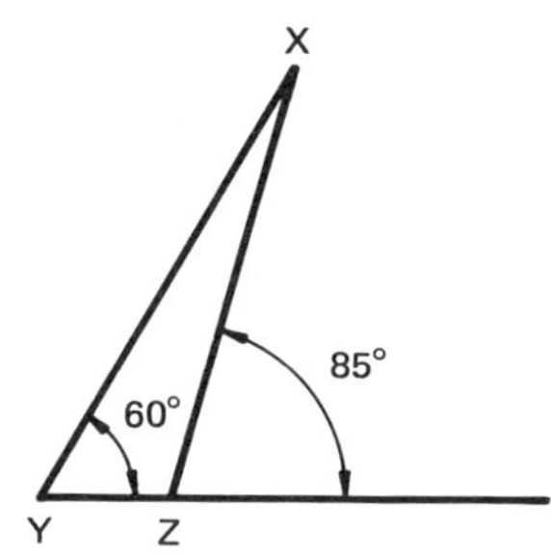

Fig. 28.27

3. Calculate the value of x in Fig. 28.28.

A 10° **B** 50° **C** 60° **D** 80°

E 100° (NW)

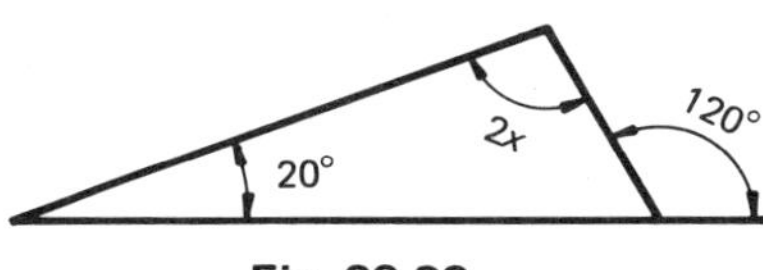

Fig. 28.28

4. The size of the angle marked b in Fig. 28.29 is

A 88° **B** 92° **C** 108° **D** 258° (EA)

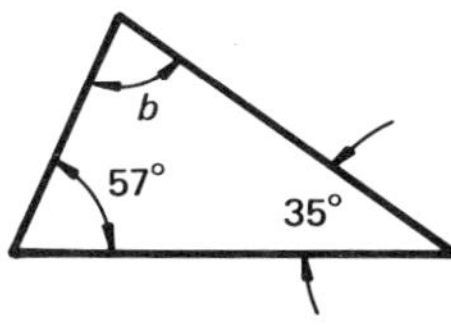

Fig. 28.29

5. In Fig. 28.30, WX = WY. Find ∠XWY.

A 40° **B** 70° **C** 100° **D** 140°

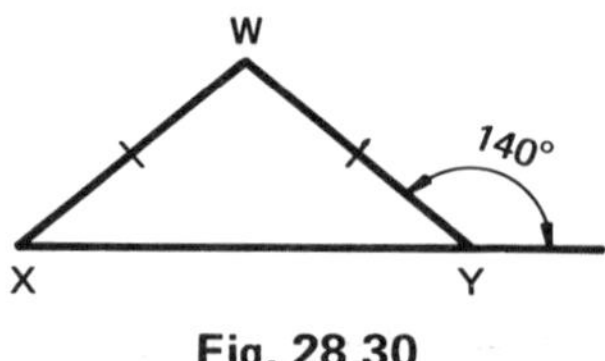

Fig. 28.30

6. In Fig. 28.31, the size of ∠BAC is

A 30° **B** 45° **C** 60° **D** 90°

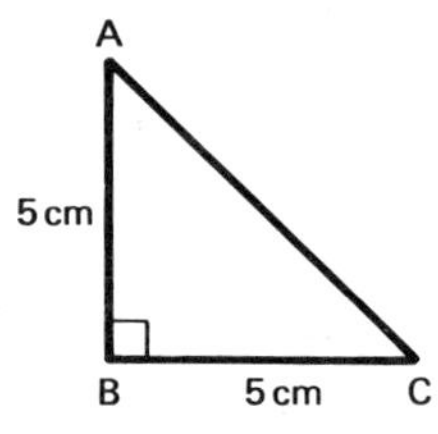

Fig. 28.31

7. In Fig. 28.32, find the size of the angle QRS.

A 70° **B** 60° **C** 50° **D** 30°

E 20°

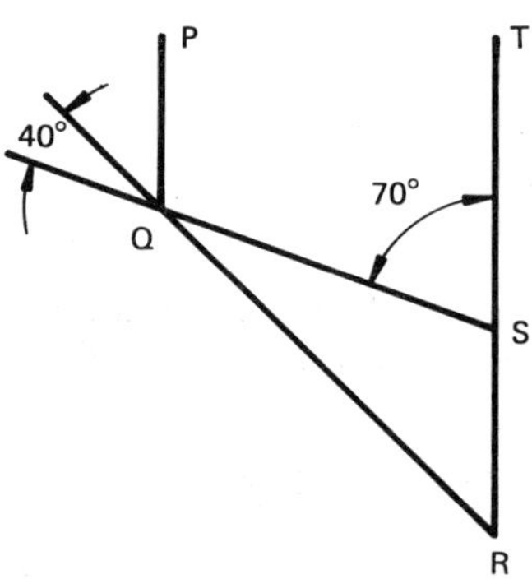

Fig. 28.32

8. In Fig. 28.33 the size of angle x is

A 30° **B** 45° **C** 75° **D** 105° (AL)

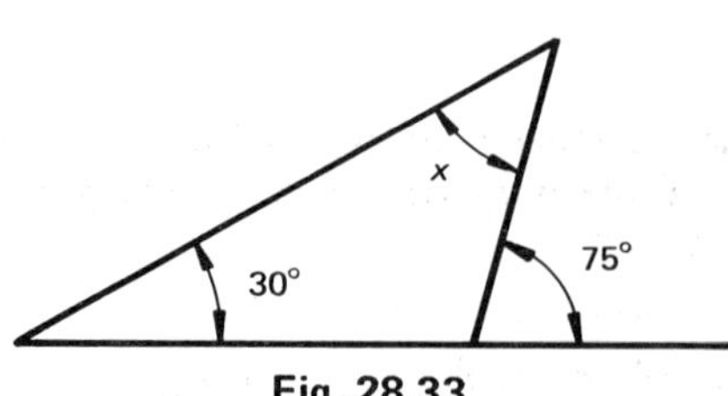

Fig. 28.33

29 QUADRILATERALS

- A QUADRILATERAL is any four-sided figure (Fig. 29.1).

- The SUM OF THE ANGLES OF A QUADRILATERAL is 360°. Thus in Fig. 29.1, $\angle A+\angle B+\angle C+\angle D=360°$.

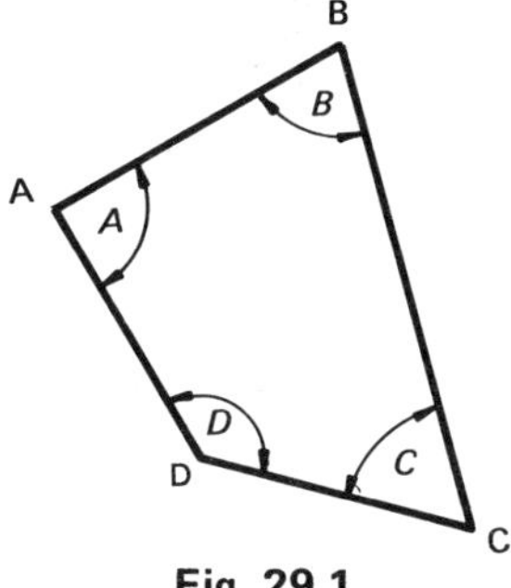

Fig. 29.1

- A PARALLELOGRAM has both pairs of opposite sides parallel (Fig. 29.2). It has the following properties:
  - (i) The sides which are opposite each other are equal in length.
  - (ii) The angles which are opposite to each other are equal.
  - (iii) The diagonals bisect each other and bisect the parallelogram so that two equal (congruent) triangles are formed.

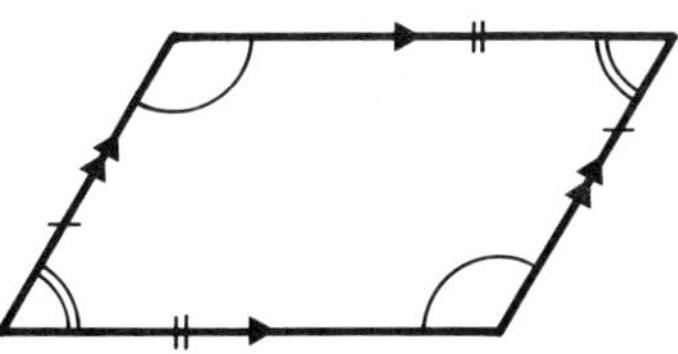
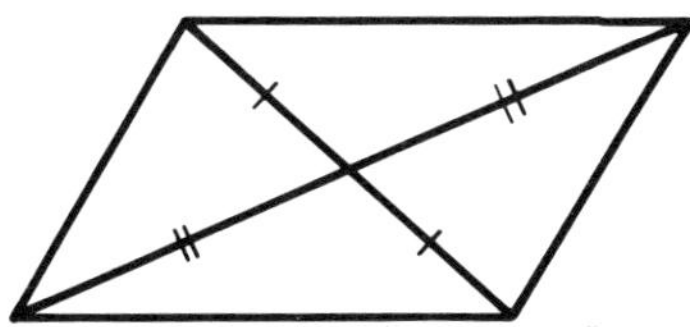

Fig. 29.2

- A RHOMBUS is a parallelogram with all its sides equal in length (Fig. 29.3, p. 160). It has all the properties of a parallelogram but, in addition, it has the following properties:
  - (i) The diagonals bisect at right angles
  - (ii) The diagonal bisects the angle through which it passes.

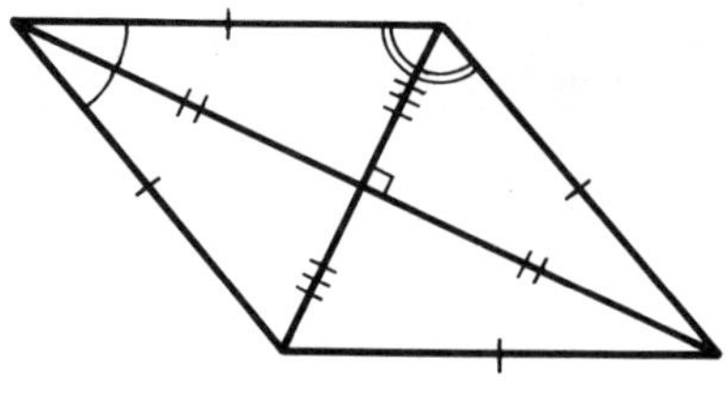

Fig. 29.3

Example

Fig. 29.4 shows a rhombus ABCD with ∠ABD = 65°. Find the sizes of ∠BDC, ∠ADB, ∠BAD and ∠AOB.

Since AB and CD are parallel

$$\angle BDC = \angle ABD = 65°$$

Since the diagonal bisects the angle through which it passes

$$\angle ADB = \angle BDC = 65°$$

Since AD and BC are parallel

$$\angle BAD = 180° - ABC = \angle ABC = 180° - 130° = 50°$$

Since the diagonals bisect at right angles

$$\angle AOB = 90°$$

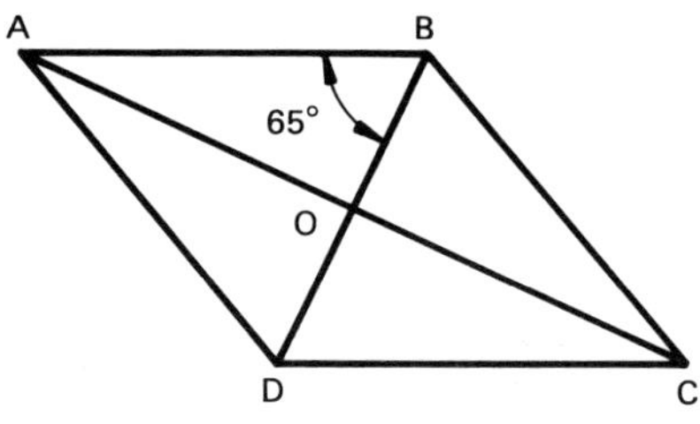

Fig. 29.4

- A RECTANGLE is a parallelogram with each of its angles equal to 90° (Fig. 29.5). A rectangle has all the properties of a parallelogram but, in addition, its diagonals are equal in length.

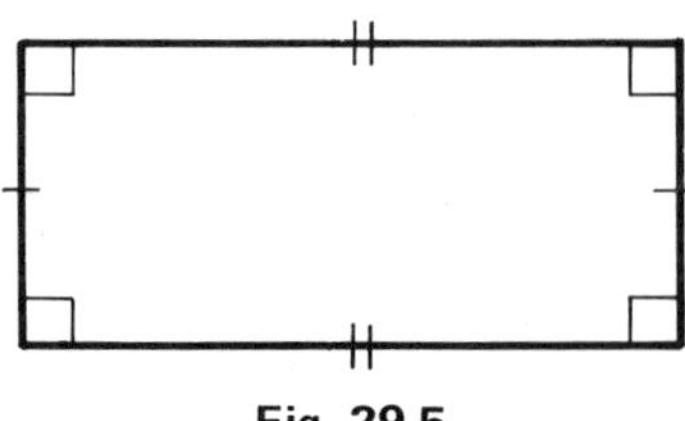

Fig. 29.5

- A SQUARE is a rectangle with all its sides equal (Fig. 29.6). It has all the properties of a parallelogram, rhombus and rectangle.

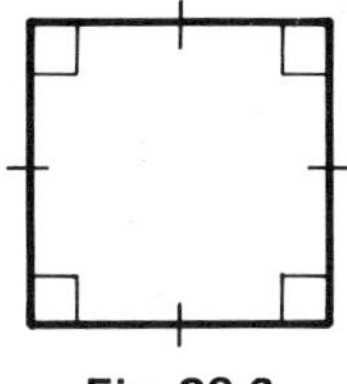

Fig. 29.6

- A TRAPEZIUM is a quadrilateral with one pair of sides parallel (Fig. 29.7).

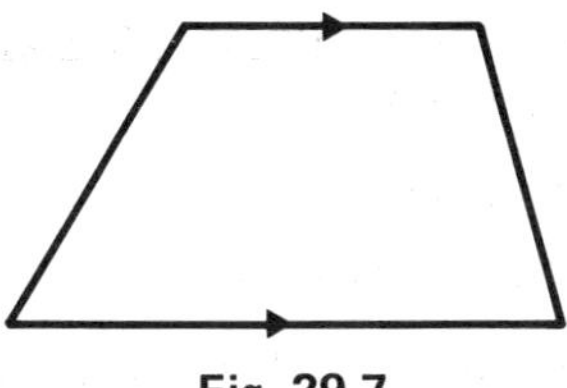

Fig. 29.7

- A KITE is a quadrilateral having two pairs of adjacent sides equal in length (Fig. 29.8).

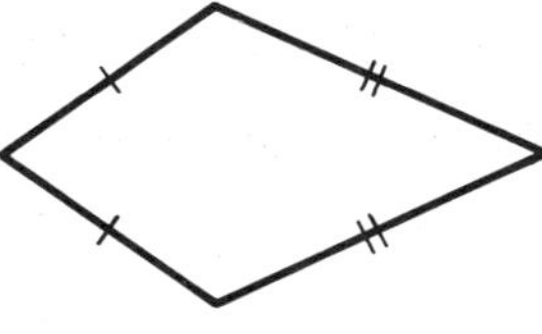

Fig. 29.8

Exercise 29.1

1. Find the angles marked x and y in Fig. 29.9.

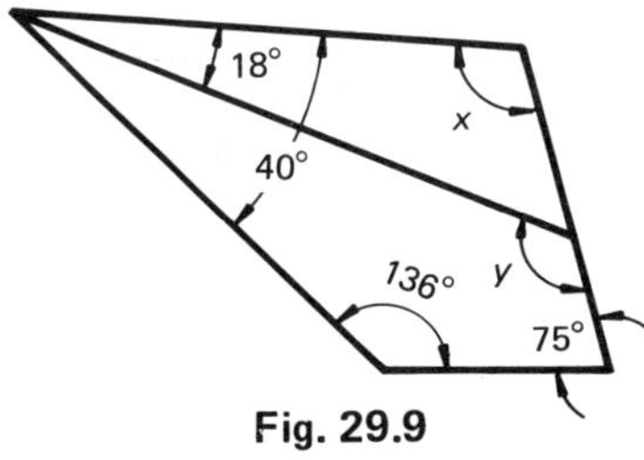

Fig. 29.9

2. Fig. 29.10 shows a parallelogram. Calculate the angles marked a and b.

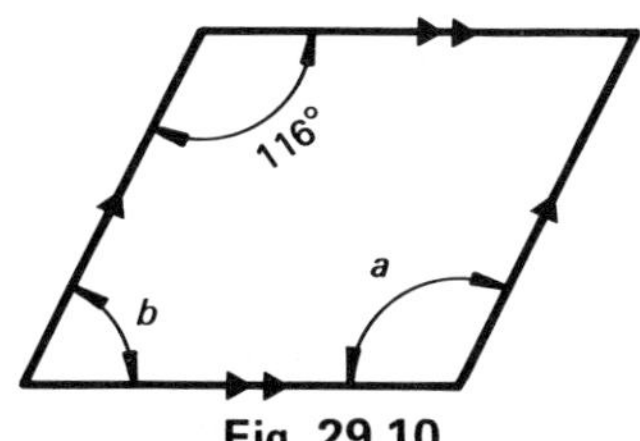

Fig. 29.10

3. In the rhombus ABCD (Fig. 29.11) write down the values of $\angle DAC$, $\angle BCD$, $\angle ABC$ and $\angle DOC$, given that $\angle ABD = 25°$.

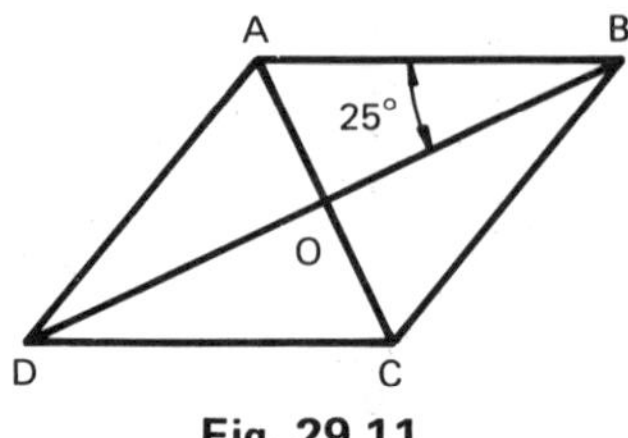

Fig. 29.11

4. In Fig. 29.12, the diagonal BD of the rhombus is 20 cm and the diagonal AC is 12 cm long.
   (a) What is the size of $\angle AOB$?
   (b) What is the length of AO?
   (c) What is the area of the rhombus ABCD?

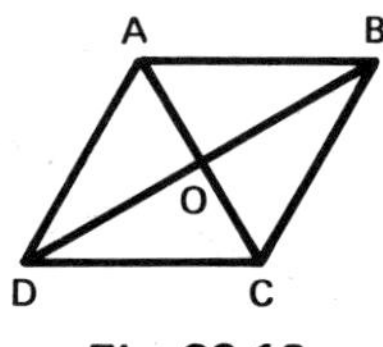

Fig. 29.12

5. A square has sides which are 8 cm long. Calculate the length of its diagonal.

6. The rectangle ABCD (Fig. 29.13) has its diagonal AC = 15 cm and its side AB = 11 cm. Calculate the length of BC.

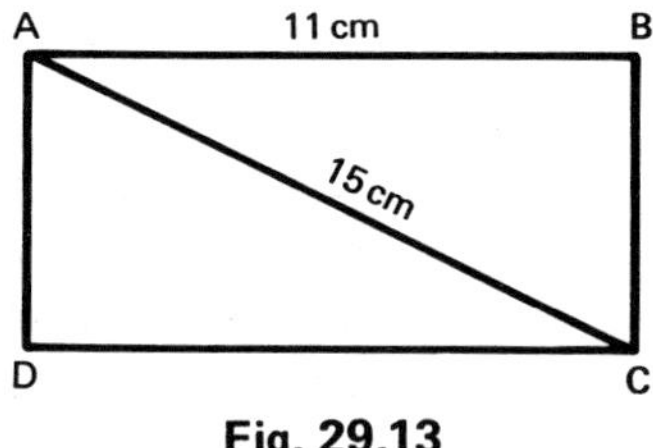

Fig. 29.13

7. In Fig. 29.14, find the size of the angle marked x.

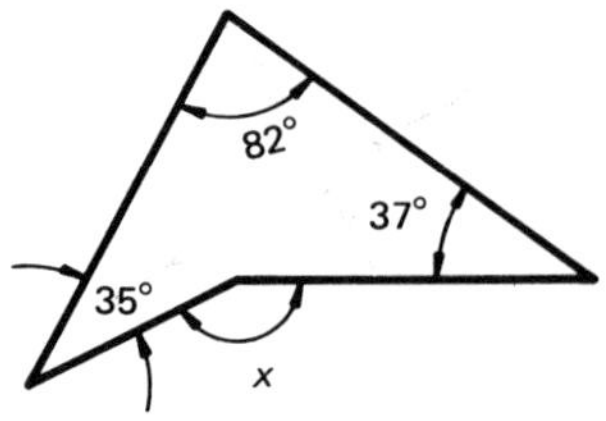

Fig. 29.14

8. Fig. 29.15 shows a kite. Given that ∠ABD = 40° and ∠BCD = 38°, find ∠ADB, ∠BAD and ∠BDC.

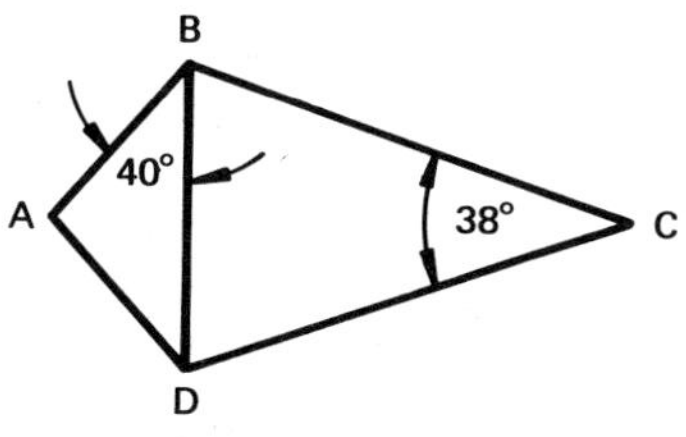

Fig. 29.15

Exercise 29.2 (All of the type found in CSE examination papers)

1. Fig. 29.16 shows an iron gate made with thin tubing. What length of tube is needed to make it? (S)

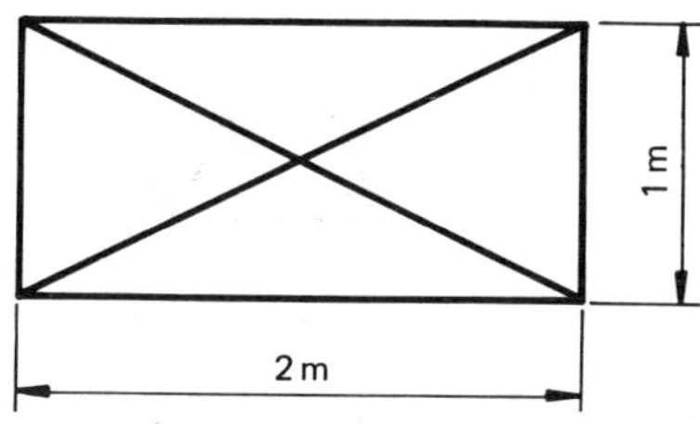

Fig. 29.16

2. In which two of the following quadrilaterals are the diagonals always equal in length?
   (a) a parallelogram (b) a rectangle
   (c) a rhombus (d) a square. (NW)

3. Fig. 29.17 shows two rods AB and CD which intersect at X. The points A, B, C and D are then joined to form a quadrilateral.
   (a) If CX = XD but AX is less than XB and the rods cross at right angles, what kind of quadrilateral is ACBD?
   (b) If CX = XD and AX = XB and the rods cross at 70°, what kind of quadrilateral is now formed by ACBD?
   (c) When AX = XB and CX = XD and the rods cross at right angles then the new quadrilateral will have what special name?

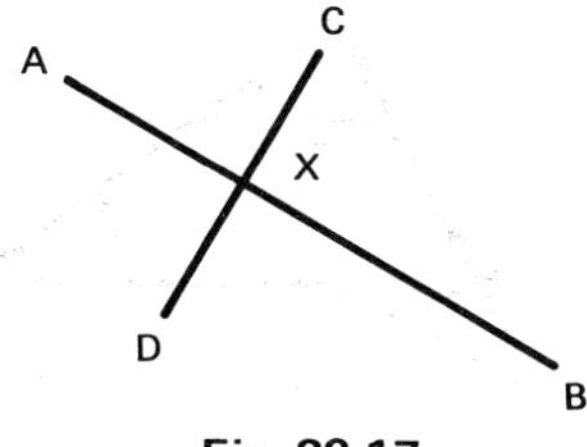

Fig. 29.17

4. In Fig. 29.18, find the size of the angle marked x. (EA)

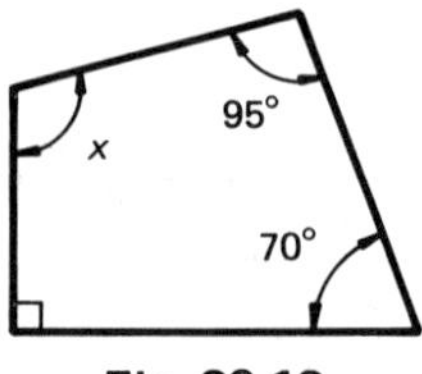

Fig. 29.18

5. Fig. 29.19 shows four different shapes.
   (a) Calculate the angle A in shape ABCD.
   (b) What is the name of the figure EFGH?
   (c) Which of the shapes is a trapezium? (EA)

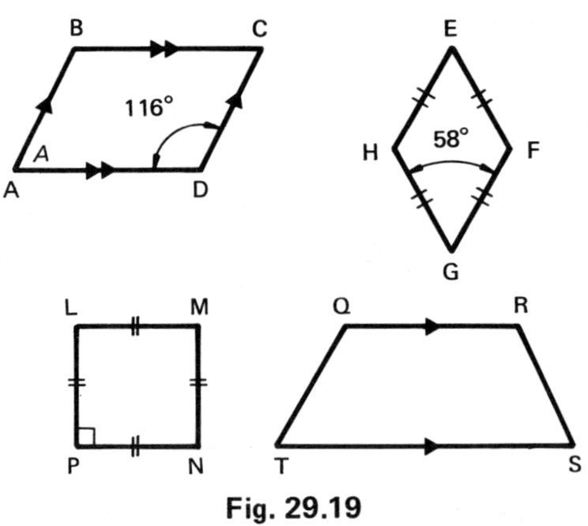

Fig. 29.19

6. In Fig. 29.20, ABCD is a rhombus. BD = 10 cm, AC = 6 cm and XE is a straight line parallel to BC.
   (a) (i) What is the size of ∠AXD?
   (ii) What is the length, in centimetres, of AX?
   (b) Calculate the area of the rhombus ABCD in square centimetres.
   (c) What is the name given to the quadrilateral EXCB? (S)

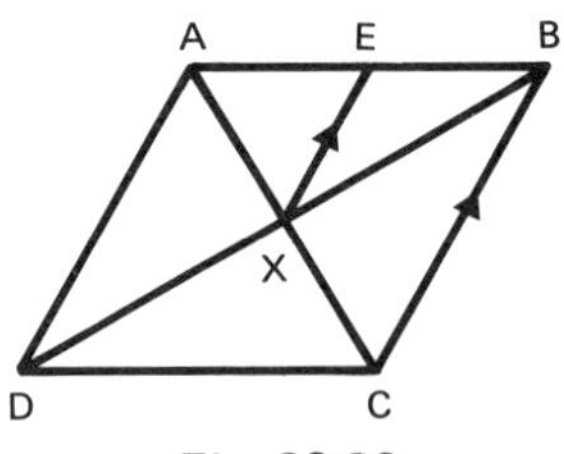

Fig. 29.20

7. In Fig. 29.21, ABCD is a parallelogram.
   (a) What is the name of the quadrilateral BCDX?
   (b) Calculate the size of the angle XDC.
   (c) Calculate the size of the angle DXB. (EA)

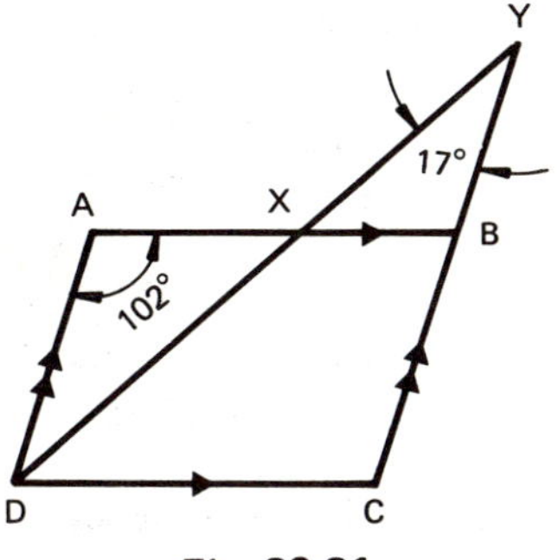

Fig. 29.21

8. For the parallelogram shown in Fig. 29.22, find:
   (a) the value of x
   (b) the value of y
   (c) the angle ACB
   (d) State the name of the triangle ABD. (EA)

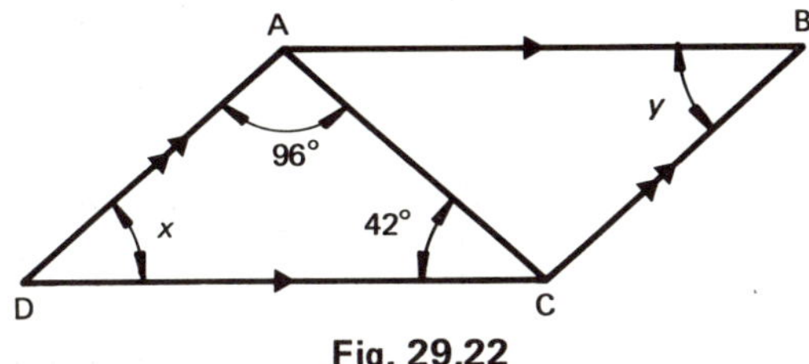

Fig. 29.22

9. In the rhombus ABCD (Fig. 29.23), diagonal AC = 24 cm and diagonal BD = 10 cm. Find:
   (a) AO and BO
   (b) AB
   (c) the area of the thombus ABCD. (EM)

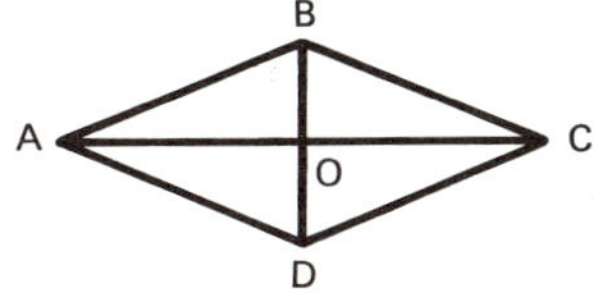

Fig. 29.23

10. In Fig. 29.24, ABCD is a rhombus. What is the size of the angle A? (AL)

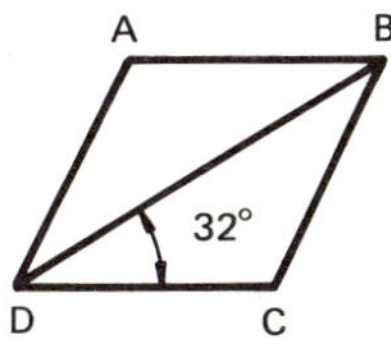

Fig. 29.24

Multi-choice questions 29

1. The opposite angles of a parallelogram are

   A supplementary B sometimes equal
   C complementary D never equal
   E always equal (NW)

2. In the trapezium in Fig. 29.25, the angles marked a and b are

   A $124°, 55°$ B $125°, 56°$
   C $125°, 46°$ D $125°, 67°$

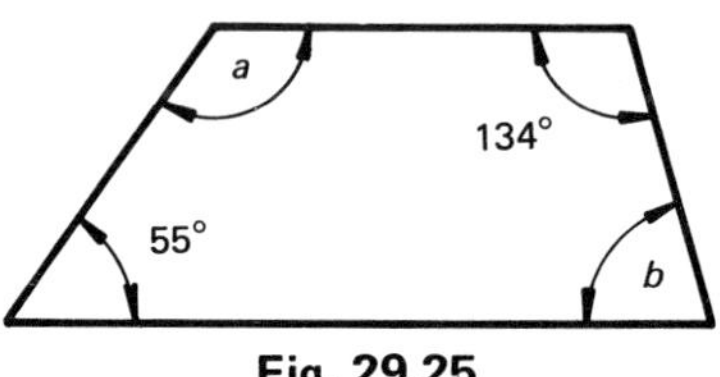

Fig. 29.25

3. Calculate the value of x in Fig. 29.26.

   A $75°$ B $85°$ C $95°$ D $105°$
   E $115°$ (NW)

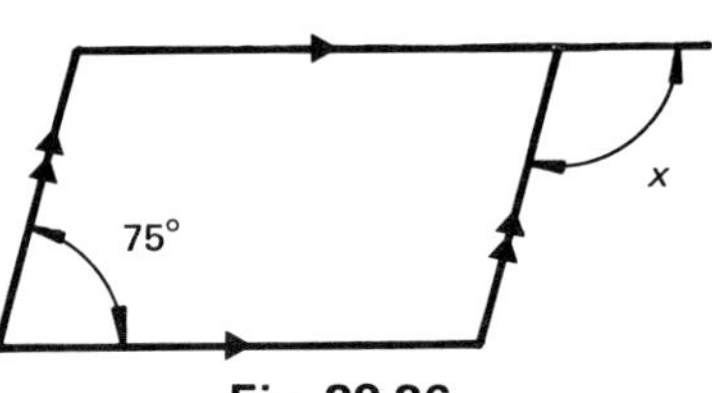

Fig. 29.26

4. Fig. 29.27 shows the rhombus WXYZ. The lengths of the diagonals WY and XZ are 12 cm and 16 cm respectively. What is the area of the rhombus?

   A 48 cm^2 B 96 cm^2 C 100 cm^2
   D 120 cm^2 E 192 cm^2 (AL)

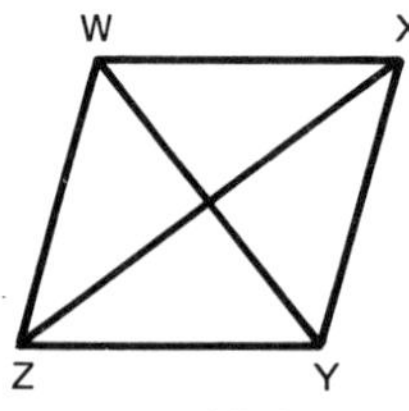

Fig. 29.27

5. The name given to the shape shown in Fig. 29.28 is a

   A kite B parallelogram C rhombus
   D trapezium E none of these (WY)

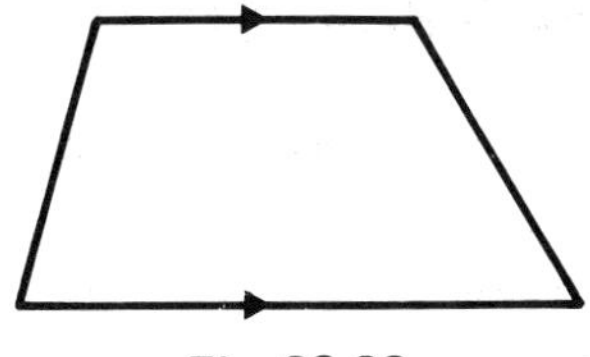

Fig. 29.28

6. A quadrilateral has three angles of 80°, 90° and 100°. The size of the fourth angle is

A 30° **B** 40° **C** 90° **D** 120°

30 POLYGONS

- A POLYGON is a plane figure bounded only by straight lines. Thus a triangle is a polygon having three sides and a quadrilateral is a polygon having four sides.

- A CONVEX polygon has no interior angle greater than 180° (Fig. 30.1).

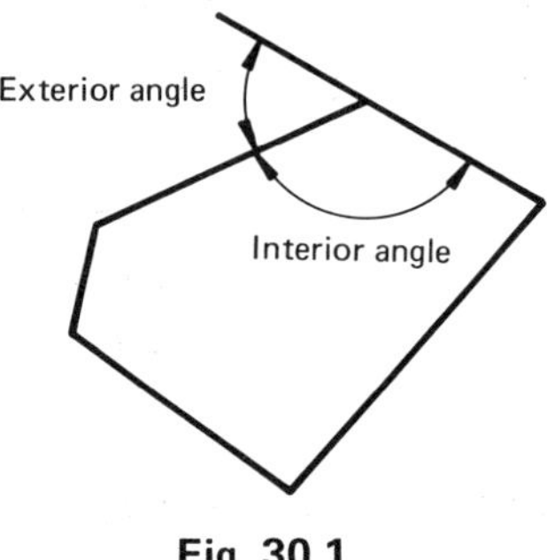

Fig. 30.1

- In a convex polygon the SUM OF THE INTERIOR ANGLES is $(2n-4)$ right angles, where n is the number of sides possessed by the polygon.

- The SUM OF THE EXTERIOR ANGLES of a polygon is 360°, no matter how many sides the polygon has.

- A REGULAR polygon has all its sides and all its angles equal.

- A RE-ENTRANT polygon has at least one angle greater than 180° (Fig. 30.2).

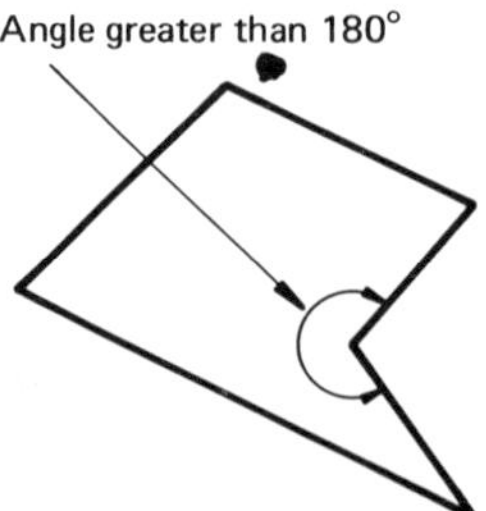

Fig. 30.2

- The table below gives the NAMES OF POLYGONS.

Name	*Number of Sides*	*Name*	*Number of Sides*
Pentagon	5	Nonagon	9
Hexagon	6	Decagon	10
Heptagon	7	Undecagon	11
Octagon	8	Dodecagon	12

Example

(a) Find the sum of the interior angles of a hexagon.

Since a hexagon has six sides, $n = 6$.

$$\begin{aligned}\text{Sum of interior angles} &= (2n-4) \text{ right angles}\\ &= (2\times6-4) \text{ right angles}\\ &= 8 \text{ right angles}\\ &= (8\times90)^\circ\\ &= 720^\circ\end{aligned}$$

(b) Each interior angle of a regular polygon is 150°. How many sides has it?

Each exterior angle $= 180^\circ - 150^\circ = 30^\circ$ (Fig. 30.3).

Since the sum of the exterior angles is 360°.

$$\text{Number of sides} = \frac{360}{30} = 12.$$

Hence the polygon has 12 sides, i.e. it is a dodecagon.

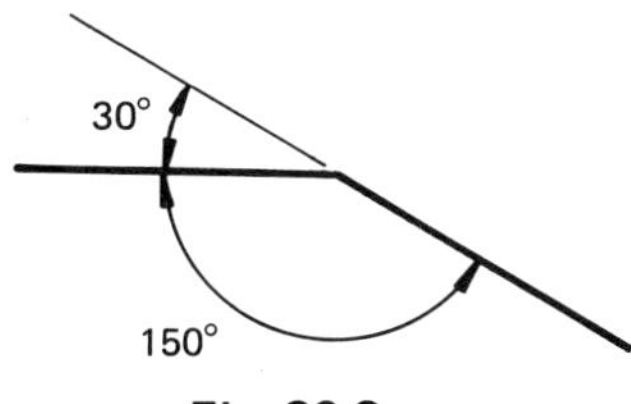

Fig. 30.3

(c) A regular polygon has 15 sides. What is the size of each interior angle?

$$\text{Each exterior angle} = \frac{360}{15} = 24^\circ$$

$$\text{Each interior angle} = 180^\circ - 24^\circ = 156^\circ$$

Exercise 30.1

1. Find the sum of the interior angles of a convex polygon with:
   (a) 6 sides (b) 9 sides (c) 11 sides.

2. A regular polygon has 8 sides. What is the size of each interior angle?

3. A regular polygon has 12 sides. What is the size of:
   (a) each exterior angle (b) each interior angle?

4. A pentagon has interior angles of 104°, 108° and 112°. If the remaining two angles are equal, what is their size?

5. Each interior angle of a regular polygon is 120°. How many sides has it?

Exercise 30.2 (All of the type found in CSE examination papers)

1. Find the value of y used to represent the two equal angles in a hexagon (Fig. 30.4). (EA)

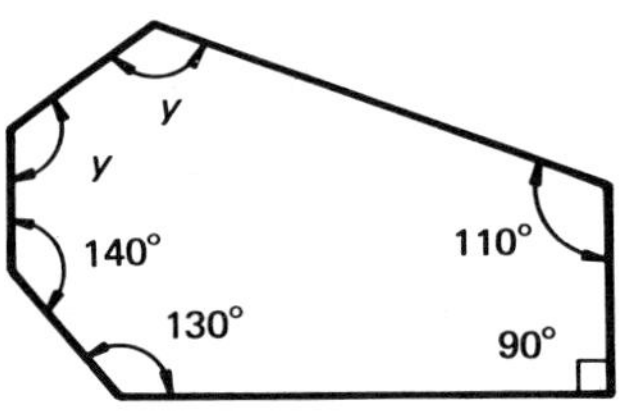

Fig. 30.4

2. A regular polygon has an interior angle of 140°. Calculate:
   (a) An exterior angle
   (b) The number of sides of the polygon.

3. The diagram in Fig. 30.5 shows a regular pentagon ABCDE inscribed in a circle with centre O. Calculate the size of:
   (a) angle AOB (b) angle ABC. (Y)

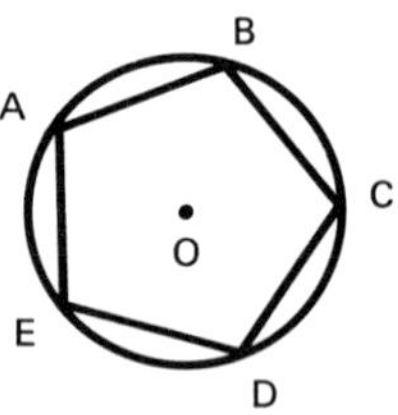

Fig. 30.5

4. The diagram in Fig. 30.6 shows a regular hexagon. What are the sizes of the angles marked x and y? (W)

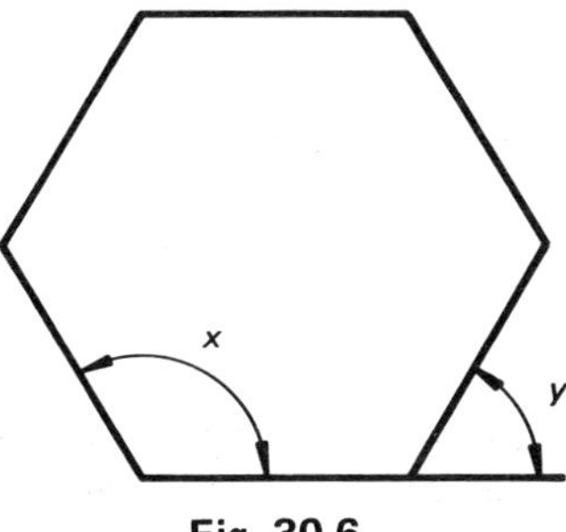

Fig. 30.6

5. The exterior angle of a regular polygon is 24°. Calculate the number of sides of the polygon. (Y)

6. UVWXYZ is a regular hexagon (Fig. 30.7) with centre O. What is the size of angle UVO?

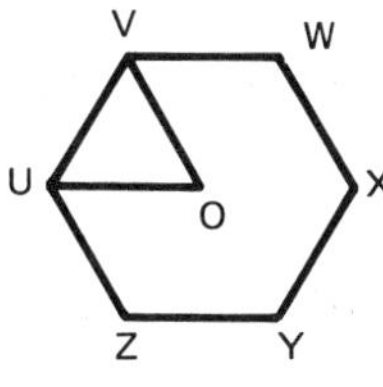

Fig. 30.7

7. Each interior angle of a regular polygon is 156°. How many sides does the polygon have?

8. Which of the shapes shown in Fig. 30.8 is a hexagon?

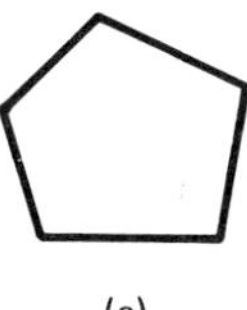
(a)

(b)

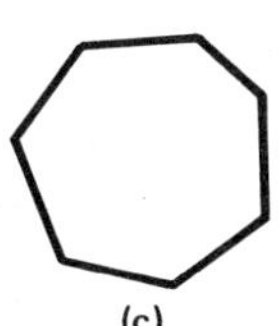
(c)

(d)

Fig. 30.8

31 THE CIRCLE

- The angle which an ARC OF A CIRCLE SUBTENDS AT THE CENTRE is twice the angle which the arc subtends at the circumference (Fig. 31.1).

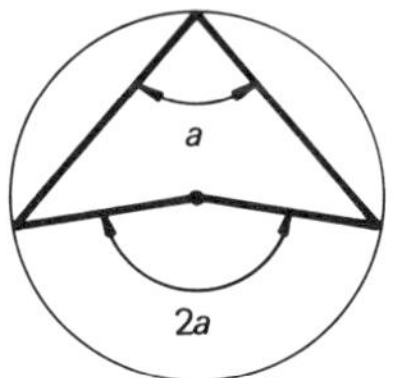

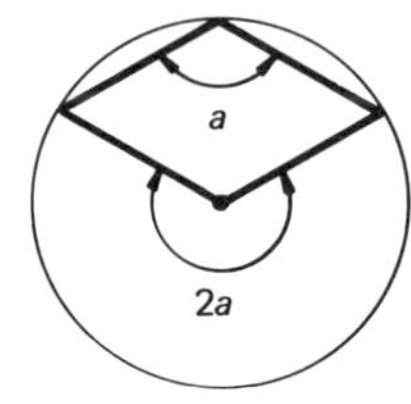

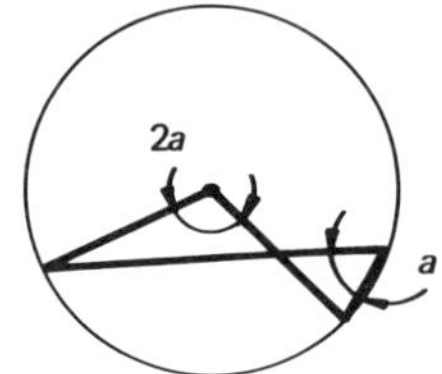

Fig. 31.1

- If a TRIANGLE IS INSCRIBED IN A SEMI-CIRCLE, the angle opposite the diameter is a right angle (Fig. 31.2).

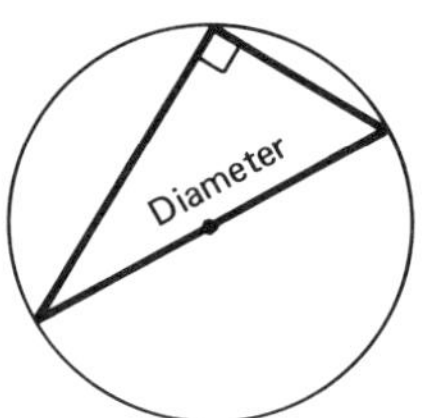

Fig. 31.2

- Angles in the SAME SEGMENT OF A CIRCLE are equal (Fig. 31.3).

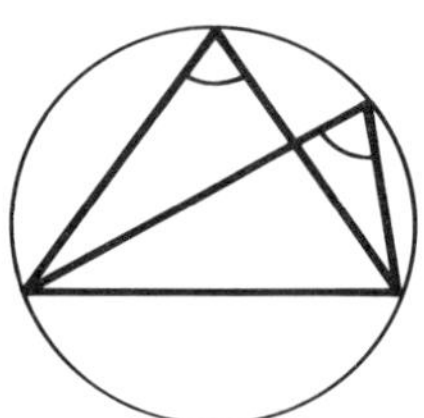

Fig. 31.3

- The opposite angles of a QUADRILATERAL INSCRIBED IN A CIRCLE (i.e. a cyclic quadrilateral) are equal to 180°. Thus in Fig. 31.4, $\angle A + \angle C = 180°$ and $\angle B + \angle D = 180°$.

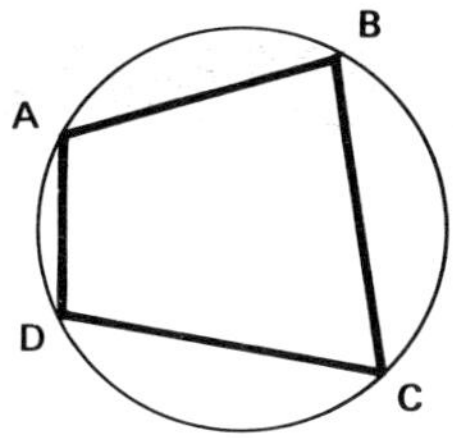

Fig. 31.4

Exercise 31.1 (All of the type found in CSE examination papers)

1. In Fig. 31.5, AC is a diameter of a circle. Calculate:
   (a) Angle BDC (b) Angle ACB. (SW)

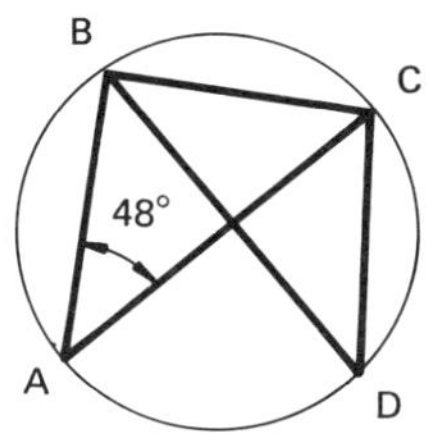

Fig. 31.5

2. P, S and R are points on the circumference of a circle with centre O (Fig. 31.6). Calculate the size of angle POR.

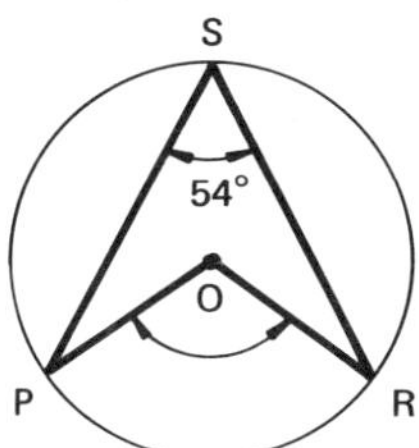

Fig. 31.6

3. In Fig. 31.7, O is the centre of the circle. Calculate the size of the obtuse angle AOB. (SW)

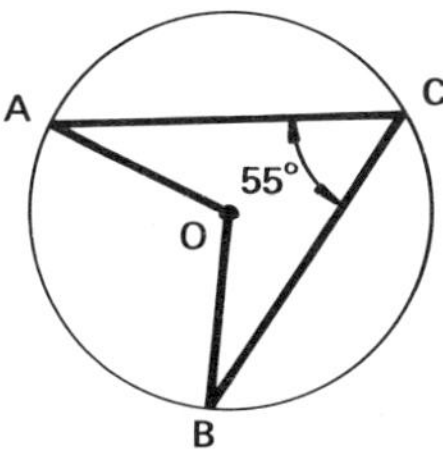

Fig. 31.7

4. A, B, C and D are points on the circumference of a circle (Fig. 31.8). If O is the centre of the circle, calculate the size of:

(a) angle BAD (b) angle BCD. (S)

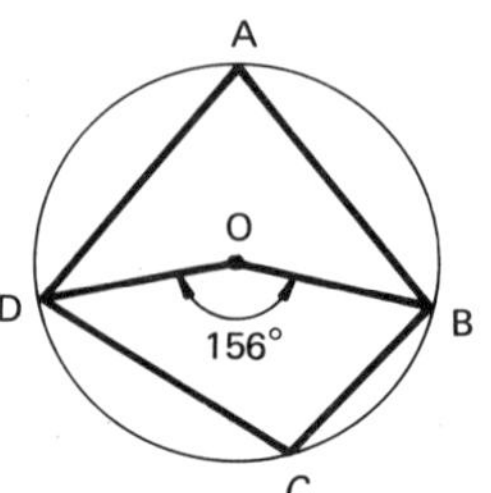

Fig. 31.8

5. The diagram in Fig. 31.9 shows a circle with centre O. Calculate the angles marked:

(a) x (b) y. (SW)

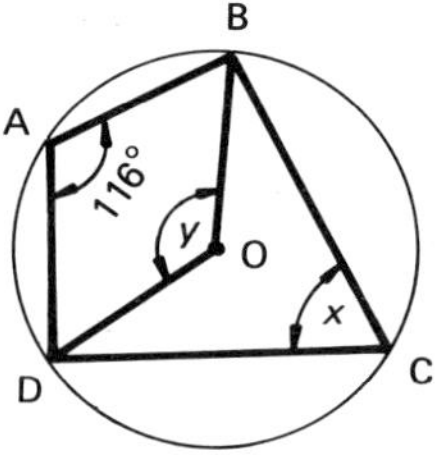

Fig. 31.9

6. In Fig. 31.10, O is the centre of the circle and AOC is a straight line. Calculate:

(a) the length of AC (b) the area of triangle ABC. (S)

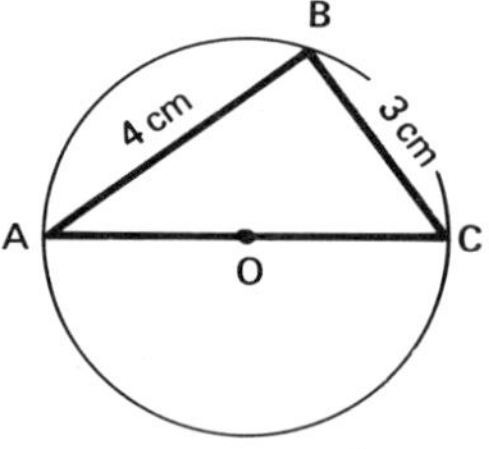

Fig. 31.10

7. In Fig. 31.11, ABC is an isosceles triangle with AB = AC. P is the centre of the circle and BPY is a straight line. Calculate:

(a) ∠BAC (b) ∠BYC (c) ∠ACY. (EA)

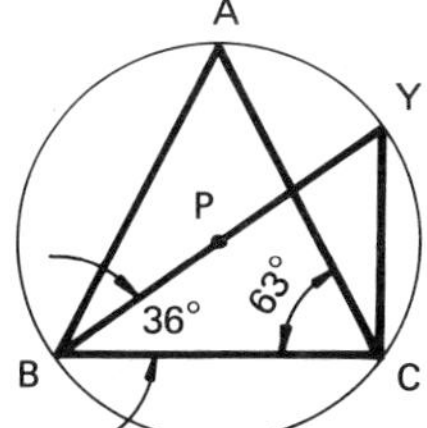

Fig. 31.11

8. A circle ABC (Fig. 31.12) has a centre O and ∠AOB = 50°. Calculate the angles ABO and OAC.

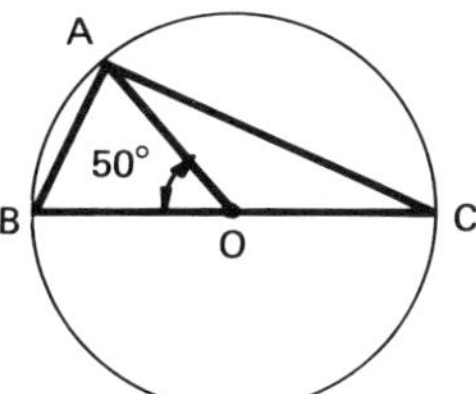

Fig. 31.12

9. In Fig. 31.13, K is the centre of the circle and AC is parallel to BD. Calculate the angles:

(a) CBA (b) ACB (c) DAB.  (EA)

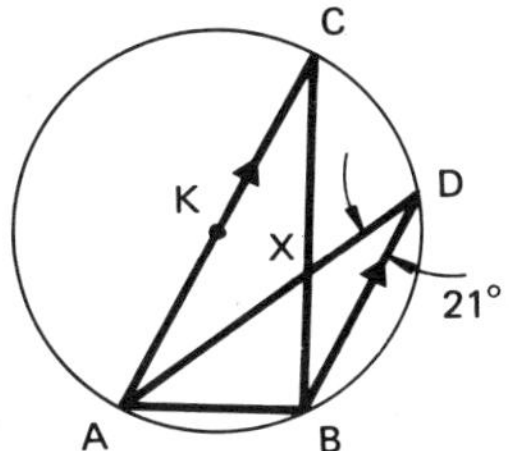

Fig. 31.13

10. In Fig. 31.14, O is the centre of the circle. Find the size of the angles.

(a) AOB (b) ADB.

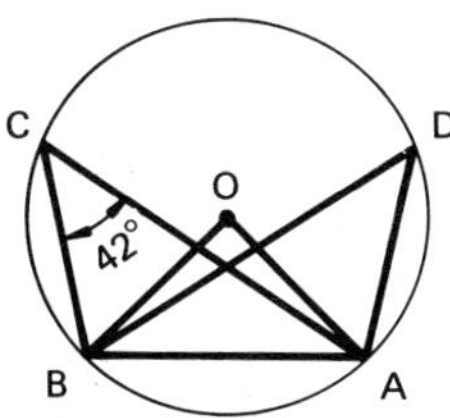

Fig. 31.14

Multi-choice questions 31

1. PQRS is a cyclic quadrilateral (Fig. 31.15) and PQ is produced to T. The size of angle RQT is

   A 40° **B** 70° **C** 90° **D** 100°
   E 110° (AL)

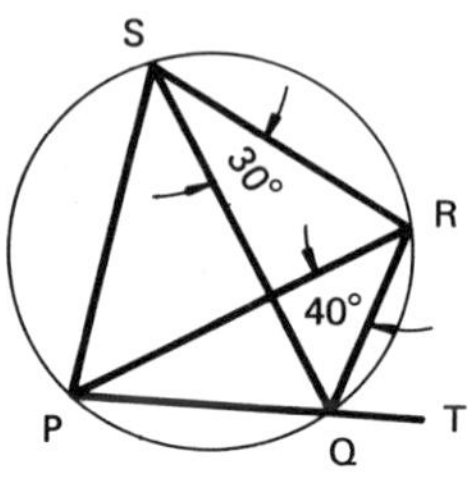

Fig. 31.15

The circle in Fig. 31.16 has a centre O and COD is a straight line. Use this diagram to answer questions 2, 3 and 4.

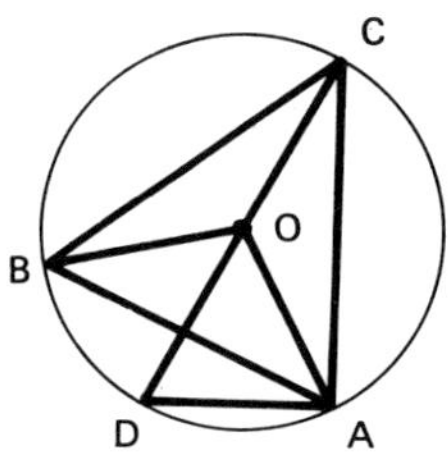

Fig. 31.16

2. If angle ACB is 50° then the obtuse angle AOB will be

   A 40° **B** 50° **C** 90° **D** 100° (EA)

3. Angle DAC will be

   A 40° **B** 50° **C** 90° **D** 100° (EA)

4. Angle ADC will be the same size as the angle

   A ABC **B** ACD **C** CAB **D** AOD (EA)

5. In Fig. 31.17, O is the centre of the circle. Calculate the size of the angle A.

   A 20° **B** 40° **C** 50° **D** 60°
   E 80° (NW)

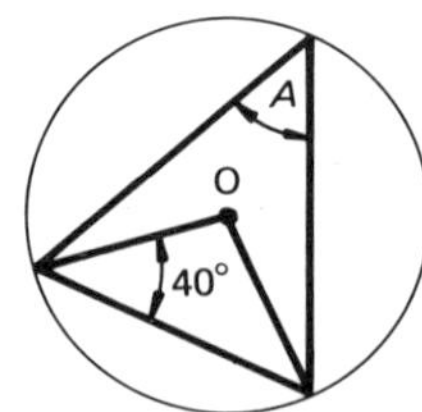

Fig. 31.17

6. In Fig. 31.18, O is the centre of the circle. Calculate x.

A 40° **B** 45° **C** 80° **D** 90°
E 160° (NW)

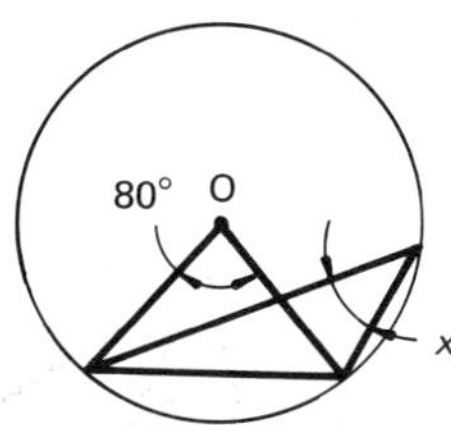

Fig. 31.18

7. In Fig. 31.19, AB is a diameter of the circle of centre O. When the angle CAD = 68°, the size of angle CBD is

A 68° **B** 112° **C** 136° **D** 158° (AL)

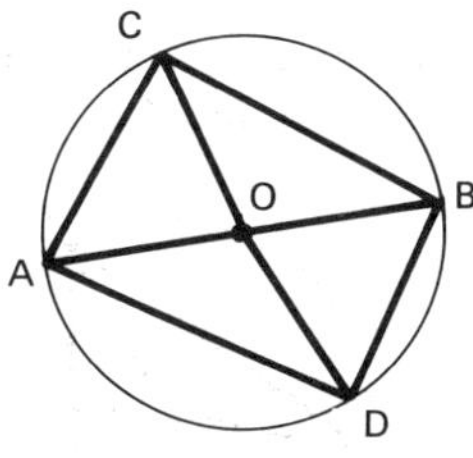

Fig. 31.19

In Fig. 31.20, ABCD is a cyclic quadrilateral. Use this diagram to answer questions 8, 9 and 10.

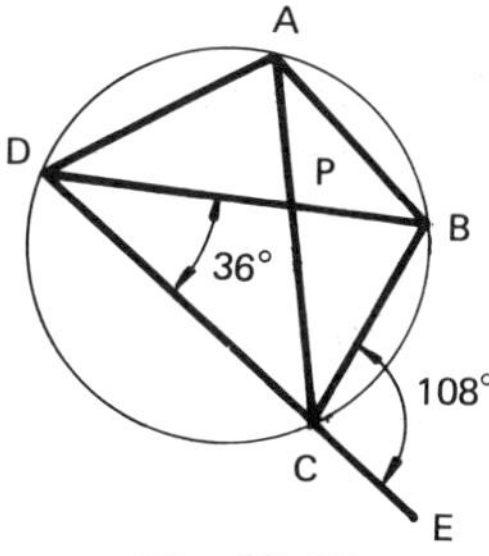

Fig. 31.20

8. What is the size of angle DAB?

A 18° **B** 54° **C** 72° **D** 108°
E 144° (AL)

9. What is the size of angle DBC?

A 18° **B** 54° **C** 72° **D** 108°
E 144° (AL)

10. What is the size of angle CAD?

A 18° **B** 54° **C** 72° **D** 108°
E 144° (AL)

32 SYMMETRY

- A LINE OR AXIS OF SYMMETRY on a shape is that line which can be used as a fold so that one half of the shape covers the other half exactly. Thus in Fig. 32.1, the trapezium shown has one line of symmetry while the rectangle has two lines of symmetry.

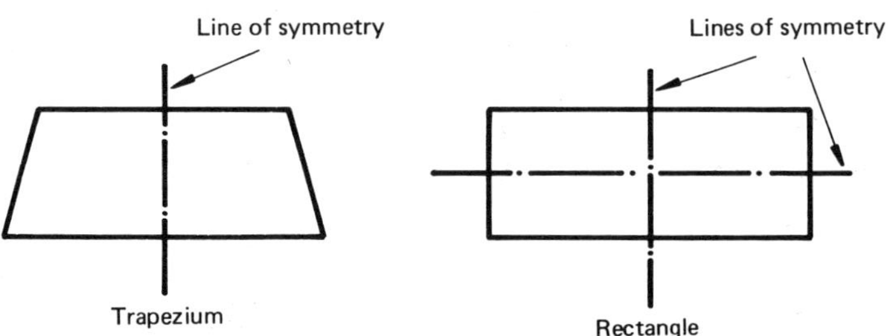

Fig. 32.1

- Fig. 32.2 shows a square with O the point where its diagonals intersect. If the square is rotated about O through one of the angles 90°, 180°, 270° and 360°, it will appear not to have moved. We say that a square has ROTATIONAL SYMMETRY of order 4. Similarly a rectangle has rotational symmetry of order 2 because when it is rotated about 180° and 360° its position will appear to be the same.

  Every shape has rotational symmetry of 1 since a rotation through 360° will bring it back to its original position.

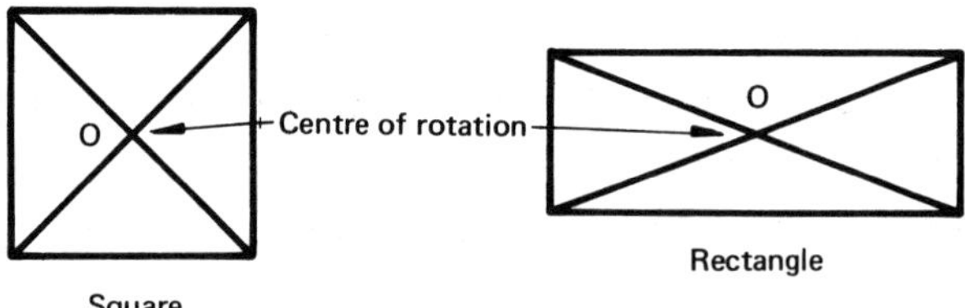

Fig. 32.2

- A plane shape has POINT SYMMETRY if it appears to be in the same position after a rotation through 180°. Thus the parallelogram shown in Fig. 32.3 has no lines of symmetry but it possesses point symmetry.

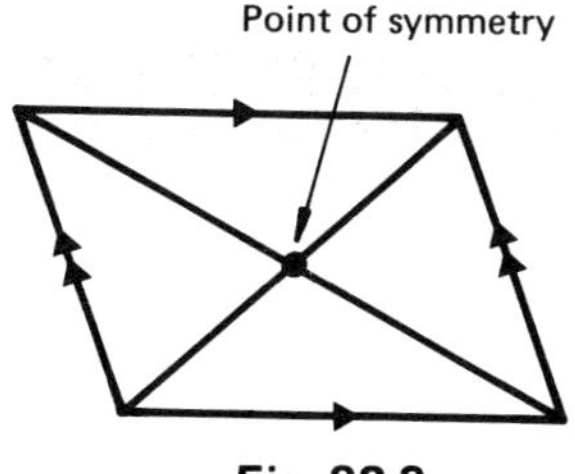

Fig. 32.3

- The table below gives details of the SYMMETRY OF VARIOUS QUADRILATERALS.

	No. of axes of symmetry	*Order of rotational symmetry*	*Point symmetry*
Square	4	4	Yes
Rectangle	2	2	Yes
Parallelogram	0	2	Yes
Rhombus	2	2	Yes
Trapezium (isosceles)	1	1	No
Kite	1	1	No

Exercise 32.1 (All of the type found in CSE examination papers)

1. Fig. 32.4 shows a rhombus drawn inside a rectangle. How many lines of symmetry has the figure? (EA)

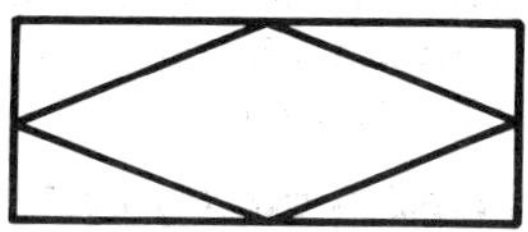

Fig. 32.4

2. Draw:

   (a) a rectangle (b) a kite

   and mark on your diagram the number of lines of symmetry that each figure has. (EM)

3. What is the order of rotational symmetry of the figure shown in Fig. 32.5?

Fig. 32.5

4. (a) Draw a triangle with three axes of symmetry.

(b) Draw a shape which has just one axis of symmetry and show that axis of symmetry as a broken line. (EM)

5. Mark the axes of symmetry (if any) for the three triangles shown in Fig. 32.6. (EM)

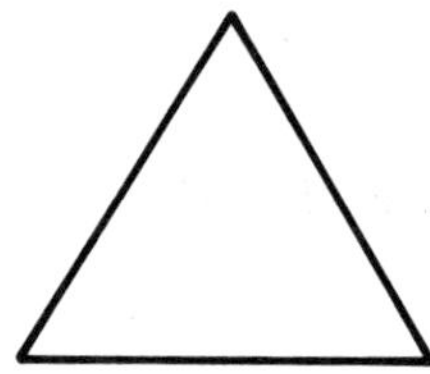

Fig. 32.6

6. (a) On a diagram of a regular pentagon draw one axis of symmetry.

(b) What order of rotational symmetry does a regular pentagon possess? (Y)

7. Copy each of the quadrilaterals shown in Fig. 32.7 and on each of them mark:

(a) the axes of symmetry (if any)

(b) the point of symmetry. (EM)

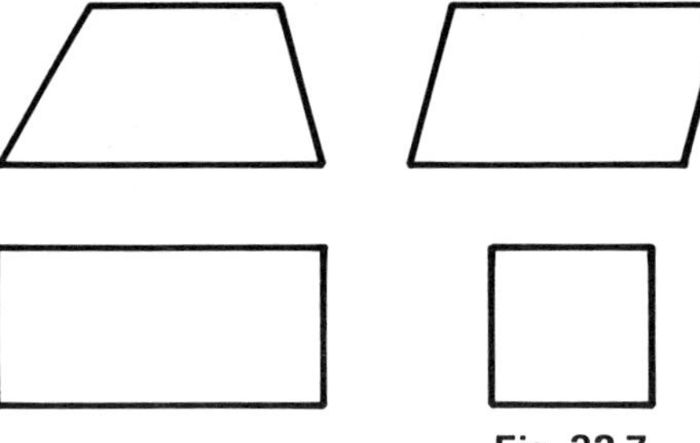
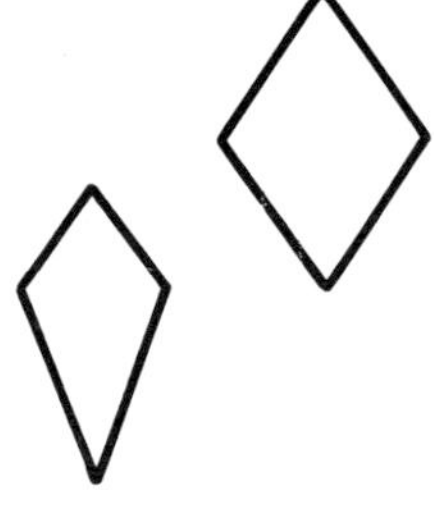

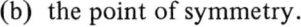

Fig. 32.7

8. How many axes of symmetry has:

(a) a regular octagon (b) a regular hexagon? (W)

Multi-choice questions 32

1. How many lines of symmetry can be drawn on the letter H shown in Fig. 32.8?

A 0 **B** 1 **C** 2 **D** 3 (EA)

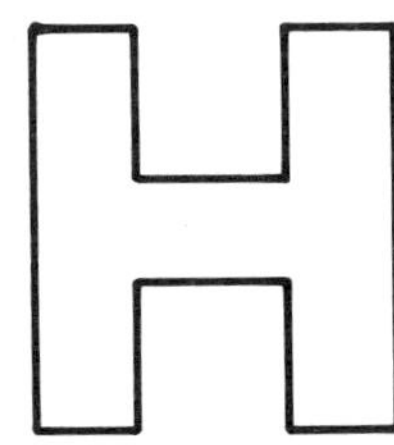

Fig. 32.8

2. Which one of the following statements is not correct?

   A Any square has four axes of symmetry.
   B Any rhombus has two axes of symmetry.
   C Any kite has one axis of symmetry.
   D Any trapezium has one axis of symmetry. (AL)

3. How many axes of symmetry has a regular hexagon?

   A 3 **B** 4 **C** 5 **D** 6

4. Fig. 32.9 shows a tessalation. How many different lines of symmetry can be drawn on it?

   A 0 **B** 1 **C** 2 **D** 4 (AL)

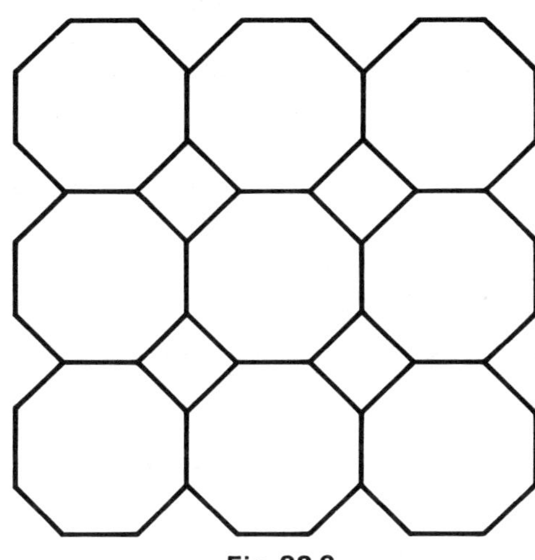

Fig. 32.9

5. Which of the following shapes has no line of symmetry but has point symmetry?

   A rhombus **B** kite **C** regular octagon
   D regular pentagon **E** none of these (WY)

6. Fig. 32.10 shows a kite. About which line is the figure symmetrical?

   A BC **B** AB **C** CD **D** AD

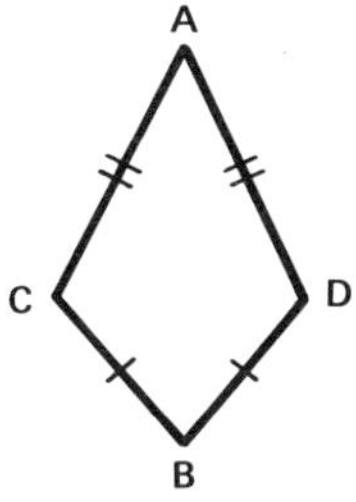

Fig. 32.10

33 GEOMETRICAL CONSTRUCTIONS

• 1. To divide a line AB into two equal parts

Construction. With A and B as centres and a radius greater than $\frac{1}{2}$AB, draw circular arcs which intersect at X and Y (Fig. 33.1). Join XY. The line XY divides AB into two equal parts and it is also perpendicular to AB.

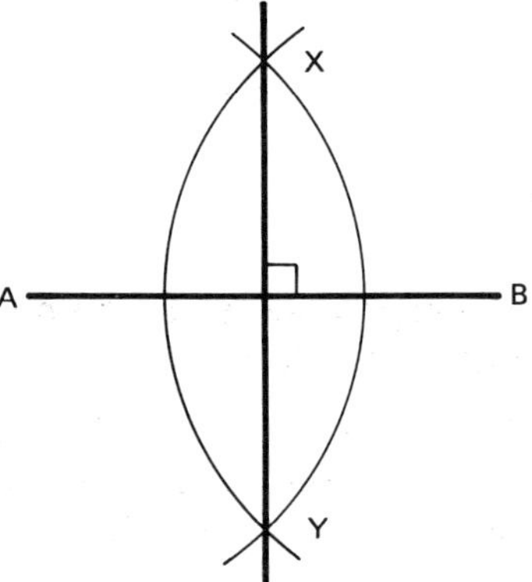

Fig. 33.1

• 2. To draw a perpendicular from a given point A on a straight line

Construction. With centre A and any radius draw a circle to cut the straight line at points P and Q (Fig. 33.2). With centres P and Q and a radius greater than AP (or AQ) draw circular arcs to intersect at X and Y. Join XY. This line will pass through A and its perpendicular to the given line.

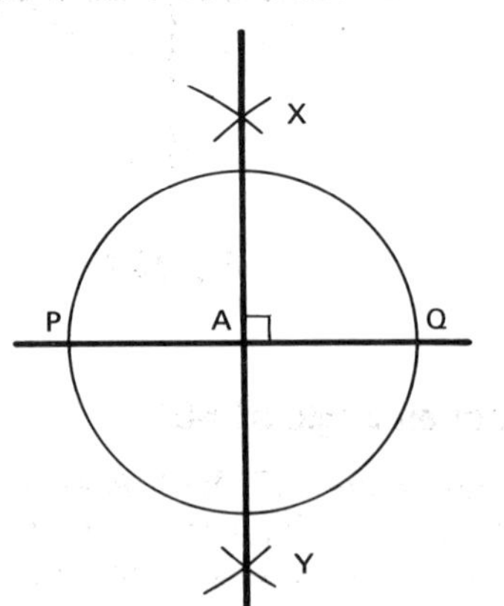

Fig. 33.2

3. To draw a perpendicular from a point A at the end of a line (Fig. 33.3)

Construction. From any point O outside the line and radius OA draw a circle to cut the line at B. Draw the diameter BC and join AC. AC is perpendicular to the straight line (because the angle in a semi-circle is 90°).

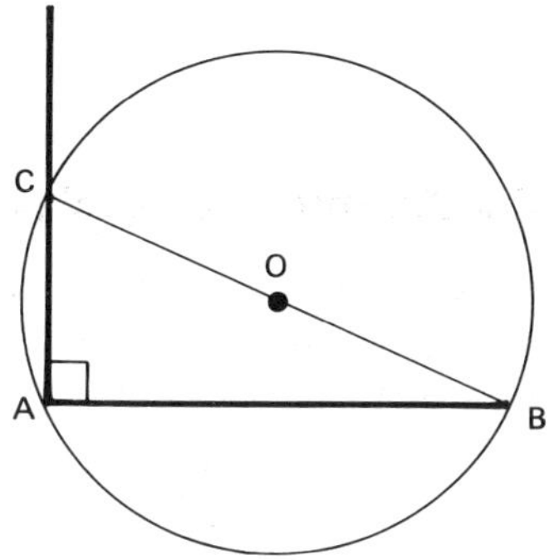

Fig. 33.3

4. To draw the perpendicular to a line AB from a given point P which is not on the line

Construction. With P as centre draw a circular arc to cut AB at points C and D. With C and D as centres and a radius greater than $\frac{1}{2}$CD, draw circular arcs to intersect at E. Join PE. The line PE is the required perpendicular (Fig. 33.4).

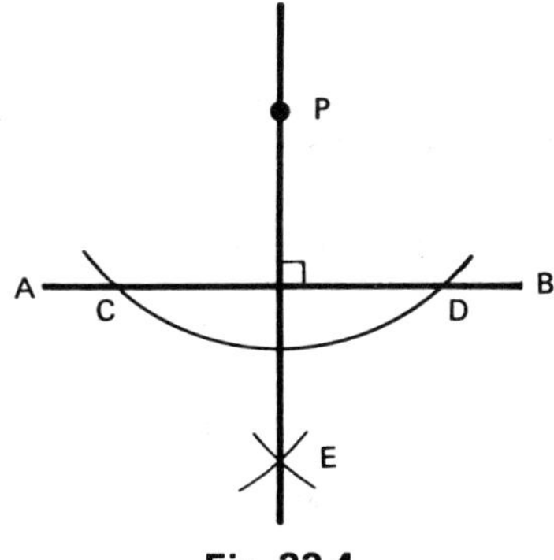

Fig. 33.4

5. To construct an angle of 60°

Construction. Draw a line AB. With A as centre and any radius draw a circular arc to cut AB at D. With D as centre and the *same* radius draw a second arc to cut the first arc at C. Join AC. The angle CAD is then 60° (Fig. 33.5, p. 184).

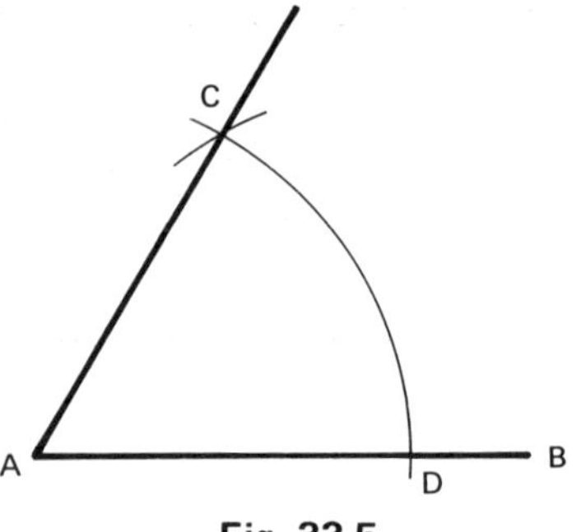

Fig. 33.5

6. To bisect a given angle BAC

Construction. With centre A and any radius draw an arc to cut AB at D and AC at E. With centres D and E and a radius greater than $\frac{1}{2}$DE draw arcs to intersect at F. Join AF, then AF bisects ∠BAC (Fig. 33.6). Note that by bisecting an angle of 60°, an angle of 30° is obtained. An angle of 45° is obtained by bisecting a right angle.

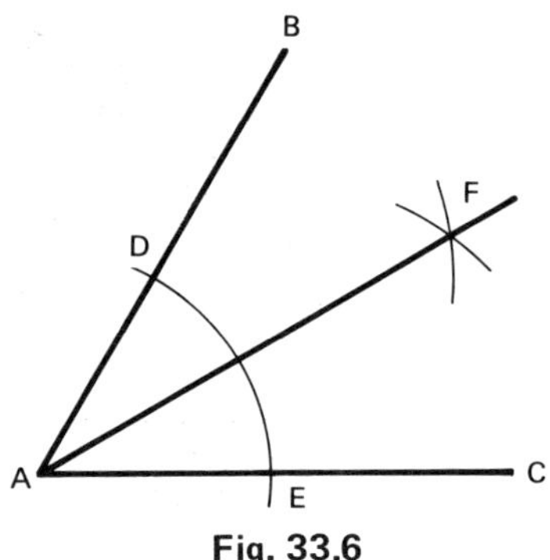

Fig. 33.6

7. To construct an angle equal to a given angle BAC

Construction. With centre A and any radius draw an arc to cut AB at D and AC at E. Draw the line XY. With centre X and the same radius draw an arc to cut XY at W. With centre W and radius equal to DE draw an arc to cut the first arc at V. Join VX, then ∠VXW = ∠BAC (Fig. 33.7).

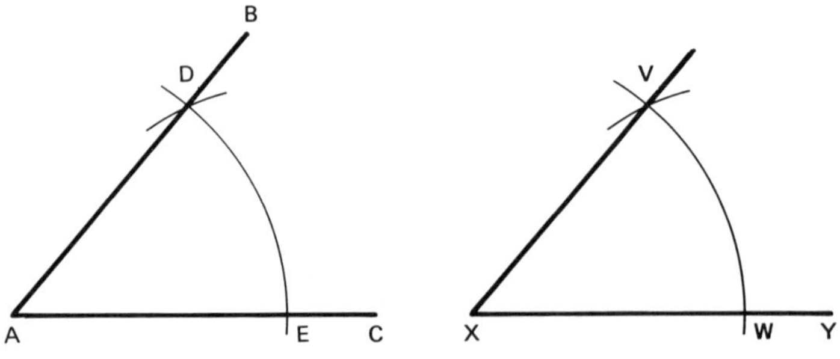

Fig. 33.7

8. To draw a line through a point P parallel to a given line AB

Construction. Mark off any two points X and Y on AB. With centre P and radius XY draw an arc. With centre Y and radius XP draw a second arc to cut the first arc at Q. Join PQ, then PQ is parallel to AB (Fig. 33.8).

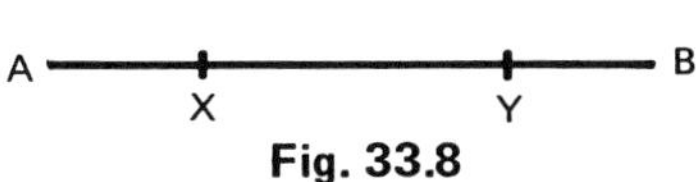

Fig. 33.8

9. To divide a straight line AB into a number of equal parts

Construction. Suppose that AB has to be divided into four equal parts. Draw AC at any angle to AB. Set off on AC, four equal parts AP, PQ, QR, RS of any convenient length. Join SB. Draw RV, QW and PX each parallel to SB. Then AX = XW = WV = VB (Fig. 33.9).

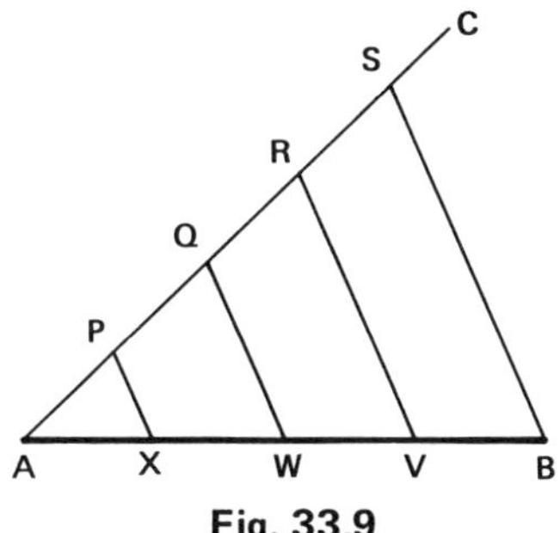

Fig. 33.9

10. To draw the circumscribed circle of a given triangle ABC

Construction. Construct the perpendicular bisectors of the sides AB and BC (using construction 1) so that they intersect at O. With centre O and radius AO draw a circle which is the required circumscribed circle (Fig. 33.10).

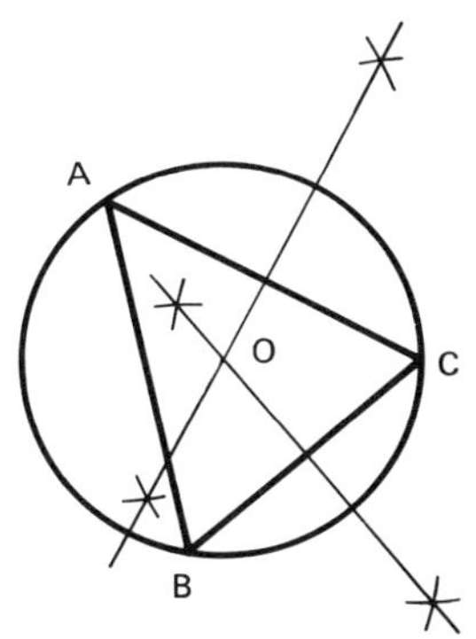

Fig. 33.10

11. To draw the inscribed circle of a given triangle ABC

Construction. Construct the internal bisectors of $\angle B$ and $\angle C$ (using construction 6) to intersect at O. With centre O draw the inscribed circle of the triangle ABC (Fig. 33.11).

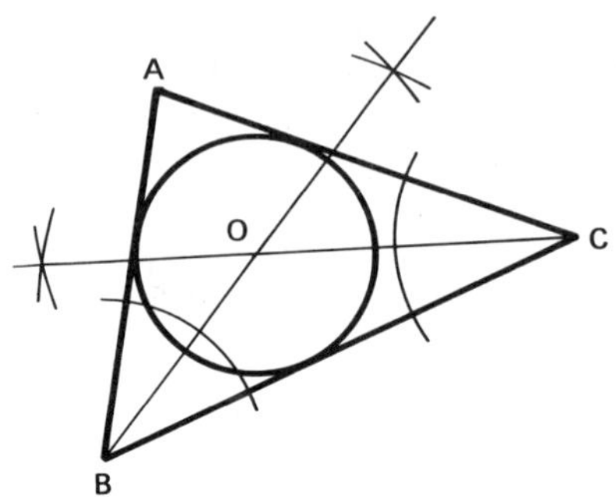

Fig. 33.11

12. To construct a triangle given the lengths of each of the three sides

Construction. Suppose $a = 6$ cm, $b = 3$ cm and $c = 4$ cm. Draw BC = 6 cm. With centre B and radius 4 cm draw a circular arc. With centre C and radius 3 cm draw a circular arc to cut the first arc at A. Join AB and AC. Then ABC is the required triangle (Fig. 33.12).

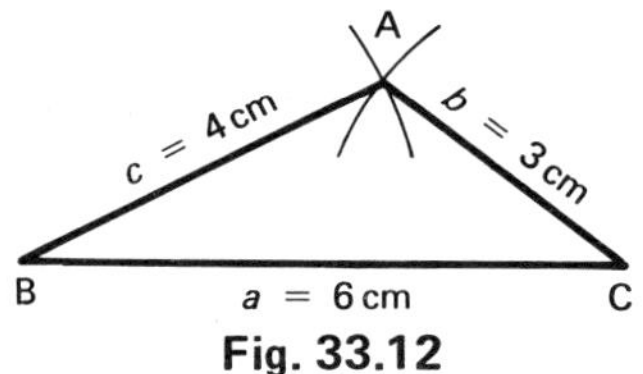

Fig. 33.12

13. To construct a triangle given two sides and the included angle between the two sides

Construction. Suppose $b = 5$ cm, $c = 6$ cm and $\angle A = 60°$. Draw AB = 6 cm and draw AX such that $\angle BAX = 60°$. Along AX mark off AC = 5 cm. Then ABC is the required triangle (Fig. 33.13).

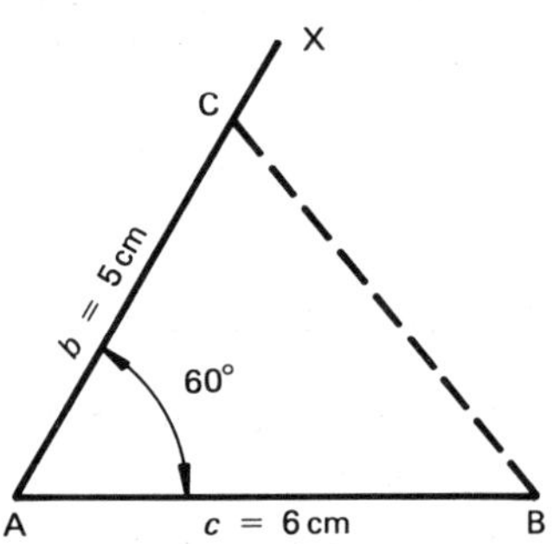

Fig. 33.13

14. To construct a triangle (or triangles) given the lengths of two of the sides and an angle which is not the included angle between the two given sides

Construction.

(a) Suppose $a = 5$ cm, $b = 6$ cm and $\angle B = 60°$. Draw BC = 5 cm and draw BX such that $\angle$CBX = 60°. With centre C and radius of 6 cm describe a circular arc to cut BX at A. Join CA, then ABC is the required triangle ABC (Fig. 33.14).

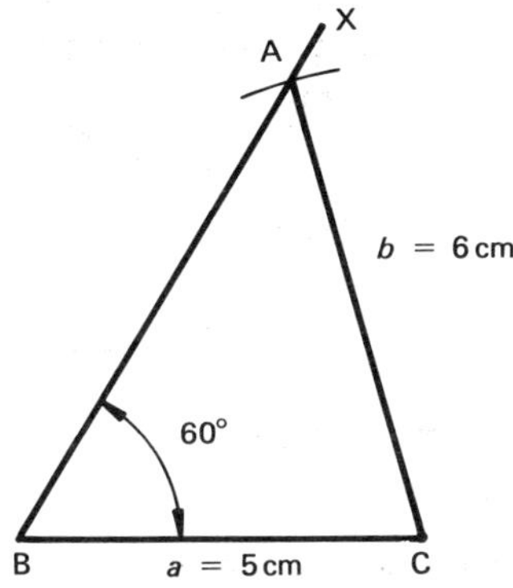

Fig. 33.14

(b) Suppose that $a = 5$ cm, $b = 4.5$ cm and $\angle B = 60°$. The construction is the same as before but the circular arc drawn with C as centre now cuts BX at two points A and A_1. This means that there are two triangles which meet the given conditions, i.e. Δs ABC and A_1BC (Fig. 33.15). For this reason this case is often called the ***ambiguous case***.

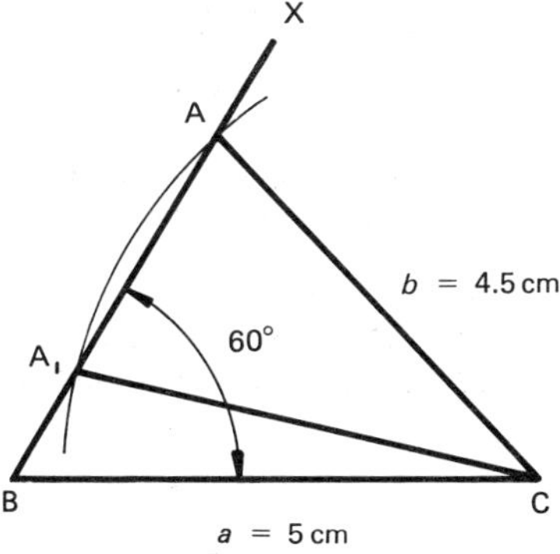

Fig. 33.15

Exercise 33.1 (All of the type found in CSE examination papers)

1. Draw a line PQ 8 cm long and construct its perpendicular bisector using ruler and compasses only.

2. Draw a line XY 9 cm long and mark off the point A such that AX = 4 cm. Through the point A draw a line perpendicular to XY.

3. Draw the line PQ 7 cm long and through P draw a line perpendicular to PQ.

4. Draw a horizontal line AB which is 9 cm long. Then mark off the point P such that it is 6 cm from A and 5 cm from B. Now construct a line perpendicular to AB which also passes through the point P.

5. (a) Construct an angle of 60°.
   (b) By bisecting your 60° angle construct an angle of 30°.

6. Draw a line AB 8 cm long and construct a line perpendicular to AB which passes through the point A. By bisecting the right angle so formed, construct an angle of 45°.

7. Draw a line 9 cm long and divide it into 5 equal parts without measuring each division.

8. (a) Construct the triangle ABC given that AB = 7 cm, BC = 8 cm and AC = 5 cm.
   (b) Draw the circumscribed circle for triangle ABC and measure its diameter.

9. (a) Construct triangle XYZ given XY = 5 cm, YZ = 7 cm and ∠XYZ = 50°.
   (b) Draw the inscribed circle for ΔXYZ and state its diameter.

10. (a) Draw the triangle ABC given that AB = 6.3 cm, AC = 7.7 cm and BC = 8.4 cm.
    (b) Draw the circle which passes through A, B and C. Measure the radius of this circle. (WM)

11. Construct the quadrilateral ABCD from the following data: AB = 10 cm, BC = 6 cm, AC = 14 cm, ∠DAC = 60° and ∠DCA = 45°. Measure and record the length DB. (W)

12. Draw the line AB 8 cm long. Now construct an equilateral triangle on the line segment AB. (Y)

34 TRIGONOMETRY

- The side opposite the right angle in Fig. 34.1 is called the HYPOTENUSE, i.e. side AB. The side BC lies opposite the angle A and hence it is called SIDE OPPOSITE A. The side AC is then called SIDE ADJACENT TO A.

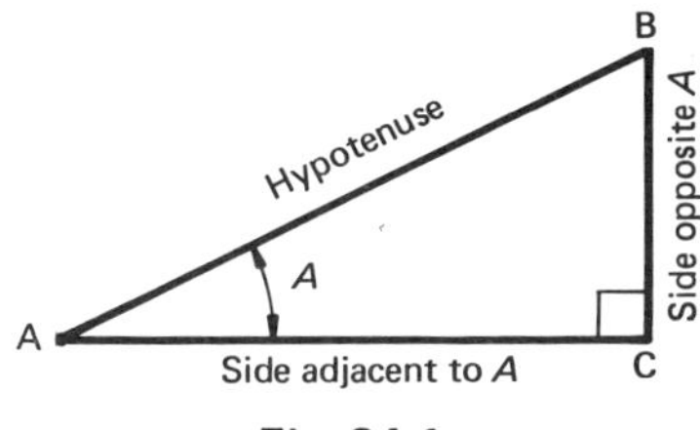

Fig. 34.1

- The three TRIGONOMETRICAL RATIOS are

$$\text{sine}\,A = \frac{\text{side opposite } A}{\text{hypotenuse}} = \frac{\text{BC}}{\text{AB}}$$

$$\text{cosine}\,A = \frac{\text{side adjacent to } A}{\text{hypotenuse}} = \frac{\text{AC}}{\text{AB}}$$

$$\text{tangent}\,A = \frac{\text{side opposite } A}{\text{side adjacent to } A} = \frac{\text{BC}}{\text{AC}}$$

The abbreviations sin, cos and tan are generally used.

- The values of the sine, cosine and tangents of angles may be found by using TRIGONOMETRICAL TABLES. Note the following
  (i) The sine of an angle goes from 0 (for 0°) to 1 (for 90°). The mean differences are ADDED.
  (ii) The cosine of an angle goes from 1 (for 0°) to 0 (for 90°). The mean differences are SUBTRACTED.
  (iii) The tangent of an angle goes from 0 (for 0°) to infinity (for 90°). The mean differences are ADDED.

Example

(a) In Fig. 34.2, find the length of AB.

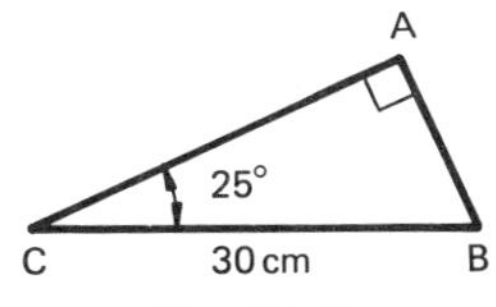

Fig. 34.2

AB is the side opposite C and BC is the hypotenuse.

$$\frac{AB}{BC} = \sin C$$

$$\begin{aligned} AB &= BC \times \sin C \\ &= 30 \sin 25^\circ \\ &= 30 \times 0.4226 \\ &= 12.68 \text{ cm} \end{aligned}$$

(b) Find the angle C (Fig. 34.3).

BC is the side adjacent to C.

$$\cos C = \frac{BC}{AC} = \frac{16}{22} = 0.7273$$

$$C = 43^\circ 21'$$

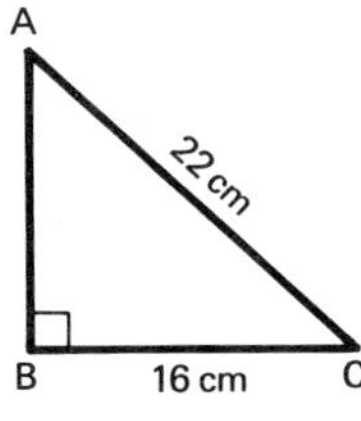

Fig. 34.3

(c) In Fig. 34.4, find BC.

AB is the side opposite C,

BC is the side adjacent to C.

$$\frac{AB}{BC} = \tan C$$

$$BC = \frac{AB}{\tan C} = \frac{25}{\tan 68^\circ}$$

$$= \frac{25}{2.4750} = 10.10 \text{ cm}$$

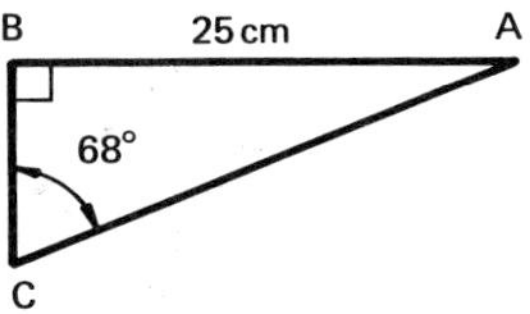

Fig. 34.4

Exercise 34.1

1. Use sine tables to find the sines of the following angles:
   (a) $14°$ (b) $19°18'$ (c) $76°52'$.
2. Use cosine tables to find:
   (a) $\cos 18°$ (b) $\cos 37°12'$ (c) $\cos 85°49'$.
3. Use the tangent tables to find:
   (a) $\tan 25°$ (b) $\tan 49°15'$ (c) $\tan 79°$.
4. (a) If $\sin A = 0.7624$, find the angle A.
   (b) If $\cos X = 0.5731$, find the angle X.
   (c) If $\tan Y = 2.4951$, find the angle Y.
5. Find the lengths of the sides marked x in Fig. 34.5.

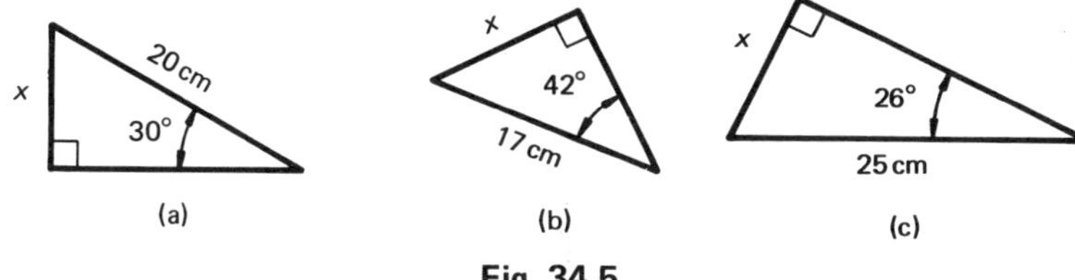

Fig. 34.5

6. Find the angles marked y in Fig. 34.6.

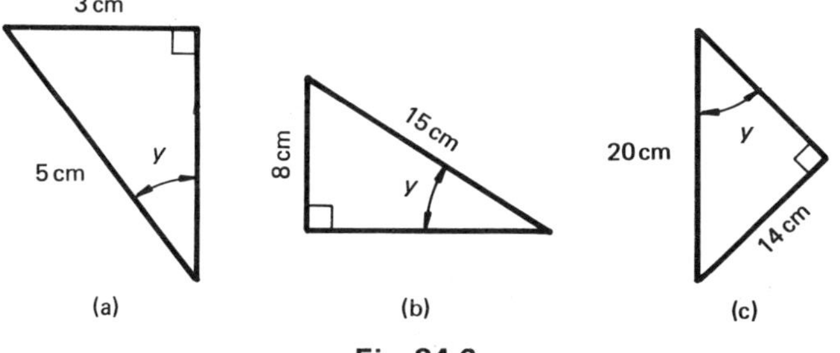

Fig. 34.6

7. An equilateral triangle has a vertical height of 20 cm. Calculate the lengths of the equal sides.
8. Calculate the vertical height of an isosceles triangle whose vertex angle is $42°$ and whose equal sides are 15 cm long.
9. In Fig. 34.7, find the angles marked x.

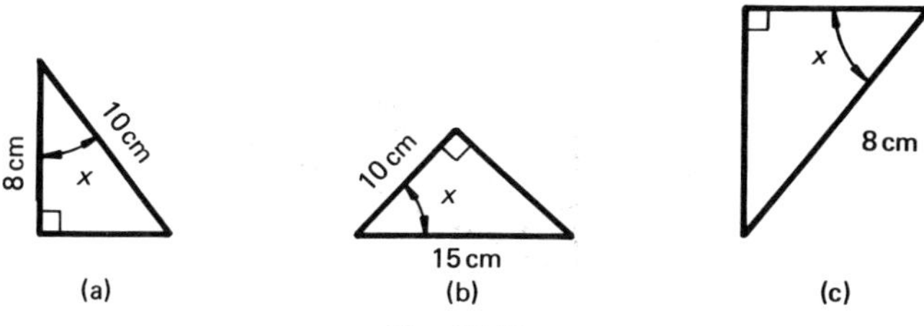

Fig. 34.7

10. Find the lengths of the sides marked p in Fig. 34.8.

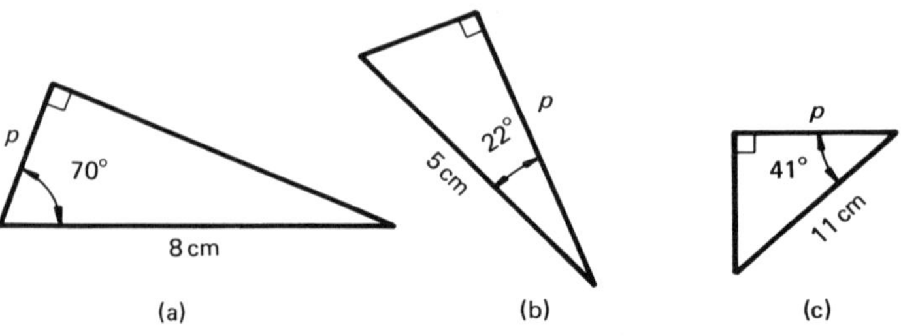

Fig. 34.8

11. The base of an isosceles triangle is 8 cm long and the equal sides are 12 cm long. Determine the angles of the triangle.

12. The vertex angle of an isosceles triangle is 56°. If the equal sides are each 14 cm long, find:
   (a) The length of the base
   (b) The vertical height of the triangle.

13. Find the lengths of the sides marked x in Fig. 34.9.

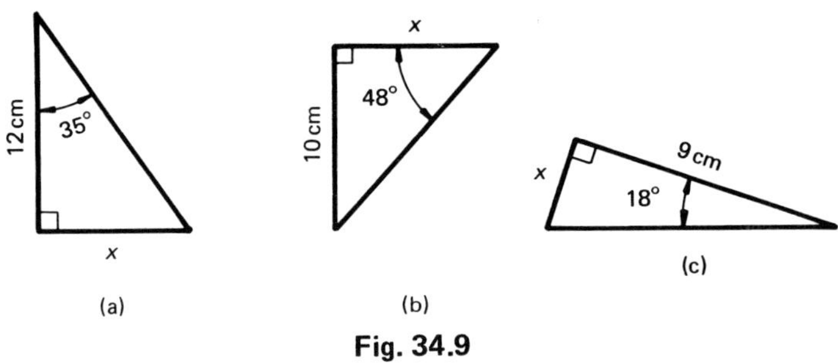

Fig. 34.9

14. Find each of the angles marked A in Fig. 34.10.

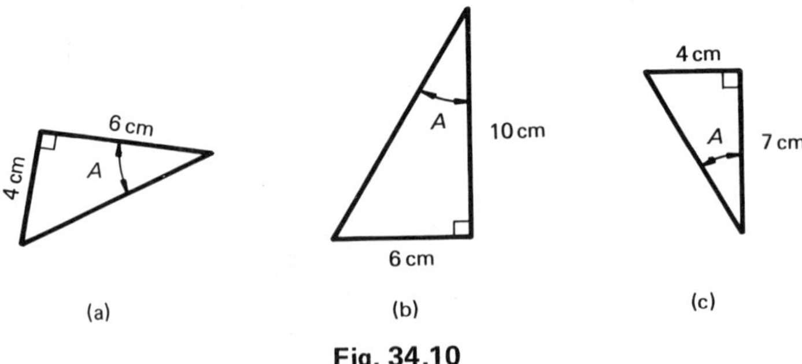

Fig. 34.10

15. An isosceles triangle has a base 12 cm long and its two equal angles are 74°. Find the vertical height of the triangle.

Exercise 34.2 (All of the type found in CSE examination papers)

1. Fig. 34.11 represents a ramp. The distance AC is 5 m and the distance AB is 1 m. Use four figure tables to find:

   (a) The angle ACB

   (b) The distance BC giving your answer correct to one decimal place. (Y)

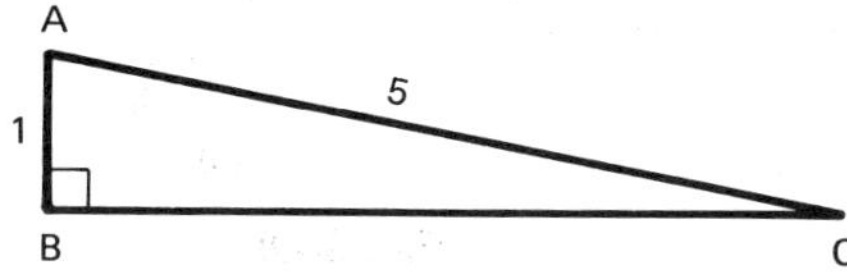

Fig. 34.11

2. In the triangle ABC (Fig. 34.12), AB = AC = 8 cm, ∠ABC = 50° and AN is perpendicular to BC. Calculate:

   (a) ∠BAC (b) the length of BC

   (c) the length of AN.

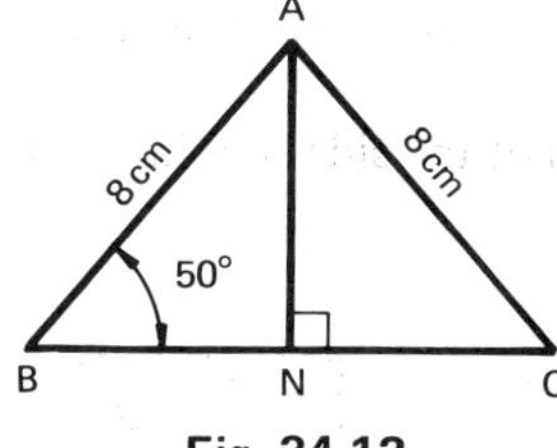

Fig. 34.12

3. Given Fig. 34.13, calculate:

   (a) the length of AD (b) the length of CD

   (c) tan ∠ABD (d) the size of ∠ABD. (W)

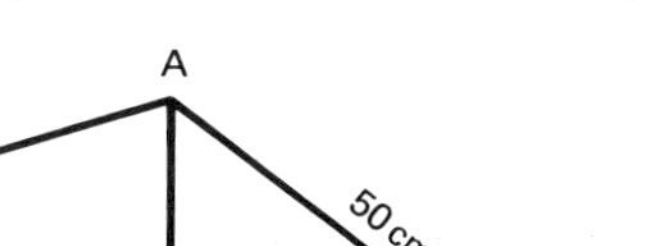

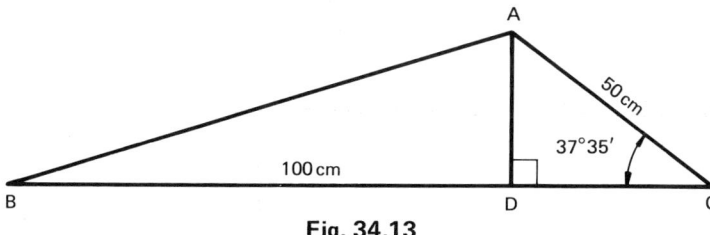

Fig. 34.13

4. Given Fig. 34.14, calculate, by using trigonometrical tables, the length of CD and AD and the size of ∠ABC. (W)

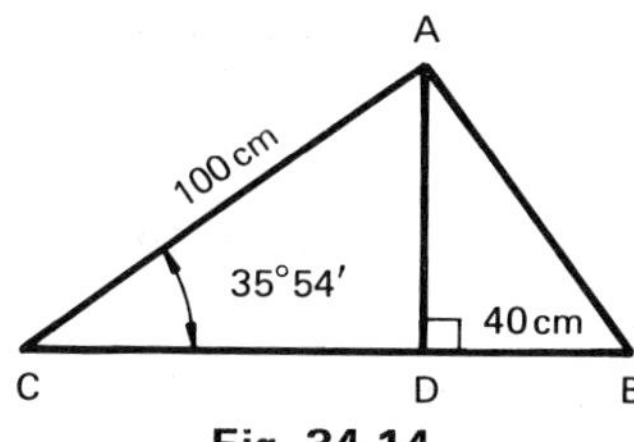

Fig. 34.14

5. In Fig. 34.15, area of rectangle ABCD = 62 cm^2, AD = 8 cm, ED = 12.9 cm and ∠ADE is a right angle.
   (a) Find the length of AB.
   (b) Calculate the area of ΔADE.
   (c) Find the size of ∠AED. (EA)

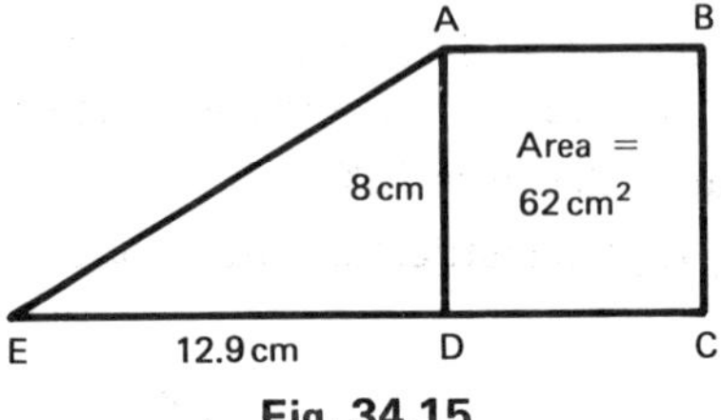

Fig. 34.15

6. (a) Find x if $\sin x = 0.4673$.
   (b) For what angle is 0.3994 the cosine?

7. If $\tan x = 0.5821$:
   (a) Find x (b) Calculate the value of $(1 - \cos x)$. (Y)

8. Write down the values of:
   (a) $\sin 26.3°$ (b) $\cos 63.7°$ (c) $\tan 26.3°$. (EA)

9. In the right-angled triangle ABC (Fig. 34.16):
   (a) Find the value of $\sin A$
   (b) Find the size of the angle A
   (c) Calculate the length of AC. (S)

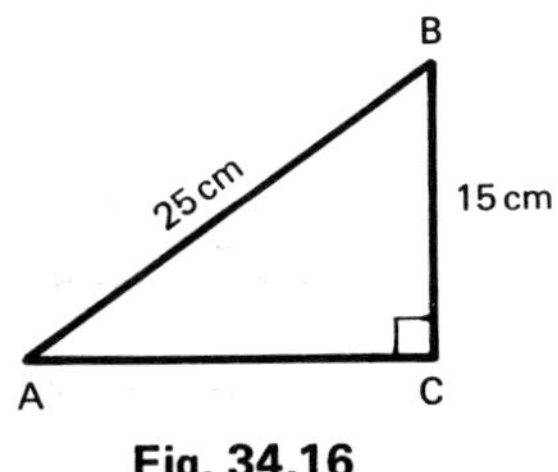

Fig. 34.16

10. Using Fig. 34.17:
   (a) Write sin ∠PRQ as a fraction
   (b) Calculate the size of ∠RPQ. (EA)

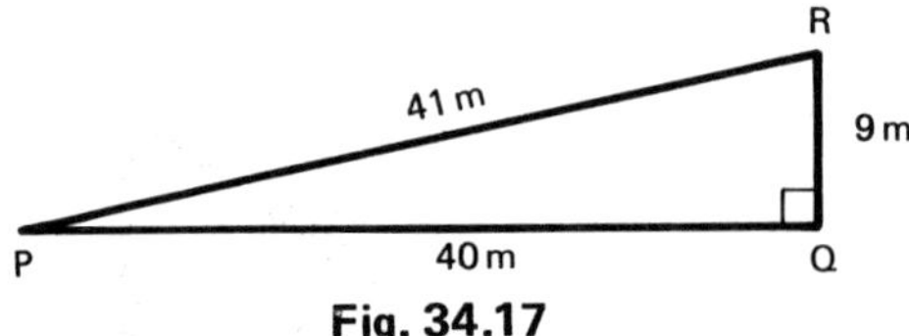

Fig. 34.17

11. Using Fig. 34.18, write down, as a fraction, the value of:
   (a) $\tan A$ (b) $\sin A$

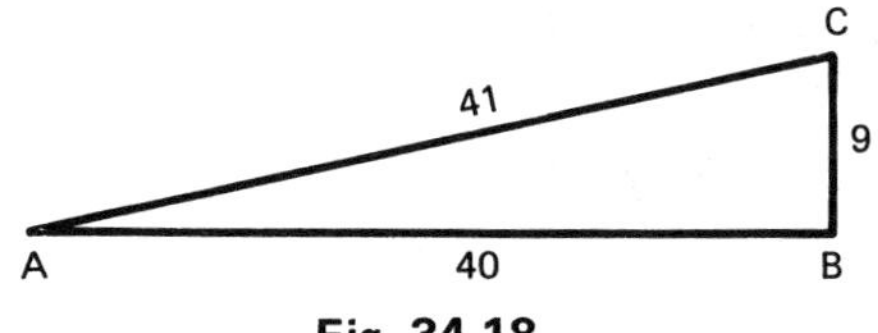

Fig. 34.18

12. CD (Fig. 34.19) is the diameter of a circle and P is a point on the circumference. CD = 8 cm and angle PCD = 31°.
    (a) Calculate the angle PDC.
    (b) Using trigonometrical tables, calculate the length of CP. (SW)

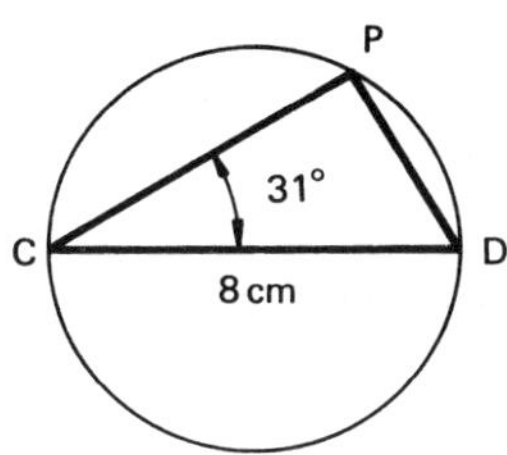

Fig. 34.19

Multi-choice questions 34

1. The sine of the angle X in Fig. 34.20 is

   A $\frac{5}{12}$ B $\frac{12}{13}$ C $\frac{5}{13}$ D $\frac{12}{13}$
   E $\frac{13}{12}$

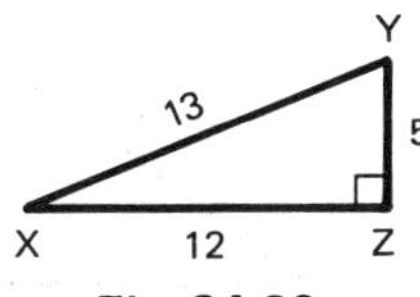

Fig. 34.20

2. Fig. 34.21 shows a road sloping at 37° to the horizontal. If a man walks 50 m up the road from A to B, the distance BC through which he rises, in metres, is

   A 30 B $37\frac{1}{2}$ C 40 D 66
   E 83

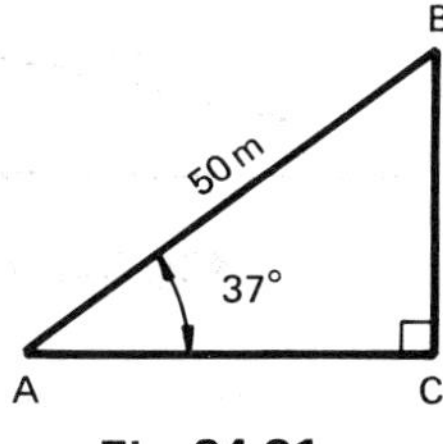

Fig. 34.21

3. Triangle ABC (Fig. 34.22) is right-angled at B. What is the value of $\cos A$?

A $\frac{5}{12}$ B $\frac{5}{13}$ C $\frac{12}{13}$ D $\frac{12}{5}$ (AL)

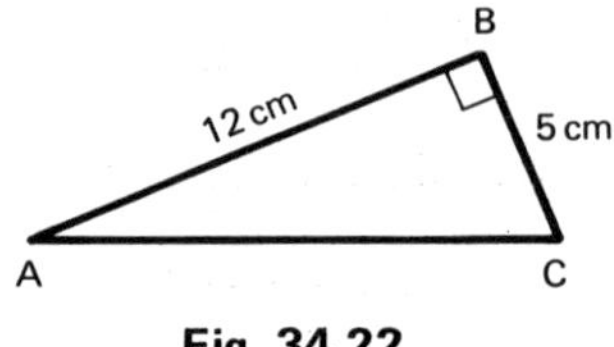

Fig. 34.22

4. In Fig. 34.23, $\tan A$ is equal to

A $\frac{9}{15}$ B $\frac{9}{12}$ C $\frac{12}{15}$ D $\frac{12}{9}$
E $\frac{15}{9}$

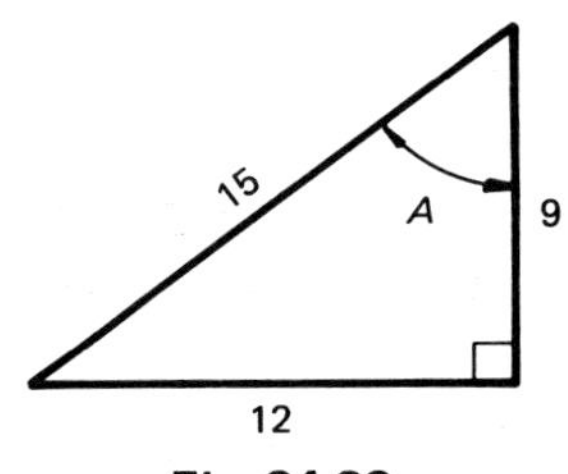

Fig. 34.23

Use Fig. 34.24 to answer questions 5, 6 and 7.

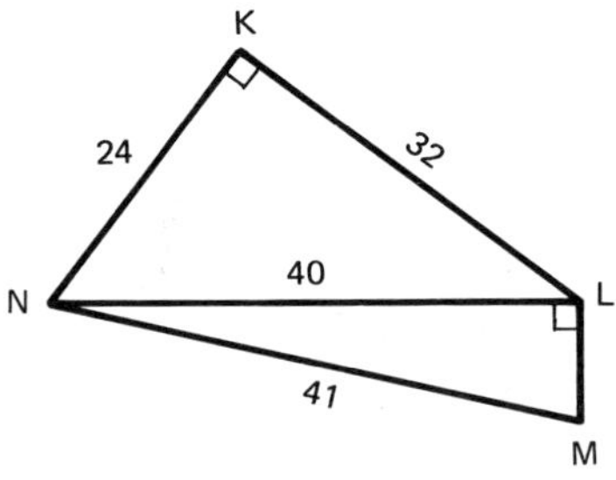

Fig. 34.24

5. Sin ∠KNL expressed as a fraction in its lowest terms is

A $\frac{3}{5}$ B $\frac{32}{40}$ C $\frac{3}{4}$ D $\frac{4}{5}$ (WY)

6. Express tan ∠KLN as a decimal fraction.

A 0.6 B 0.75 C 0.8 D 1.25 (WY)

7. Calculate the size of ∠LNM to the nearest degree.

A 13° B 44° C 46° D 77° (WY)

Questions 8, 9 and 10 refer to Fig. 34.25.

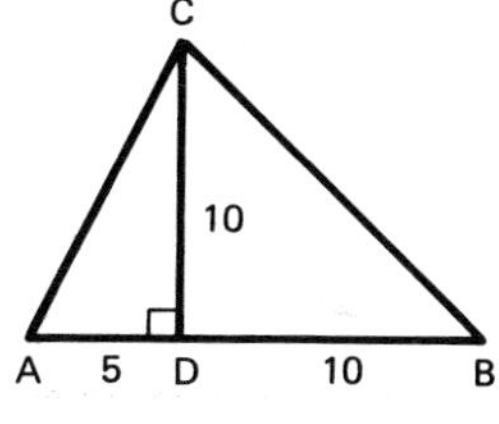

Fig. 34.25

8. What is the sine ratio for angle A?

A $\frac{DC}{AC}$ **B** $\frac{BC}{AC}$ **C** $\frac{BC}{AC}$ **D** $\frac{AD}{AC}$ (EA)

9. What is the size of the angle B?

A 10° **B** 20° **C** 45° **D** 60° (EA)

10. Which angle in the diagram has a tangent of 2?

A CAD **B** ABC **C** ACD **D** CDB (EA)

Questions 11, 12 and 13 refer to Fig. 34.26.

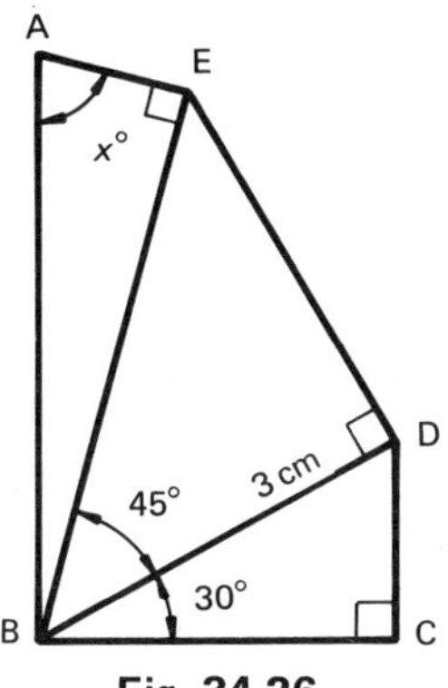

Fig. 34.26

11. If ∠ABC is a right angle, what is the value of x?

A 135 **B** 105 **C** 75 **D** 45 (AL)

12. The length of DE in centimetres is

A $\frac{3}{\sin 45°}$ **B** 3 **C** $3 \cos 45°$ **D** $3 \tan 75°$ (AL)

13. The length of CD in centimetres is

A $3 \tan 30°$ **B** $3 \tan 60°$ **C** $3 \sin 30°$ **D** $3 \cos 30°$ (AL)

14. Fig. 34.27 shows a rhombus with edges 20 cm long and ∠PSR = 60°. The height *h* cm to the nearest centimetre is

A 10 **B** 14 **C** 17 **D** 18

E 20 (AL)

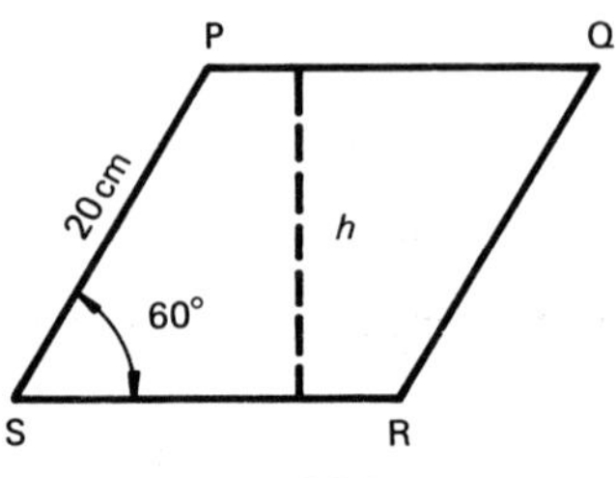

Fig. 34.27

35 SETS

- A SET is a collection of objects, numbers, ideas, etc.

  The different objects, etc., are called the *elements* or *members* of the set.

- A SET MAY BE DESCRIBED by using one of the following methods.
  (i) By listing all of the elements, for instance $A = \{1,3,5,7,9,11\}$. The order of the elements does not matter and each element is listed once only.
  (ii) By listing only enough of the elements to indicate the pattern and showing that the pattern continues by using dots. Thus $B = \{2,4,6,8,\ldots\}$.
  (iii) By a description such as {all even numbers}.
  (iv) By using an algebraic expression, such as $C = \{x : 0 \leqslant x \leqslant 8$, x is an integer$\}$ which is used as 'the set of elements x such that x is an integer whose value lies between 0 and 8, i.e. $C = \{0,1,2,3,4,5,6,7,8\}$.

- There are several TYPES OF SETS.
  (i) A *finite set* is one in which all the elements are listed, such as $\{2,3,4,5,6\}$.
  (ii) An *infinite set* in which it is impossible to list all the members. Thus if B = {all even numbers}, it is not possible to list all of the elements and so we write $B = \{2,4,6,8,\ldots\}$ where the dots mean 'and so on'.
  (iii) The *null set* which contains no elements. It is denoted by $\emptyset$ or by $\{\,\}$.

- The symbol $\in$ means 'is a MEMBER OF THE SET'. Thus the fact that 7 is a member of $A = \{3,5,7,9\}$ is denoted by writing $7 \in A$. The symbol $\notin$ means 'is not a member of'. Because 6 is not a member of A we write $6 \notin A$.

- If all the members of a set A are also members of the set B then A is a SUBSET of B and we write $A \subset B$. Thus if $S = \{a,b,c\}$ and $T = \{a,b,c,d,e\}$, then $S \subset T$.

Example

List all of the subsets of $\{a,b,c\}$.

The subsets are: $\emptyset$, $\{a\}$, $\{b\}$, $\{c\}$, $\{a,b\}$, $\{a,c\}$, $\{b,c\}$ and $\{a,b,c\}$.

Every possible subset of a given set, except the set itself is called a *proper subset*.

- If there are n elements in a set, then the NUMBER OF SUBSETS is given by

$$N = 2^n$$

Thus if a set has 5 elements then the number of subsets that can be formed is

$$N = 2^5 = 32$$

- The order in which the elements of a set are written does not matter. Thus $\{1, 3, 5, 7\}$ is the same as $\{3, 5, 7, 1\}$. Two sets are said to be EQUAL if their elements are identical. Thus if $A = \{3, 5, 7, 9\}$ and $B = \{7, 5, 9, 3\}$, then $A = B$.

- The UNIVERSAL SET for any particular problem is the set which contains all the available elements for that problem. Thus if the universal set is all the odd numbers up to 16, then we write

$$\mathcal{E} = \{1, 3, 5, 7, 9, 11, 13, 15\}$$

- The COMPLEMENT of a set A is the set of elements of $\mathcal{E}$ which do not belong to A. Thus if $\mathcal{E} = \{1, 3, 5, 7, 9\}$ and $A = \{3, 5, 7\}$, then the complement of A is $A' = \{1, 9\}$.

- Sets and set problems may be represented by diagrams called VENN DIAGRAMS.

The universal set is shown by a rectangle and subsets of this set are shown by closed curves. Thus in Fig. 35.1, the shaded region represents A' (i.e. the complement of A). Subsets are represented by a curve within a curve and Fig. 35.2 shows $A \subset B \subset \mathcal{E}$.

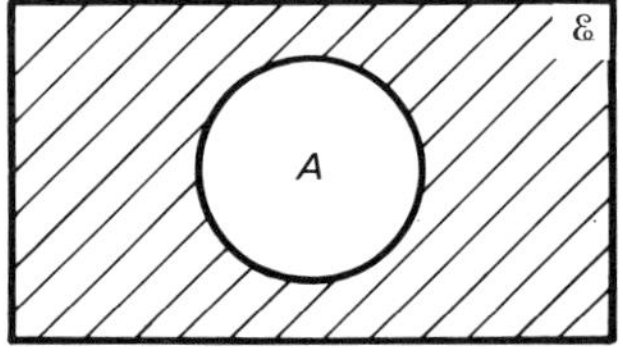

A' is represented by the shaded region

Fig. 35.1

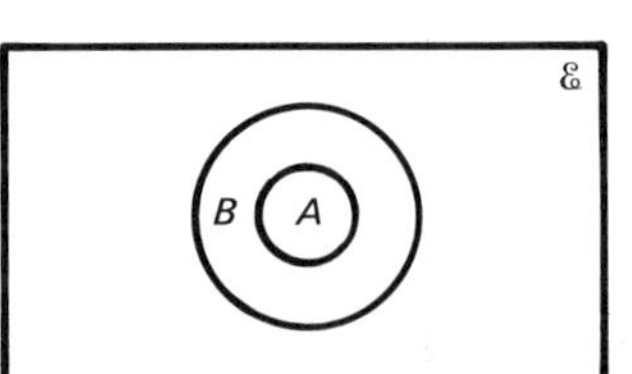

Venn diagram representing $A \subset B \subset \mathcal{E}$

Fig. 35.2

- The INTERSECTION of two sets A and B is the set of elements which are members of *both* A and B. Thus if $A = \{3, 5, 8\}$ and $B = \{2, 3, 4, 5\}$, then the intersection of A and B is $\{3, 5\}$. We write $A \cap B = \{3, 5\}$.

The shaded portion of the Venn diagram in Fig. 35.3 represents $X \cap Y$.

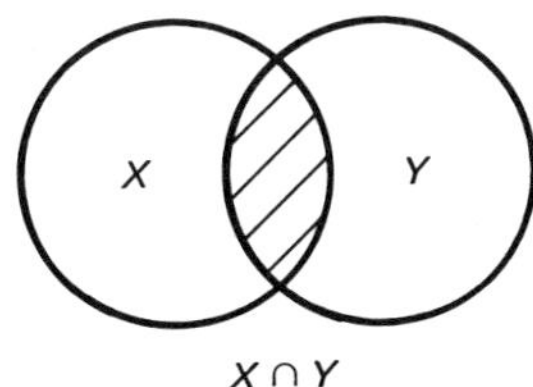

Fig. 35.3

- The UNION of the sets A and B is the set of all the elements contained in A and B. If $A = \{1,3,5\}$ and $B = \{2,3,5,6,8\}$, then the union of A and B is $\{1,2,3,5,6,8\}$ and we write $A \cup B = \{1,2,3,5,6,8\}$. The shaded portion of the Venn diagram shown in Fig. 35.4 represents the union of the sets A and B.

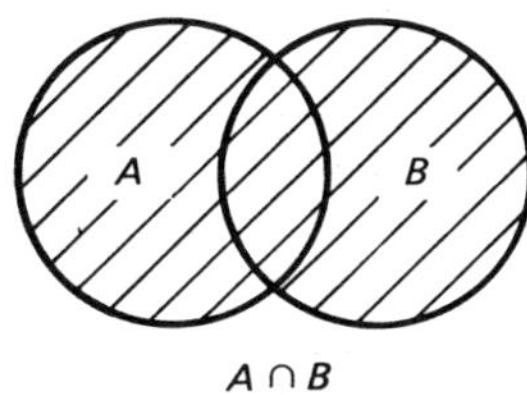

Fig. 35.4

Example

If $A = \{2,3,4,5\}$, $B = \{2,4,6,8,10\}$ and $\mathscr{E} = \{1,2,3,4,5,6,7,8,9,10,11,12\}$, draw a Venn diagram to represent this information. Hence write down the elements of:

(a) A' (b) $A \cap B$ (c) $A \cup B$.

The Venn diagram is shown in Fig. 35.5.

(a) $A' = \{1,6,7,8,9,10,11,12\}$.

(b) $A \cap B = \{2,4\}$.

(c) $A \cup B = \{2,3,4,5,6,8,10\}$.

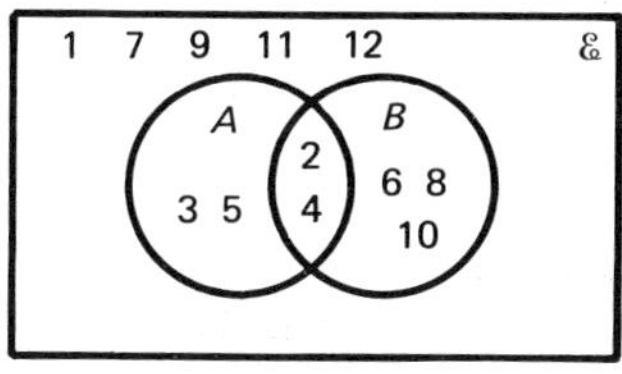

Fig. 35.5

- The NUMBER OF ELEMENTS in a set A is denoted by $n(A)$. Thus if $S = \{2,4,6,8\}$ then $n(S) = 4$ because the set S contains 4 elements.

Example

Fig. 35.6 shows two intersecting sets A and B. The figures give the number of elements in each region. Determine:

(a) $n(A)$ (b) $n(B)$ (c) $n(A \cap B)$ (d) $n(A \cup B)$.

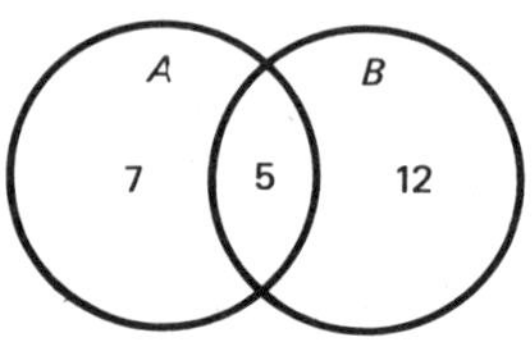

Fig. 35.6

(a) $n(A) = 7 + 5 = 12$.

(b) $n(B) = 5 + 12 = 17$.

(c) $n(A \cap B) = 5$.

(d) $n(A \cup B) = 7 + 5 + 12 = 24$.

Exercise 35.1

1. Write down the members of the following sets:

(a) {prime numbers less than 20}
(b) {multiples of 5 up to 40}
(c) {odd numbers between 8 and 16}
(d) $\{x : 2 < x < 8, x \text{ an integer}\}$
(e) $\{x : 1 \leqslant x \leqslant 9, x \text{ an integer}\}$.

2. State which of the following sets are finite, infinite or null:

(a) $A =$ {even numbers}
(b) $B = \{1, 3, 5, 7, 9\}$
(c) $C =$ {letters of the alphabet}
(d) $D =$ {even numbers divisible by 3}
(e) $E =$ {people who have jumped 5 m high unaided}
(f) $F =$ {points on a straight line}.

3. State which of the following statements are true:

(a) $3 \in$ {prime factors of 30}
(b) eel $\in$ {fish}
(c) rectangle $\in$ {quadrilaterals}
(d) octagon $\in$ {triangles}
(e) polygon $\in$ {solid figures}
(f) $5 \in$ {odd numbers}.

4. If $A = \{3, 5, 7, 8, 9, 10, 12, 13, 15, 18, 19, 22\}$, list the subsets whose members are:

   (a) All the odd numbers of A
   (b) All the even numbers of A
   (c) All the prime numbers of A
   (d) All the factors of 30.

5. If $X = \{2, 3, 4, 5, 7, 9\}$, how many subsets are there?

6. If $A = \{3, 5, 7, 8, 9\}$, $B = \{5, 7, 9\}$ and $C = \{7, 10\}$, which of the following statements is/are correct?

   $A \subset B \qquad B \subset C \qquad C \subset A \qquad C \subset B.$

7. & = {all quadrilaterals}. Write down the subset of quadrilaterals with diagonals that bisect at right angles.

8. & = {squares of natural numbers}. Write down the subset of the squares of the first four natural numbers.

9. Fig. 35.7 is a Venn diagram which shows the universal set and two subsets A and B.

   (a) Write down all of the elements of &.
   (b) Write down the elements of A.
   (c) Write down the elements of B.
   (d) Write down the elements of A'.
   (e) Write down the elements of $A \cap B$.
   (f) Write down the elements of $A \cup B$.
   (g) Write down the elements of $(A \cup B)'$.

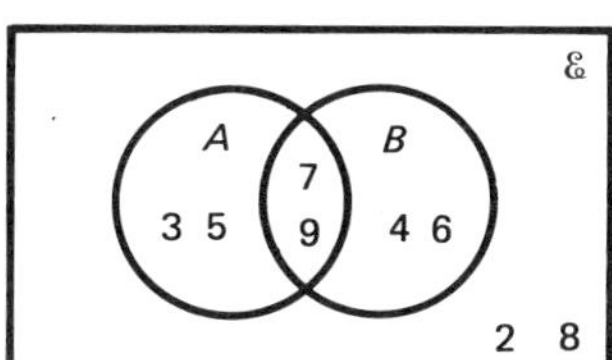

Fig. 35.7

10. Fig. 35.8 shows a Venn diagram representing the universal set & and two subsets X and Y. Draw a similar diagram and insert on it the elements to represent:

    & = {1, 2, 3, 4, 5, 6, 7, 8, 9, 10}, $X = \{2, 4, 6, 7, 8\}$ and $Y = \{1, 3, 4, 5, 7, 10\}$.

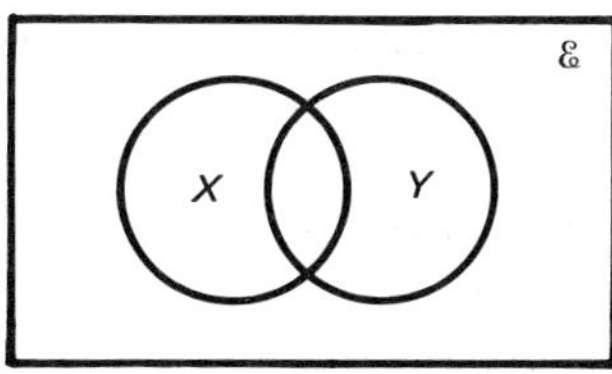

Fig. 35.8

11. Use set notation to describe the shaded portions of the Venn diagrams shown in Fig. 35.9

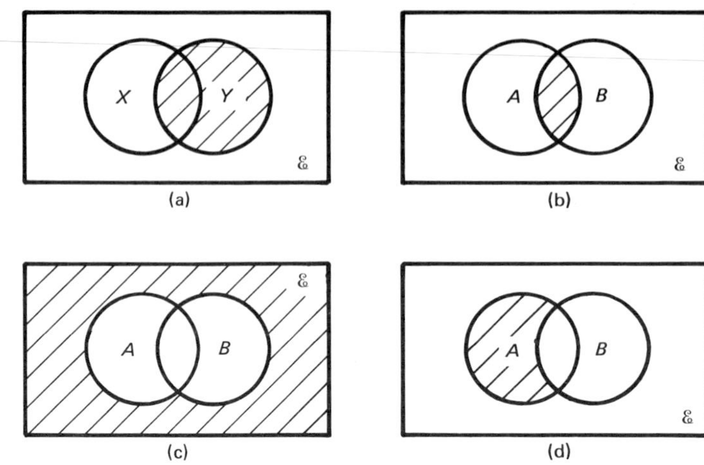

Fig. 35.9

12. Fig. 35.10 shows two intersecting sets A and B. The entries give the number of elements in each region. Find:

(a) $n(A)$ (b) $n(B)$ (c) $n(A \cup B)$ (d) $n(A \cap B)$.

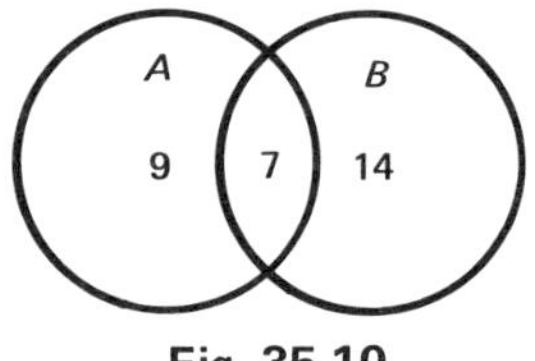

Fig. 35.10

Exercise 35.2 (All of the type found in CSE examination papers)

1. Given that:

$$\mathcal{E} = \{\text{prime numbers}\}$$
$$T = \{\text{prime factors of } 30\}$$
$$S = \{\text{prime factors of } 70\}$$

(a) List the members of the sets T and S.

(b) Draw a Venn diagram similar to Fig. 35.11 and shade $S \cap T'$. Write down the only member of this set. (SW)

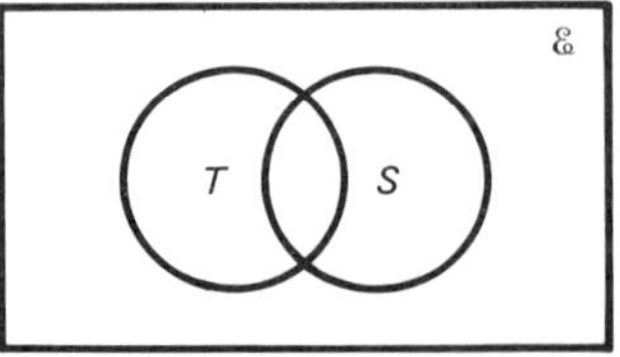

Fig. 35.11

2. Copy the Venn diagram shown in Fig. 35.12 and shade A'. (SW)

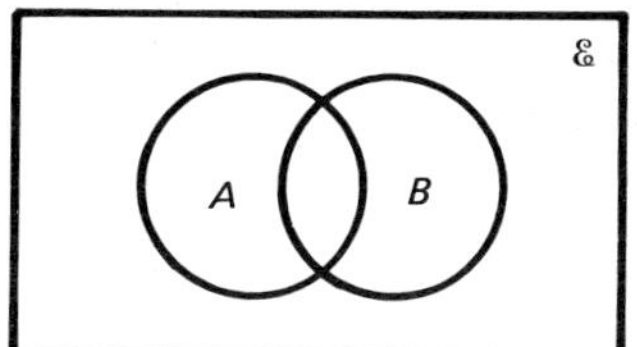

Fig. 35.12

3. State which of the diagrams of Fig. 35.13 show:

(a) $B \subset A$ (b) $A \cap B = \emptyset$

(c) $A \cap B' = A$ (d) $A \cap B = A$. (SW)

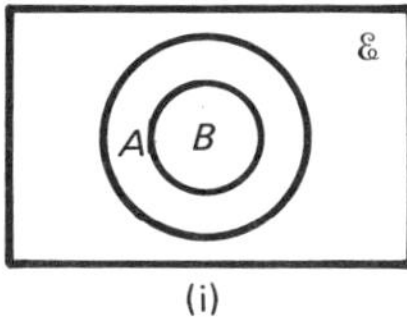

(i)

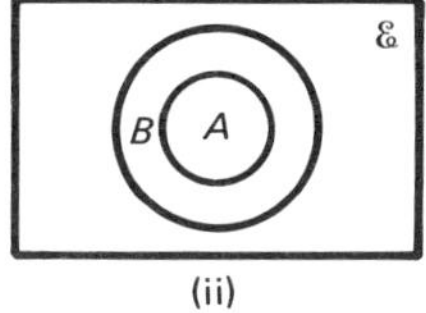

(ii)

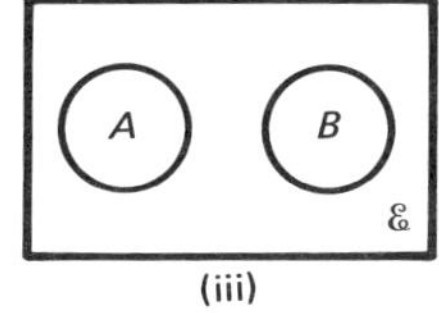

(iii)

Fig. 35.13

4. Given that:

$$\mathscr{E} = \{1,2,3,4,5,6,7,8,9,10\}$$
$$A = \{2,4,6,8,10\}$$
$$B = \{1,2,3,5,7\}$$

(a) Write down the elements of $A \cap B$.

(b) Write down the elements of A'.

(c) Find the number of elements in $A' \cup B'$, i.e. find $n(A' \cup B')$. (EA)

5. Copy the Venn diagram of Fig. 35.14 and shade the region $A \cap B'$. (NW)

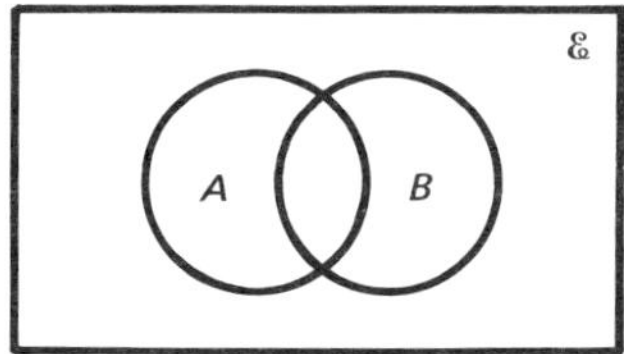

Fig. 35.14

6. If $A = \{G, E, O, F\}$ and $B = \{B, O, Y, C, T\}$, list the elements of the set $A \cup B$. (Y)

7. If $P = \{a, b, c, d, e, f, g,\}$ and $Q = \{b, x, y, d\}$, write down the set $P \cap Q$. (NW)

8. If $A = \{\text{positive multiples of } 3\}$ and $B = \{\text{all factors of } 36\}$, list the members of:

(a) B (b) $A \cap B$. (Y)

9. T is the set of factors of 12, F is the set of multiples of 5 less than 28. List the elements of:

(a) T (b) F (c) $T \cap F$. (Y)

10. The Venn diagram in Fig. 35.15 shows the number of elements in each of the two sets X and Y which are subsets of the universal set &. Use this diagram to solve the following:

(a) Write down $n(X)$ (b) Write down $n(X \cap Y)$

(c) Write down $n(X \cap Y')$ (d) Write down $n(X \cup Y)'$.

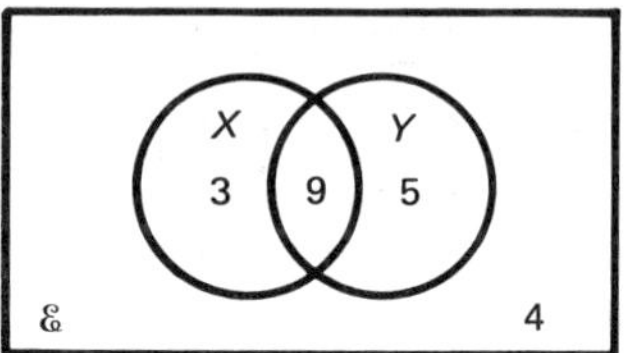

Fig. 35.15

11. The Venn diagram in Fig. 35.16 represents the subsets A and B of the universal set &. List the members of:

(a) $A \cap B$ (b) $(A \cup B)'$. (Y)

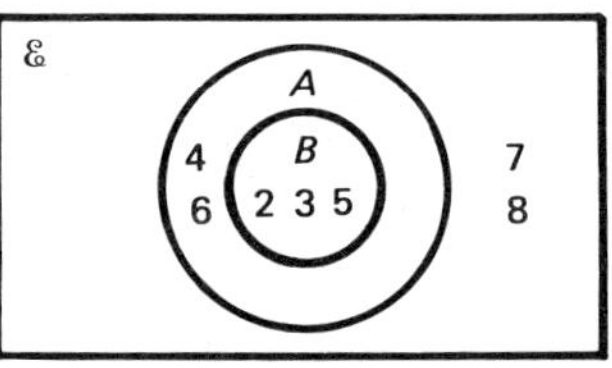

Fig. 35.16

12. If & $= \{1, 2, 3, 4, 5, 6, 7, 8, 9, 10\}$, $A = \{2, 4, 6, 8, 10\}$ and $B = \{1, 2, 5, 10\}$. Write down:

(a) The value of $n(B)$,

(b) The elements of $A \cap B$,

(c) The elements of A',

(d) The elements of $(A \cup B)'$. (EA)

Multi-choice questions 35

The following information is to be used in questions 1, 2, 3 and 4. & $= \{1, 2, 3, 4, 5, 6, 7\}$, $X = \{1, 2, 6, 7\}$, $Y = \{2, 3, 4, 5, 6\}$ and $Z = \{4, 5, 6, 7\}$.

1. $X \cap Y$ is

A $\{2, 6\}$ **B** $\emptyset$

C $\{3, 4, 5\}$ **D** $\{1, 2, 4, 5, 6, 7\}$ (WY)

2. $X \cup Z$ is

A $\{2, 6\}$ **B** $\emptyset$

C $\{3, 4, 5\}$ **D** $\{1, 2, 4, 5, 6, 7\}$ (WY)

3. X' is

A $\{2,6\}$ B $\emptyset$

C $\{3,4,5\}$ D $\{1,2,4,5,6,7\}$ (WY)

4. $Y \cap Y'$ is

A $\{2,6\}$ B $\emptyset$

C $\{3,4,5\}$ D $\{1,2,4,5,6,7\}$ (WY)

5. Which of the diagrams in Fig. 35.17 shows the relation $P \subset Q$? (SE)

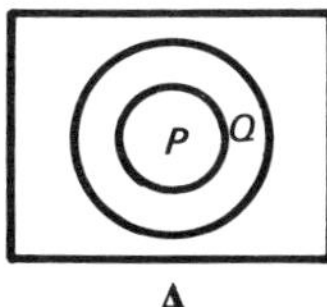

A

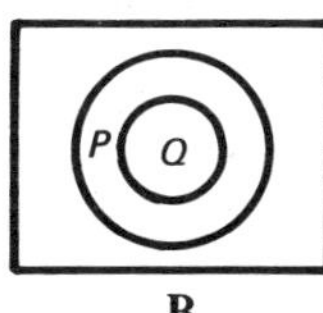

B

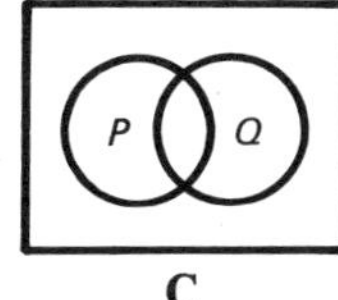

C

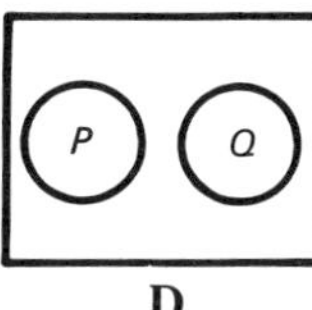

D

Fig. 35.17

6. The shaded region in the Venn diagram of Fig. 35.18 may be described in set notation by

A $A \cup B$ B $A \cap B$ C $A' \cup B'$ D $(A \cup B)'$

E $(A \cap B)'$

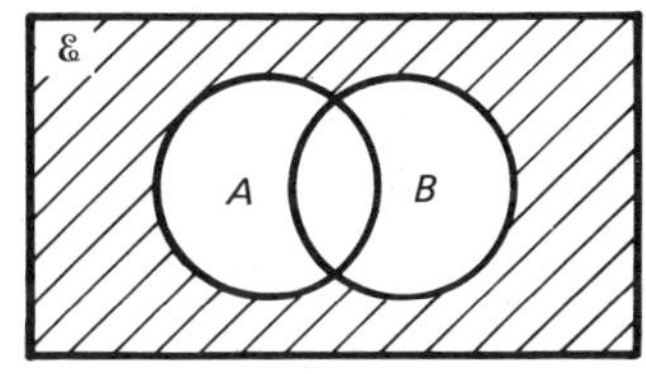

Fig. 35.18

7. The shaded part of Fig. 35.19 represents

A $P \cup Q$ B $P \cap Q$ C P' D Q' (WY)

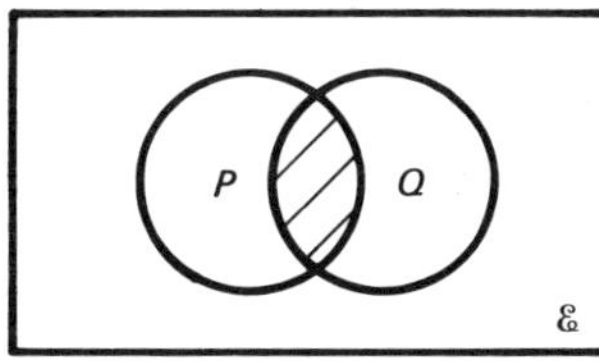

Fig. 35.19

8. The shaded part of Fig. 35.20 represents

A $P \cup Q$ B $P \cap Q$ C Q D Q' (WY)

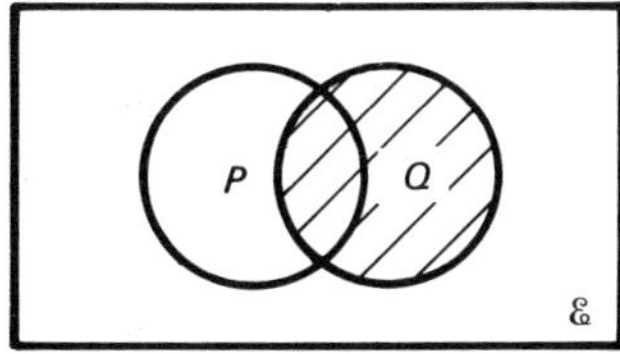

Fig. 35.20

9. The shaded part of Fig. 35.21 represents

A $R \cup S$ **B** $R \cap S$ **C** R' **D** S' (WY)

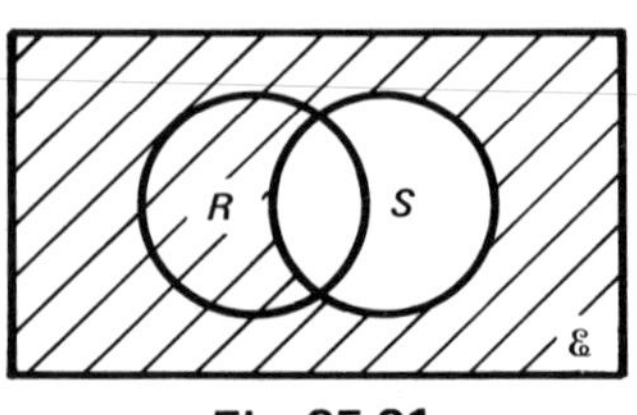

Fig. 35.21

36 NUMBER BASES

- In the DECIMAL SYSTEM (sometimes called the denary system) the digits 0 to 9 are used. The value of each digit in the number depends upon its position in the number. Thus the number

$$\begin{aligned} 4958 &= (4 \times 1000) + (9 \times 100) + (5 \times 10) + (8 \times 1) \\ &= (4 \times 10^3) + (9 \times 10^2) + (5 \times 10^1) + (8 \times 10^0) \end{aligned}$$

It will be noticed that only powers of 10 are used and so we say that the number base is 10.

We can show a number in base 10 by using columns as shown below.

Base 10 headings	10^3	10^2	10^1	10^0
	4	9	5	8

- Number scales OTHER THAN 10 may be used, but it must be remembered that the greatest digit that can be used is one less than the order of the base. Thus in base 3 the greatest digit used is 2 (which is one less than 3) and in base 7 the greatest digit used is 6 (since this is one less than 7).

Example

(a) Convert 73_{10} into a number in base 5.

In base 5 the column headings are powers of 5.

Base 5 headings	5^2	5^1	5^0
	2	4	3

Hence $73_{10} = 243_5$ (since $243_5 = (2 \times 5^2) + (4 \times 5^1) + (3 \times 5^0)$
$= (2 \times 25) + (4 \times 5) + (3 \times 1)$
$= 50 + 20 + 3 = 73$).

(b) Convert 354_6 into decimal notation.

Base 6 headings	6^2	6^1	6^0
	3	5	4

Hence

$$\begin{aligned} 354_6 &= (3 \times 6^2) + (5 \times 6^1) + (4 \times 6^0) \\ &= (3 \times 36) + (5 \times 6) + (4 \times 1) \\ &= 108 + 30 + 4 = 142_{10} \end{aligned}$$

Exercise 36.1

Convert the following into decimal notation:

1. 21_3
2. 374_8
3. 10111_2
4. 234_5
5. 4351_6.

Convert the following decimal numbers into numbers in the base stated:

6. 35 in base 5
7. 243 in base 6
8. 27 in base 2
9. 739 in base 7
10. 9483 in base 8.

- ADDITION AND SUBTRACTION may be carried out in any number base as shown in the following example.

Example

(a) Find the value of $234_5 + 421_5$, stating the sum in base 5.

```
 234
 421
----
1210
111
```

In the first column $4+1=10$ (i.e. 0 and carry 1). In the second column we have $3+2+1=11$ (i.e. 1 and carry 1). In the third column $2+4+1=12$ (i.e. 2 and carry 1).

Hence $$234_5 + 421_5 = 1210_5$$

(b) Find the value of $212_3 - 121_3$ stating the difference in base 3.

```
212
121
---
 21
```

In the first column $2-1=1$. In the second column, 2 from 1 will not go so borrow 1 from the third column. The 1 then becomes $1+3=4$ and $4-2=2$. Since we borrowed 1 from the third column the 2 becomes 1 and $1-1=0$.

Hence $$212_3 - 121_3 = 21_3$$

- MULTIPLICATION can also be carried out in any number base.

Example

Find the value of $122_3 \times 21_3$ stating the product in base 3.

```
  122
   21
  ---
  122
 1021
-----
11102
```

$122 \times 1 = 122$
$122 \times 2 = 1021$
Adding gives 11102

Hence $$122_3 \times 21_3 = 11102_3$$

Exercise 36.2

Find values for each of the following in the base stated:

1. 110_2+1111_2
2. 12_3+212_3
3. 414_5+314_5
4. 425_6+314_6
5. $37_8+46_8+527_8$
6. 202_3-111_3
7. 33_4-21_4
8. 2122_3-212_3
9. 311_4-32_4
10. 402_5-343_5
11. $11_2\times10_2$
12. $111_2\times11_2$
13. $210_3\times12_3$
14. $1221_3\times111_3$
15. $321_4\times210_4$.

Exercise 36.3 (All of the type found in CSE examination papers)

1. Write the number 11010_2 in base 10. (NW)
2. Work out $101_2\times11_2$ giving the answer in base 2. (SW)
3. Work out 615_8-236_8 giving the answer in base 8. (SW)
4. Write the number 46_8 as a number in base 10. (Y)
5. Find the value of $103_5+24_5+1011_5$ in base 5. (EA)
6. 49_{10} is equivalent to 144_n. What is the value of n? (SW)
7. Write 81_{10} as a number in base 8. (SW)
8. Convert 134_5 to a number in base 10. (EA)
9. Convert 26_{10} to binary form (i.e. to a number in base 2). (Y)
10. What is the base of the number system in which $3\times15=47$? (EA)
11. Calculate $23_4\times12_4$ leaving your answer in base 4. (Y)
12. $X=\{10111_2, 210_3, 42_5, 30_8\}$.
    (a) Write down the least member of X
    (b) Write down the greatest member of X. (EA)
13. Calculate, leaving your answer in base 3:
    (a) 121_3+222_3 (b) $21_3\times12_3$. (EA)
14. Calculate 1000_2-1_2 leaving the answer as a binary number. (EA)
15. Find the value of n when $14_5\times n_5=102_5$. (EA)

Multi-choice questions 36

1. Express the number 1011_2 as a base 10 number.
   A 7 B 9 C 10 D 11
   E 12 (NW)

2. Express 3213_4 as a base 10 number.

A 9 **B** 37 **C** 71 **D** 119

E 231 (NW)

3. In base 4, $121_4 + 323_4$, is equal to

A 444_4 **B** 510_4 **C** 1000_4 **D** 1010_4

E 1110_4 (NW)

4. The binary number 1 0 0 when changed to base 10 is

A 2 **B** 4 **C** 7 **D** 10

E 200 (AL)

5. The binary number 1 1 1 0 1 is equal to

A 44_5 **B** 31_8 **C** 25_{12} **D** 51_6 (AL)

6. 63_8 as a number in base 2 is

A 1 1 0 1 1 **B** 1 1 0 0 0 1

C 1 1 0 0 1 1 **D** 1 1 0 0 0 1 1 (Y)

7. $434_5 + 321_5$ is equal to

A 755_5 **B** 810_5 **C** 1200_5 **D** 1255_5

E 1310_5 (SE)

37 MATRICES

- A MATRIX is an array of numbers or letters in the form of a square or a rectangle. Each number or letter is called an *element* of the matrix. The order of a matrix is the number of rows × the number of columns. Thus

$$\begin{pmatrix} 1 & 3 & 4 \\ 2 & 5 & 6 \end{pmatrix} \quad \text{is of order } 2\times 3$$

$$\begin{pmatrix} a & b \\ p & q \\ x & y \end{pmatrix} \quad \text{is of order } 3\times 2$$

- Matrices may be ADDED or SUBTRACTED provided that they are of the same order. The operation is performed by adding (or subtracting) corresponding elements. Thus

$$\begin{pmatrix} 3 & 5 & 6 \\ 2 & 4 & 1 \end{pmatrix} + \begin{pmatrix} 2 & 4 & 3 \\ 1 & 7 & 5 \end{pmatrix} = \begin{pmatrix} 3+2 & 5+4 & 6+3 \\ 2+1 & 4+7 & 1+5 \end{pmatrix} = \begin{pmatrix} 5 & 9 & 9 \\ 3 & 11 & 6 \end{pmatrix}$$

$$\begin{pmatrix} 7 & 2 \\ -3 & 4 \end{pmatrix} - \begin{pmatrix} -3 & 2 \\ 4 & 1 \end{pmatrix} = \begin{pmatrix} 7-(-3) & 2-2 \\ -3-4 & 4-1 \end{pmatrix} = \begin{pmatrix} 10 & 0 \\ -7 & 3 \end{pmatrix}$$

- SCALAR MULTIPLICATION may be performed as follows

$$3\begin{pmatrix} 2 & 4 \\ 5 & 1 \end{pmatrix} = \begin{pmatrix} 3\times 2 & 3\times 4 \\ 3\times 5 & 3\times 1 \end{pmatrix} = \begin{pmatrix} 6 & 12 \\ 15 & 3 \end{pmatrix}$$

- GENERAL MATRIX MULTIPLICATION allows two matrices to be multiplied together provided that the number of columns in the first matrix equals the number of rows in the second matrix. Thus

$$\begin{pmatrix} 5 & 3 & 6 \\ 2 & 7 & 8 \end{pmatrix}\begin{pmatrix} 1 & 9 \\ 4 & 2 \\ 9 & 4 \end{pmatrix} = \begin{pmatrix} 5\times 1+3\times 4+6\times 9 & 5\times 9+3\times 2+6\times 4 \\ 2\times 1+7\times 4+8\times 9 & 2\times 9+7\times 2+8\times 4 \end{pmatrix}$$

$$= \begin{pmatrix} 71 & 75 \\ 102 & 64 \end{pmatrix}$$

- It is usual to DENOTE MATRICES by bold capital letters.

Example

If $\mathbf{A} = \begin{pmatrix} 2 & 3 \\ 5 & 6 \end{pmatrix}$ and $\mathbf{B} = \begin{pmatrix} 1 & 2 \\ 4 & 7 \end{pmatrix}$, form $\mathbf{C} = \mathbf{A} + \mathbf{B}$.

$$\mathbf{C} = \begin{pmatrix} 2+1 & 3+2 \\ 5+4 & 6+7 \end{pmatrix} = \begin{pmatrix} 3 & 5 \\ 9 & 13 \end{pmatrix}$$

- Two matrices are EQUAL if their corresponding elements are equal. Thus

If $$\begin{pmatrix} a & b \\ c & d \end{pmatrix} = \begin{pmatrix} r & s \\ t & u \end{pmatrix}$$

then $a = r$, $b = s$, $c = t$ and $d = u$.

- The UNIT MATRIX of order 2×2 is

$$\begin{pmatrix} 1 & 0 \\ 0 & 1 \end{pmatrix}$$

- If $\mathbf{AB} = \mathbf{I}$, $\mathbf{I}$ being the unit matrix, then $\mathbf{B}$ is called the INVERSE of $\mathbf{A}$. The inverse of $\mathbf{A}$ is usually written as $\mathbf{A}^{-1}$.

If $$\mathbf{A} = \begin{pmatrix} a & b \\ c & d \end{pmatrix} \text{ then } \mathbf{A}^{-1} = \frac{1}{ad-bc}\begin{pmatrix} d & -b \\ -c & a \end{pmatrix}$$

Example

If $\mathbf{A} = \begin{pmatrix} 3 & 2 \\ 1 & 2 \end{pmatrix}$, form $\mathbf{A}^{-1}$.

$$\mathbf{A}^{-1} = \frac{1}{3\times 2 - 2\times 1}\begin{pmatrix} 2 & -2 \\ -1 & 3 \end{pmatrix} = \begin{pmatrix} \frac{1}{2} & -\frac{1}{2} \\ -\frac{1}{4} & \frac{3}{4} \end{pmatrix}$$

Exercise 37.1

If $\mathbf{A} = \begin{pmatrix} 2 & 1 \\ 3 & 5 \end{pmatrix}$ and $\mathbf{B} = \begin{pmatrix} 1 & 6 \\ 2 & 4 \end{pmatrix}$, form the following:

1. $\mathbf{A} + \mathbf{B}$
2. $\mathbf{A} - \mathbf{B}$
3. $\mathbf{B} - \mathbf{A}$
4. $\mathbf{AB}$
5. $\mathbf{BA}$
6. $\mathbf{A}^{-1}$
7. $\mathbf{B}^{-1}$.

8. If $\begin{pmatrix} 4 & 2 \\ 6 & 8 \end{pmatrix}\begin{pmatrix} 2 \\ 3 \end{pmatrix} = k\begin{pmatrix} 28 \\ 72 \end{pmatrix}$, find the value of k.

9. Find the values of x and y if $\begin{pmatrix} x & 4 \\ 3 & y \end{pmatrix}\begin{pmatrix} 2 \\ 5 \end{pmatrix} = \begin{pmatrix} 8 \\ 21 \end{pmatrix}$.

10. If $\mathbf{P} = \begin{pmatrix} 1 & 0 \\ 0 & 1 \end{pmatrix}$ and $\mathbf{Q} = \begin{pmatrix} 3 & -2 \\ 5 & 0 \end{pmatrix}$, form **PQ** and **QP**.

Exercise 37.2 (All of the type found in CSE examination papers)

1. Calculate $\begin{pmatrix} 3 \\ 4 \end{pmatrix} - \begin{pmatrix} -2 \\ 5 \end{pmatrix}$.

2. If matrix $\mathbf{P} = \begin{pmatrix} 2 & 1 \\ 4 & 5 \end{pmatrix}$ and matrix $\mathbf{Q} = \begin{pmatrix} 1 & -1 \\ 2 & 4 \end{pmatrix}$, find the value of the matrix $\mathbf{P} + \mathbf{Q}$. (Y)

3. $\mathbf{A} = \begin{pmatrix} 4 & -1 \\ 0 & 3 \end{pmatrix}$, $\mathbf{B} = \begin{pmatrix} -2 & -3 \\ 1 & -4 \end{pmatrix}$ and $\mathbf{C} = \begin{pmatrix} 2 & 2 \\ 1 & 2 \end{pmatrix}$. Calculate:

   (a) $\mathbf{A} + \mathbf{B}$ (b) **AC**. (Y)

4. (a) What is the order of the matrix $\begin{pmatrix} 2 & 1 & 5 \\ 1 & -1 & 3 \end{pmatrix}$?

   (b) If $\mathbf{A} = \begin{pmatrix} 3 & 2 \\ -1 & 4 \end{pmatrix}$ and $\mathbf{B} = \begin{pmatrix} 2 & 0 \\ 0 & 2 \end{pmatrix}$, calculate $2(\mathbf{A} + \mathbf{B})$.

5. If $\mathbf{P} = \begin{pmatrix} 1 & 3 \\ 2 & 6 \end{pmatrix}$ and $\mathbf{Q} = \begin{pmatrix} 3 & 1 \\ -1 & -1 \end{pmatrix}$, work out the matrix **PQ**. (Y)

6. $\mathbf{P} = \begin{pmatrix} 3 & 5 \\ 1 & 2 \end{pmatrix}$ and $\mathbf{R} = \begin{pmatrix} 4 & 16 \\ 12 & 36 \end{pmatrix}$:

   (a) Work out $\frac{1}{4}\mathbf{R}$.

   (b) Find, as a single matrix, $\mathbf{P} + \mathbf{R}$.

   (c) Write down the inverse of **P**. (EA)

7. If $\mathbf{A} = \begin{pmatrix} 2 \\ -3 \end{pmatrix}$ and $\mathbf{B} = \begin{pmatrix} 3 \\ 1 \end{pmatrix}$, find $2\mathbf{A} + 3\mathbf{B}$. (Y)

8. If $\mathbf{A} = \begin{pmatrix} -3 \\ 2 \end{pmatrix}$ and $\mathbf{B} = \begin{pmatrix} 6 \\ 1 \end{pmatrix}$, find $\mathbf{B} - \mathbf{A}$.

9. Find x and y given that $\begin{pmatrix} x \\ y \end{pmatrix} = \begin{pmatrix} 1 & 2 \\ 0 & 3 \end{pmatrix}\begin{pmatrix} 4 \\ 1 \end{pmatrix} + \begin{pmatrix} 1 \\ -3 \end{pmatrix}$. (EA)

10. $\mathbf{A} = \begin{pmatrix} 3 & 1 \\ -1 & 2 \end{pmatrix}$ and $\mathbf{B} = \begin{pmatrix} 2 & -1 \\ 3 & 0 \end{pmatrix}$.

(a) Find $\mathbf{A} - \mathbf{B}$.

(b) Calculate $\mathbf{AB}$.

(c) Calculate $\mathbf{BA}$.

(d) Calculate the inverse of $\mathbf{A}$. (EA)

38 TRANSFORMATIONS

- A TRANSFORMATION describes the relation between any point and its image point. Thus in Fig. 38.1, the point P(2,3) has been transformed into the point P′(−2,3). The point P is called the *pre-image* and the point P′ is called the *image*.

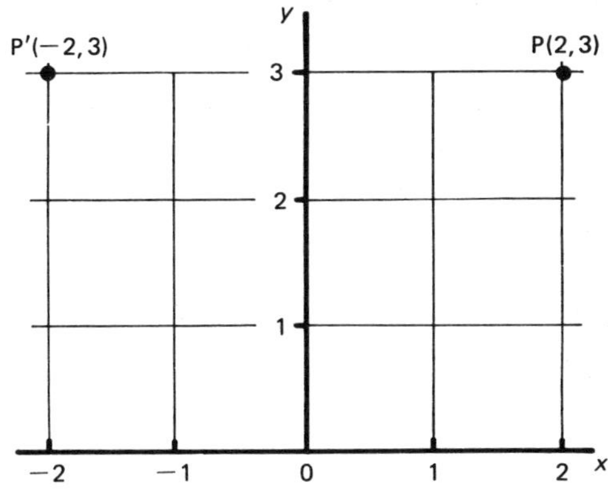

Fig. 38.1

- If every point in a line or a plane figure moves the same distance in the same direction the transformation is called a TRANSLATION. Thus in Fig. 38.2, every point in the line AB has been moved 2 units to the right and 3 units upwards. That is, A(2, 1) translates to A′(4,4) and B(6,3) translates to B′(8,6).

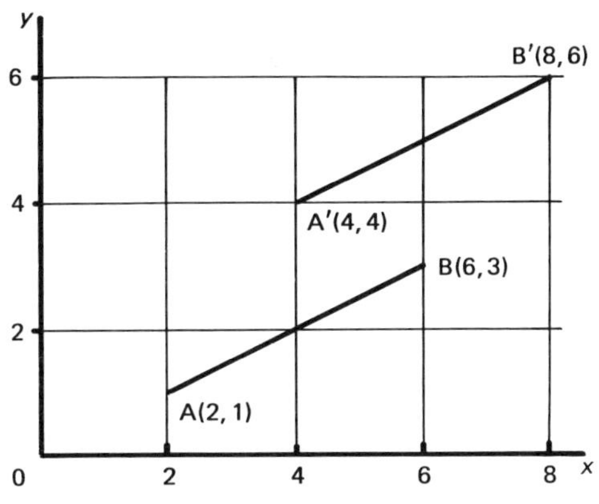

Fig. 38.2

- If a point P is REFLECTED in a mirror so that its image is P′, the mirror or line of reflection is the perpendicular bisector of PP′. Thus in Fig. 38.3, the point P has been reflected in the mirror line OM so that its image is P′.

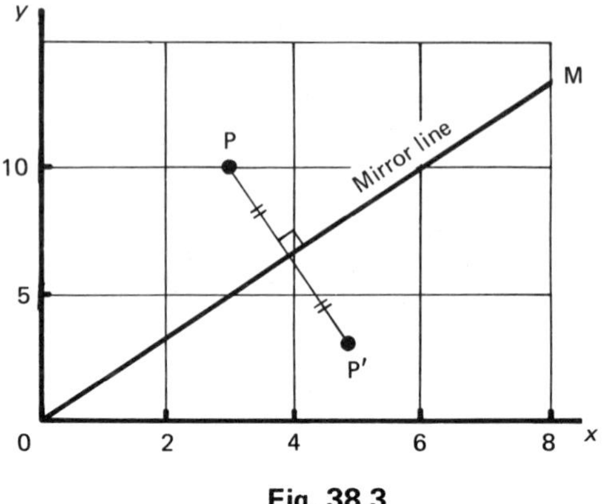

Fig. 38.3

Example

The triangle ABC is formed by joining the three points A(3, 2), B(5, 2) and C(3, 6). Draw this triangle on graph paper and reflect it in the x-axis (i.e. in the line $y = 0$). State the coordinates of the transformed points A, B and C.

The triangle ABC has been drawn in Fig. 38.4 and its image is the triangle A′B′C′ whose vertices are the points A′(3, −2), B′(5, −2) and C′(3, −6).

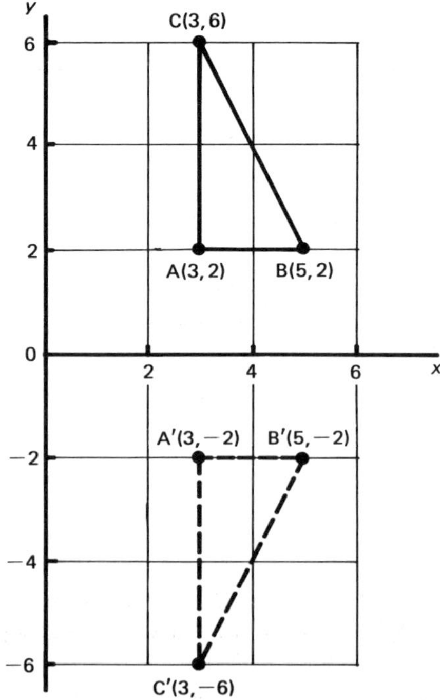

Fig. 38.4

- In Fig. 38.5, the triangle ABC has been ROTATED to the new position A′B′C′. The centre of the rotation is the point O and the angle of rotation is $\theta°$. The rotation has been made in a clockwise direction.

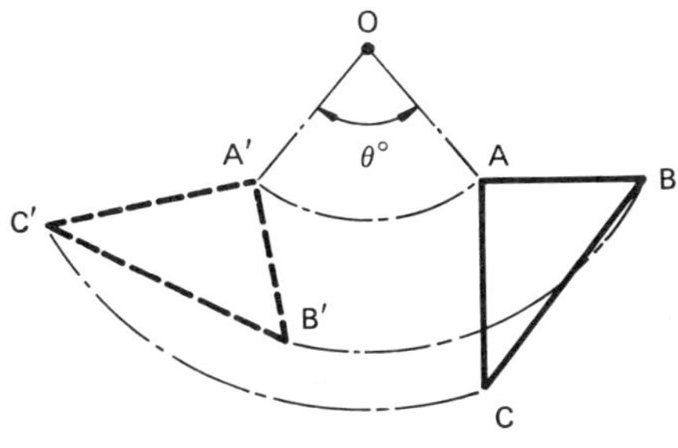

Fig. 38.5

Example

A rectangle has vertices which are A(1, 5), B(5, 5), C(5, 3) and D(1, 3). Draw this rectangle on graph paper and rotate it through 90° anticlockwise about the origin. State the coordinates of the images of the points A, B, C and D.

The rectangle has been drawn in Fig. 38.6 and its image after the rotation is the rectangle A′B′C′D′ whose vertices are the points A′(−5, 1), B′(−5, 5), C′(−3, 5) and D′(−3, 1).

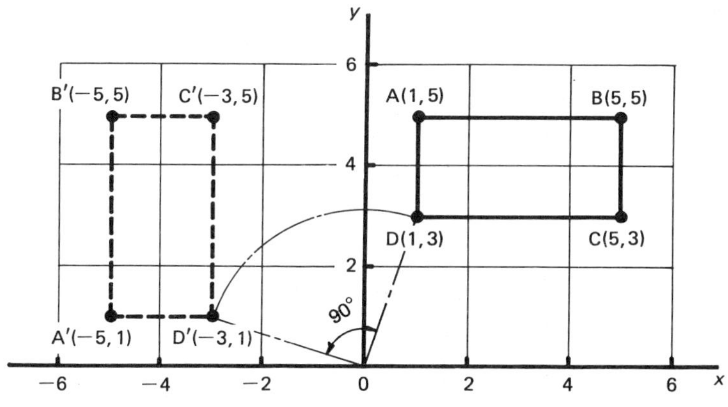

Fig. 38.6

Exercise 38.1

1. The vertices of the square ABCD are respectively A(0, 1), B(2, 1), C(2, 3) and D(0, 3). Draw the square on graph paper and translate ABCD so that it is displaced by:
   (a) 3 units to the right and 2 units upwards
   (b) 5 units to the right and 4 units upwards.

2. A rectangle has vertices which are the points A(1, 5), B(5, 5), C(5, 3) and D(1, 3). Draw this rectangle on graph paper and
   (a) Reflect it in the x-axis
   (b) Write down the coordinates of the transformed points A′B′C′D′
   (c) Reflect the rectangle in the y-axis
   (d) Write down the coordinates of the transformed points A″B″C″D″.

3. PQRS is a square such that P is the point (2, 2), Q is the point (6, 2) and R is the point (6, 6).
   (a) State the coordinates of the point S.
   (b) Draw the square on graph paper using a scale of 1 large square to represent 2 units on both axes.
   (c) Draw the line $y = x$.
   (d) Reflect PQRS in the line $y = x$.
   (e) Write down the coordinates of the transformed points P′Q′R′S′.

4. Triangle XYZ has vertices X(3, 3), Y(6, 3) and Z(6, 5). Plot these points on graph paper and join them to form ΔXYZ. Show the image of XYZ after it has been rotated 90° in a clockwise direction about the origin.

5. Using the origin as the centre of rotation, draw the image of the rectangle ABCD (Fig. 38.7) after it has been given a rotation of 45° clockwise. State the coordinates of the transformed points A′B′C′D′.

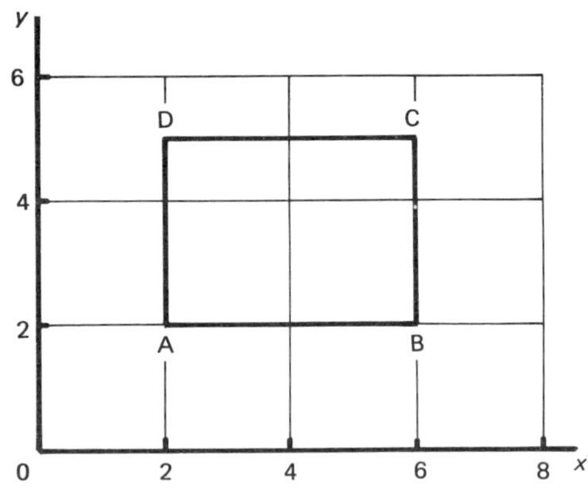

Fig. 38.7

Exercise 38.2 (All of the type found in CSE examination papers)

1. The point (4, −2) is reflected in the line $y = -x$. Write down the coordinates of the reflection. (EA)

2. The point A(2, 6) is reflected in a mirror line on to the point B(2, −2). What is the equation of the mirror line? (Y)

3. P is the point (5, 2). Plot this point on graph paper.
   (a) The image of P when reflected in the x-axis is the point A. Mark A on your graph and state its coordinates.
   (b) The image of P when reflected in the y-axis is the point B. Mark B on your diagram and give its coordinates.

4. The triangle ABC has vertices A(1, 0), B(2, 1) and C(0, 2). Draw this triangle on graph paper and draw the reflection of triangle ABC in the y-axis ($x = 0$). (NW)

5. State the coordinates (p, q) of the image of (3, 1) under a translation of 3 units to the right and 2 units upwards. (EA)

6. The point (5, 3) is reflected in the line $y = x$. Write down the coordinates of the reflection. (EA)

7. Copy the diagram in Fig. 38.8 and draw the image of the letter E when it is reflected in the y-axis. (NW)

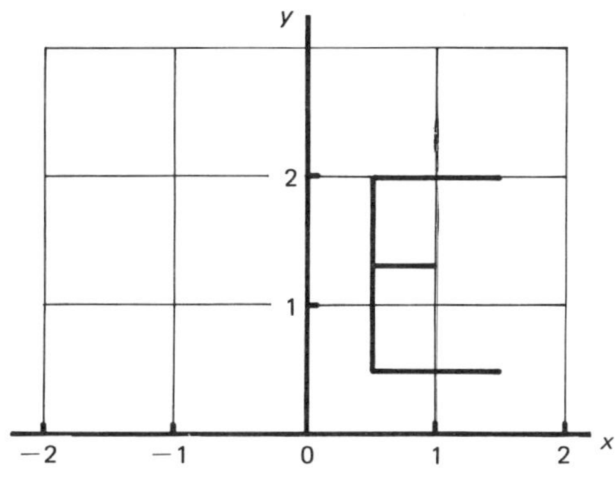

Fig. 38.8

8. The point P(4, −3) is given a rotation of 30° anticlockwise about the origin. Write down the coordinates of the image point after rotation.

9. State the coordinates (t, u) of the image of (3, 1) under an anticlockwise rotation of 90° about the origin. (EA)

10. The coordinates of a point P are (−3, 4). Find the coordinates of the point which is reached by P when OP is rotated about the origin O in an anticlockwise direction through angle of 90°. (EA)

39 STATISTICS

- In a BAR CHART it is usual to plot the frequency (i.e. the number of items) on the vertical axis and the items in the survey on the horizontal axis.

Example

The table below shows the results of a survey to find out the colours of doors on a housing estate. Draw a bar chart to represent this information.

Colour of door	White	Red	Green	Brown	Blue
Number of houses	85	17	43	70	15

The bar chart is shown in Fig. 39.1.

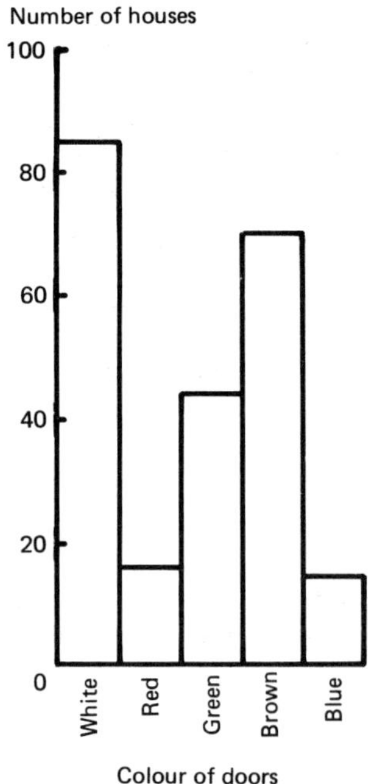

Fig. 39.1

- A PIE CHART uses sector areas to represent proportions of the total frequency.

Example

The table below shows the results of a survey to find out the favourite sports of a group of boys. Draw a pie chart to depict this information.

Sport	Athletics	Cricket	Football	Swimming
Number of boys	20	15	40	25

$$\text{Total number of boys in the survey} = 20+15+40+25 = 100$$

In the pie chart (Fig. 39.2) the full circle (i.e. 360°) represents 100 boys.

$\frac{20}{100} \times 360° = 72°$ represents athletics

$\frac{15}{100} \times 360° = 54°$ represents cricket

$\frac{40}{100} \times 360° = 144°$ represents football

$\frac{25}{100} \times 360° = 90°$ represents swimming

$\text{Total} = 360°$

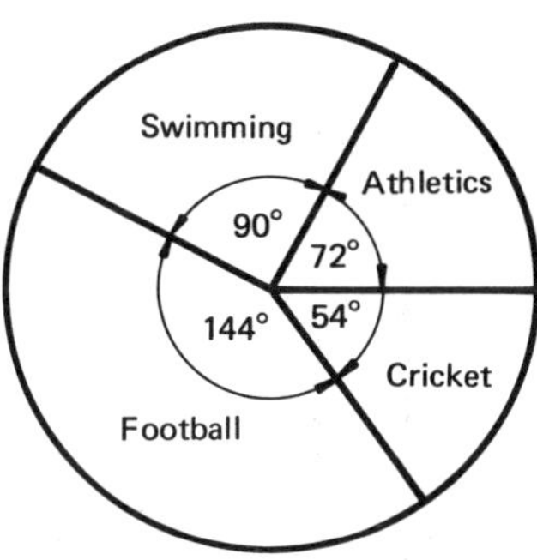

Fig. 39.2

- In a PICTOGRAM the information is represented by a series of symbols. Each symbol represents one unit and a fraction of a symbol is used to represent a fraction of a unit.

Example

The table below shows the number of bicycles manufactured for the years 1970 to 1974. Draw a pictogram to represent this information.

Year	1970	1971	1972	1973	1974
Number of bicycles	2000	4000	7000	8500	9000

The pictogram is shown in Fig. 39.3 where it will be seen that each bicycle represents an output of 2000 bicycles. Part of a bicycle as shown in 1972, 1973 and 1974 is used to represent a fraction of 2000.

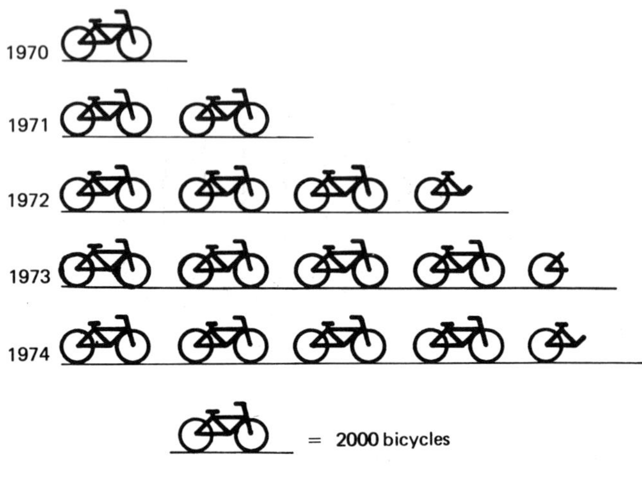

Fig. 39.3

- FREQUENCY DISTRIBUTIONS are used to organise information. Thus the table below is a frequency distribution which shows the marks obtained in a test by 60 students.

Mark	*Frequency*
2	4
3	6
4	8
5	16
6	17
7	6
8	3
Total	60

Note that 2 marks were obtained by 4 students (a frequency of 4), 3 marks were obtained by 6 students (a frequency of 6), and so on.

- A frequency distribution may be represented by a type of bar chart known as a HISTOGRAM. A histogram for the information relating to the marks of students is drawn in Fig. 39.4.

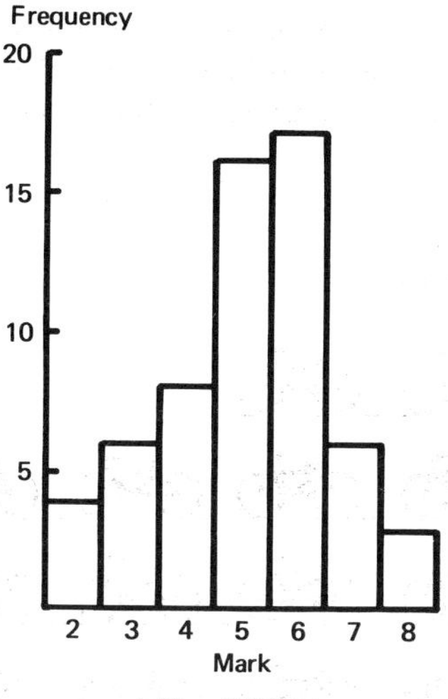

Fig. 39.4

- In statistics, three kinds of AVERAGE are used. They are the arithmetic mean, the median and the mode.

- The ARITHMETIC MEAN (often simply called the mean) is found by adding up all the values in a set of data and dividing by the number of items in the data. That is

$$\text{Mean} = \frac{\text{sum of all the values}}{\text{the number of values}}$$

Example

(a) Find the mean of 3, 5, 8, 9 and 10.

$$\text{Mean} = \frac{3+5+8+9+10}{5} = \frac{35}{5} = 7$$

(b) The values 3, 4, 6, 8, 9 and x have a mean of 7. What is the value of x?

$$\frac{3+4+6+8+9+x}{6} = 7$$

$$\frac{30+x}{6} = 7$$

$$30+x = 6\times 7$$

$$30+x = 42$$

$$x = 42-30$$

$$x = 12$$

- When a set of values is arranged in ascending (or descending) order, the MEDIAN is the value which lies half way along the series. The median of the numbers 2, 4, 5, 5, 6, 7, 7, 8, 9 is 6 because there are four numbers below this value and four numbers above it. When there are an even number of values in the set the median is found by taking the mean of the two middle values. The median of the numbers 3, 4, 6, 7, 9, 10, 12, 14 is $\frac{7+9}{2} = 8$.

Example

Find the median of 8, 5, 4, 7, 9, 10, 2, 5.

Arranging the values in ascending order we have

$$2, 4, 5, 5, 7, 8, 9, 10$$

Hence the median is $\frac{5+7}{2} = 6$.

- The MODE of a set of values is the value which occurs most frequently. The mode of 3, 4, 4, 5, 5, 6, 6, 6, 7, 7, 8, 9, 10 is 6 because this number occurs three times which is more than any of the other numbers.

Example

Find the mode of 10, 11, 11, 12, 12, 12, 13, 13, 13, 13, 14, 14, 14, 15, 15.

We see that 13 is the number which occurs most times (four). Therefore it must be the mode.

- When a fair die is rolled, each of its six faces has an equal chance of landing face upwards. The throwing of a 1, 2, 3, 4, 5 or 6 are called EQUI-PROBABLE EVENTS.

- Suppose that we want to cut an ace from a pack of 52 playing cards. Each time we cut an ace we say that there has been a FAVOURABLE OUTCOME. Since there are four aces in the pack the number of equi-probable events which produce a favourable outcome is 4.

- The PROBABILITY of obtaining a favourable outcome is calculated from

$$\text{Probability} = \frac{\text{number of equi-probable events which produce a favourable outcome}}{\text{total number of equi-probable events}}$$

- On the PROBABILITY SCALE all probabilities have a value which lies between 0 and 1. A probability of 0 occurs when an event can never happen. A probability of 1 represents an event which is certain to happen.

- The TOTAL PROBABILITY, covering all possible events is 1. That is

$$\begin{pmatrix}\text{Probability of}\\ \text{favourable outcomes}\end{pmatrix} + \begin{pmatrix}\text{probability of}\\ \text{unfavourable outcomes}\end{pmatrix} = 1$$

Example

One letter is chosen from the word 'HORIZON'. What is the probability that it is

(a) the letter R
(b) the letter O
(c) a vowel (i.e. a, e, i, o or u)
(d) not a vowel?

(a) Since there are 7 letters in the given word, there is a total of 7 equi-probable events of which only 1 is favourable. Hence

$$\text{Probability of choosing the letter R} = \tfrac{1}{7}$$

(b) In this case there are 2 favourable outcomes since there are two letter O's in the given word. Hence

$$\text{Probability of choosing the letter O} = \tfrac{2}{7}$$

(c) Since there are 3 vowels in the given word, in this case the number of favourable outcomes is 3. Hence

$$\text{Probability of choosing a vowel} = \tfrac{3}{7}$$

(d) Since the total probability is 1

Probability of choosing a vowel + probability of not choosing a vowel = 1

$\therefore$ Probability of not choosing a vowel $= 1 - \tfrac{3}{7} = \tfrac{4}{7}$

Exercise 39.1

1. A survey to establish the most liked sports of 300 boys gave the following results. Draw a bar chart to represent this information.

Sport	Athletics	Football	Cricket	Hockey	Swimming
Number of boys	30	88	64	42	76

2. The information in the table below gives details of the transport used by commuters in South-East England. Draw a pie chart to represent this information.

Type of transport	Car	Underground	Rail	Bus
Numbers using	156	84	32	28

(300 commuters took part in the survey.)

3. A pie chart was drawn to show how a family spent each pound of its income. The sector angles corresponding to the items of expenditure were as follows:

Item	*Sector angle*
Food and drink	137°
Housing	61°
Transport	43°
Clothing	47°
Other	72°

If the total family income is £100, calculate how much was spent on each item.

4. The table below gives details of the number of houses completed (in thousands) in South-West England. Represent this information by means of a pictogram.

Year	1974	1976	1978	1980	1982
Number completed	81	69	73	84	80

5. The information given below shows the output of tyres by the Treadwell Tyre Company for the first six months of 1981. Draw a pictogram to represent this information.

Month	Jan	Feb	Mar	Apr	May	June
Output (thousands)	40	43	39	38	37	45

6. A survey was made to find out the ages of 30 members of a youth club with the following results:

Age	13	14	15	16	17	18
Frequency	2	10	6	6	4	2

Draw a bar chart to represent this information.

7. The following marks were obtained by 50 students during a test:

Mark	1	2	3	4	5	6	7	8
Frequency	2	2	11	11	12	7	4	1

Draw a histogram to represent this information.

8. The heights of 5 men are: 177.8 cm, 175.3 cm, 174.8 cm, 179.1 cm and 176.5 cm. Calculate the mean height of the 5 men.

9. Calculate the mean of the numbers 5, 8, 9, 10.

10. The numbers 3, 5, 8, 10 and x have a mean of 8. Find the value of x.

11. Find the median of the numbers 5, 3, 8, 6, 4, 2, 8.

12. Find the median of the numbers 2, 4, 6, 5, 3, 1, 8, 9.

13. Find the mode of 3, 5, 4, 8, 3, 6, 5, 9, 5, 4, 7.

14. Thirteen people were asked to guess the weight of a cake to the nearest half kilogram. Their estimates were as follows: $3\frac{1}{2}$, $2\frac{1}{2}$, 2, 1, $3\frac{1}{2}$, 2, $3\frac{1}{2}$, 3, 3, 1, $1\frac{1}{2}$, $2\frac{1}{2}$, $3\frac{1}{2}$ kg. What was:
(a) the modal value
(b) the median value
(c) the mean value?

15. A die is rolled. Calculate the probability that it will result:
(a) in a 3
(b) in a score less than 4
(c) in an even number.

16. A letter is chosen from the word 'TERRIFIC'. Determine the probability that it will be:
(a) an F
(b) an R
(c) a vowel (i.e. a, e, i, o or u)
(d) not a vowel.

17. A bag contains 3 red balls, 5 blue balls and 2 green balls. A ball is chosen from the bag. Calculate the probability that it will be:
(a) green
(b) blue
(c) not red.

Exercise 39.2 (All of the type found in CSE examination papers)

1. One letter is chosen from the word 'EXCELLENT'. What is the probability that it will be the letter E?

2. Calculate the mean of the numbers 0, 3, 7, 10.

3. A pie chart represents a period of 24 hours. What is the angle at the centre of the circle for the sector which represents 8 hours of work?

4. The total number of children in an infant school for each of six years is:

Year	1976	1977	1978	1979	1980	1981
Number of children	106	84	92	80	66	58

Draw a bar chart to represent this information. (EM)

5. Given the numbers 5, 7, 3, 6 and 4, calculate:
(a) Their sum
(b) Their mean
(c) Their mean if each number is increased by 2
(d) Their mean if each number is doubled. (W)

6. The mean of 4, x, 8, x and 5 is 7. Find the value of x. (Y)

7. What additional number must be included in the following list to make the median of the six numbers equal to $3\frac{1}{2}$?

$$1, 3, 7, 1, 5$$ (Y)

8. The pie chart shown in Fig. 39.5 illustrates the grouping of people who took part in a survey. The total number of people was 264. Calculate the number of people who formed group E.

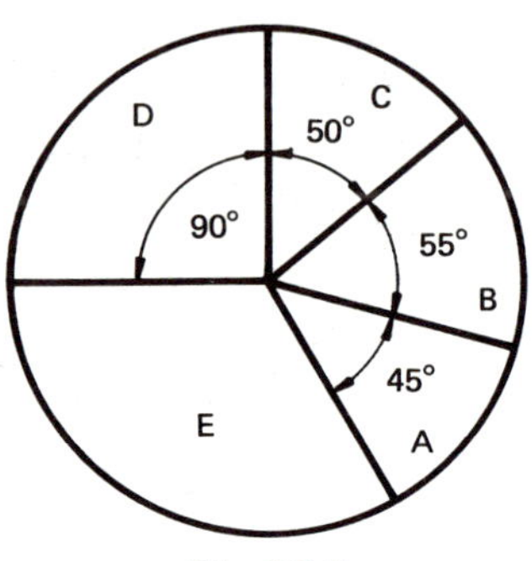

Fig. 39.5

9. There are 7 tomatoes in a bag, 4 of them green and the rest red. A tomato is picked out of the bag at random. What is the probability that it is red? (EM)

10. There are 900 pupils in a school. The pie chart in Fig. 39.6 shows the proportions of children travelling to school by private bus, public transport, bicycle and on foot.

(a) How many pupils travel to school on foot?

(b) What percentage of the pupils travel by public transport? (EA)

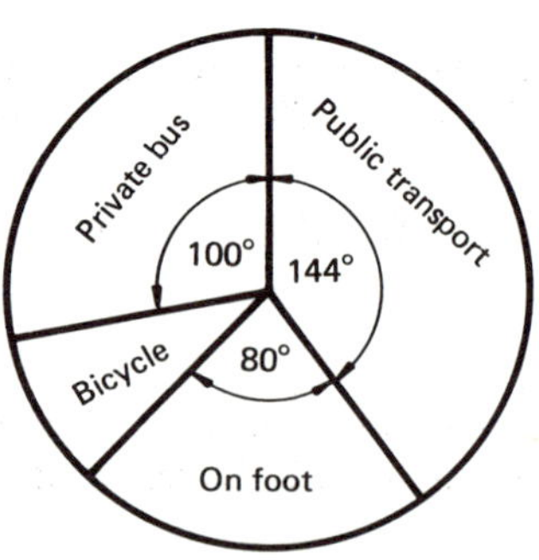

Fig. 39.6

11. The hours of sunshine recorded daily in a city during eight days in May in 1979 were as follows:

7.8, 3.9, 5.8, 2.2, 4.1, 7.2, 7.7 and 6.0

Find the median. (Y)

12. The pie chart (Fig. 39.7) shows how the Brown family spend their housekeeping money.

(a) Which item is the mode?

(b) If £60 is allowed per week, how much rent do they pay each week? (EM)

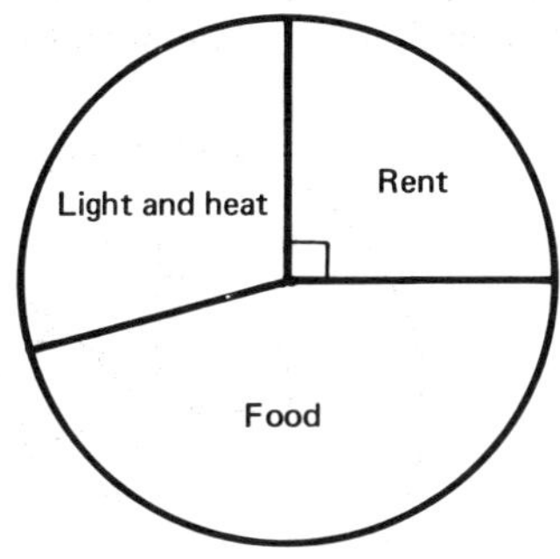

Fig. 39.7

Multi-choice questions 39

1. The mean of 11, 17, 25, 32 and 35 is

A 12 **B** 24 **C** 120 **D** 600 (WY)

2. Ten numbers have a mean of 18.5. The sum of the ten numbers is

A 1.85 **B** 8.5 **C** 18.5 **D** 185 (WY)

3. Seven numbers have a mode of 72. Six of the numbers are 46, 48, 62, 72, 72 and 84. Which of the following could be the seventh number?

A 46 **B** 48 **C** 62 **D** 72

E 84 (WY)

4. Nine numbers have a median of 15. Eight of the numbers are 4, 6, 8, 12, 15, 16, 20 and 27. Which of the following could be the ninth number?

A 8 **B** 10 **C** 12 **D** 14

E 16 (WY)

5. The bar chart (Fig. 39.8, p. 232) shows the marks gained in a school test. How many pupils took the test?

A 5 **B** 17 **C** 50 **D** 520 (WM)

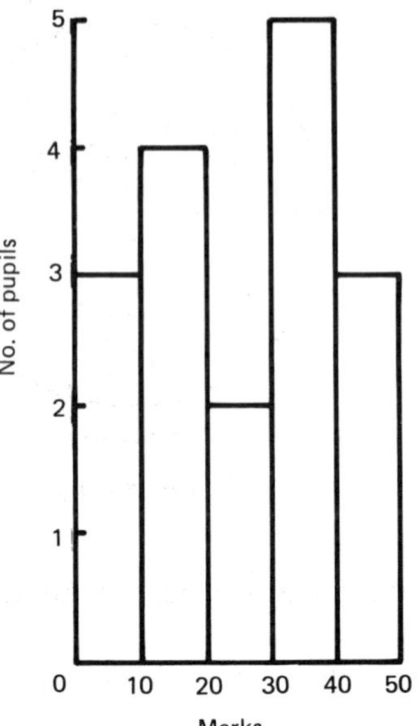

Fig. 39.8

6. 5000 students were interviewed to find out which of four sporting activities they liked most. The pie chart shown in Fig. 39.9 illustrates the information obtained. What is the angle of the sector representing those who liked hockey?

   A 95° **B** 72° **C** 65° **D** 60°

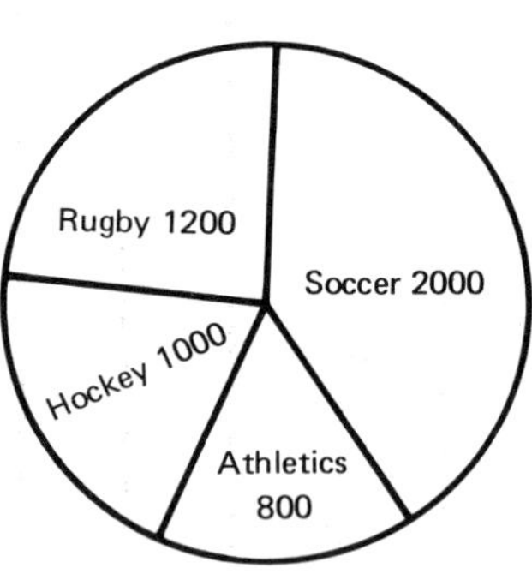

Fig. 39.9

7. The mean of the numbers 12, 24 and x is the same as the mean of the numbers 9, 12, 18 and 21. What is the value of x?

   A 9 **B** 15 **C** 18 **D** 24

8. Fig. 39.10 is a pie chart which shows the total sales for a department store during one week. The total sales were £40 000. What were the sales of clothing?

   A £12 000 **B** £10 000
   C £9000 **D** £6000

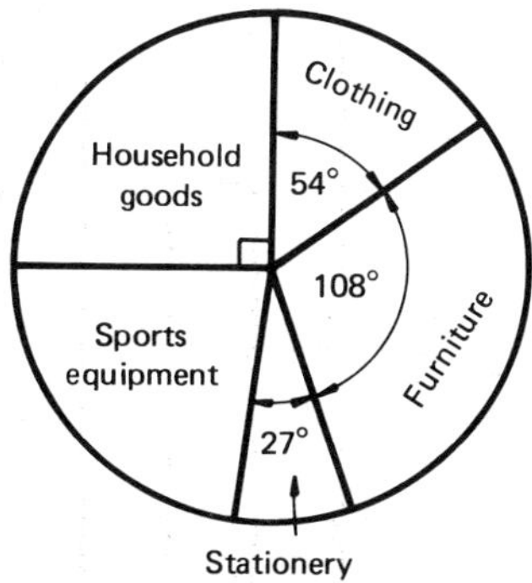

Fig. 39.10

9. Which of the following cannot be the probability of an event occurring?

A $\frac{1}{2}$ **B** 0.7 **C** 1 **D** 2

10. Fig. 39.11 illustrates the frequency distribution of marks obtained in a test. How many students took the test?

A 15 **B** 24 **C** 50 **D** 62

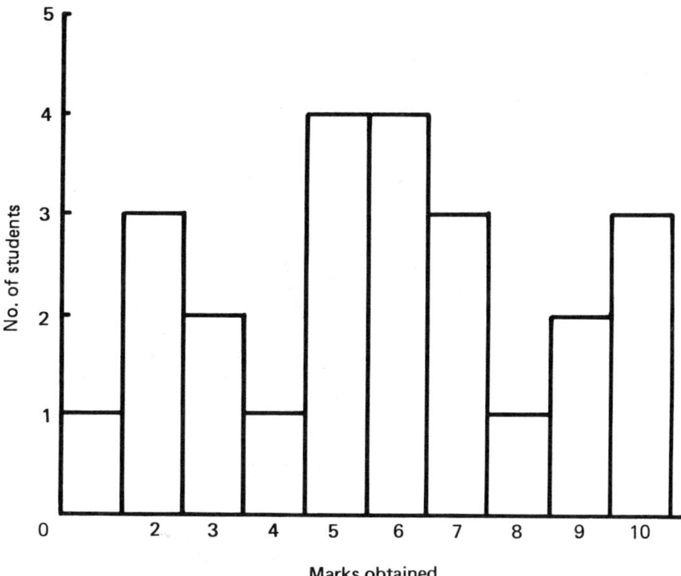

Fig. 39.11

11. Three red balls and two green balls are put into a box. One is drawn out at random. The probability of its being red is

A 0.6 **B** 2 out of 3

B 3 to 2 **D** likely

12. The main leisure interests of 150 fifth-year pupils are shown as percentages:

Football	40%
Television	30%
Reading	10%
Scouts and Guides	8%
Other	12%

If a pie chart is drawn from this information, what size of angle will be required for those pupils whose main interest is television?

A 30° **B** 50° **C** 72° **D** 108°
E 120° (NW)

13. The marks of ten students in a test were 8, 4, 5, 10, 9, 8, 6, 5, 8 and 6. What is the modal mark?

A 10 **B** $8\frac{1}{2}$ **C** 8 **D** 6.9

14. The pie chart shown in Fig. 39.12 shows the number of passes in various subjects in examinations. Out of 240 students, how many passed in Maths?

A 48 **B** 60 **C** 90 **D** 216

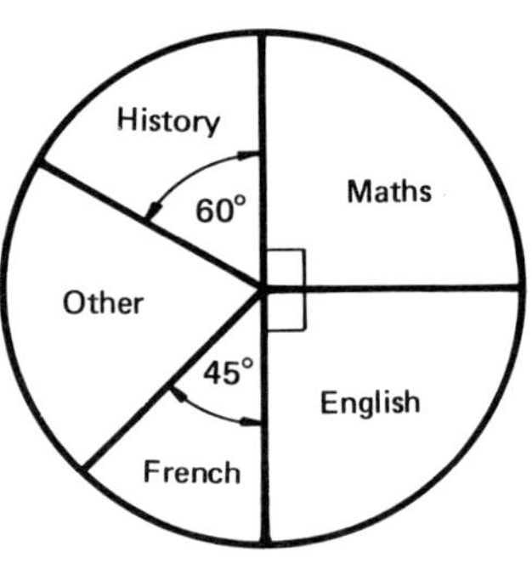

Fig. 39.12

15. The table below shows how a group of children travel to and from school:

Bus	26
Cycle	9
Walk	21
Car	3
Train	1

This information is to be represented in a pie chart. The angle representing the pupils who travel by bus will be

A 16° **B** 26° **C** 130° **D** 156° (Y)

16. The pie chart in Fig. 39.13 shows the breakdown of costs for items made by a small firm. The selling price for a certain item was £18. What is the cost of materials for this item?

A £4.50 **B** £9 **C** £12 **D** £18 (EA)

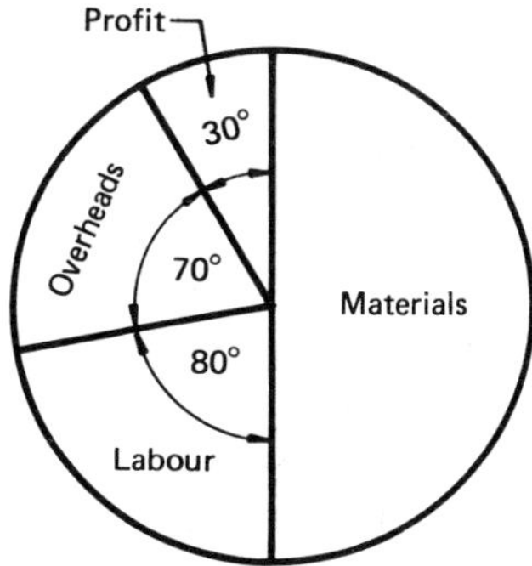

Fig. 39.13

17. In Fig. 39.13, the cost of labour to make another item was £6. What was the selling price for this item?

A £12 **B** £18 **C** £27 **D** £48 (EA)

40 COMPASS BEARINGS

- The FOUR CARDINAL DIRECTIONS are north, south, east and west. The directions north-east, north-west, south-east and south-west are also used and are as shown in Fig. 40.1.

  A bearing of N 20° E means an angle of 20° measured from north towards east as shown in Fig. 40.2. Similarly a bearing of S 50° E means an angle of 50° measured from south towards east.

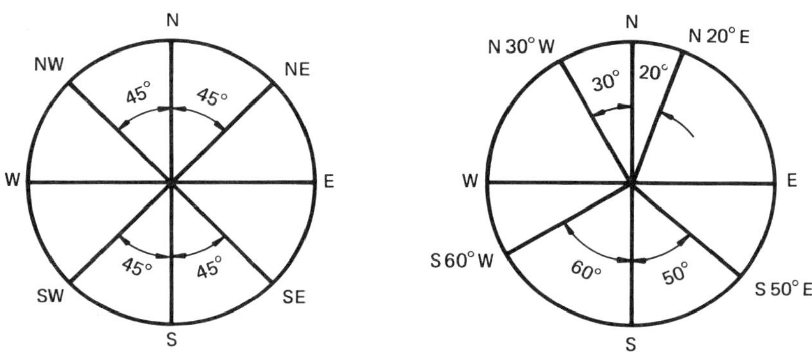

Fig. 40.1 **Fig. 40.2**

Bearings stated in this way are always measured from N and S and never from E and W.

- A second way of STATING A BEARING is to measure the angle denoting the bearing from north in a clockwise direction. Some bearings stated in this way are shown in Fig. 40.3.

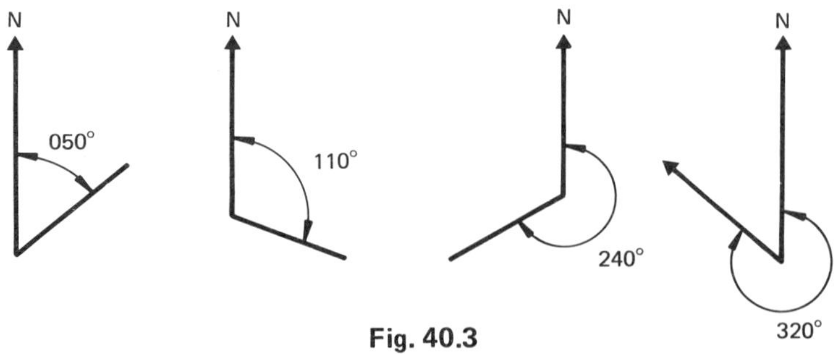

Fig. 40.3

Note that three figures are always stated. For example 005° is written instead of 5° and 073° instead of 73°. East will be 090°, south will be 180° and west will be 270°.

Example

(a) B is a point due east of a harbour A and C is a point on the coast which is 8 km due south of A, the distance BC being 9 km. Make a scale drawing and hence find the bearing of C from B.

First choose a suitable scale (in an examination the scale will usually be given) to represent the distances AC and BC. In Fig. 40.4, a scale of 1 cm = 1 km has been chosen. Remembering that east has a bearing of 090° and south a bearing of 180°, the scale drawing will look like Fig. 40.4. The reflex angle at B gives the bearing of C from B and by using a protractor this is found to be 207°.

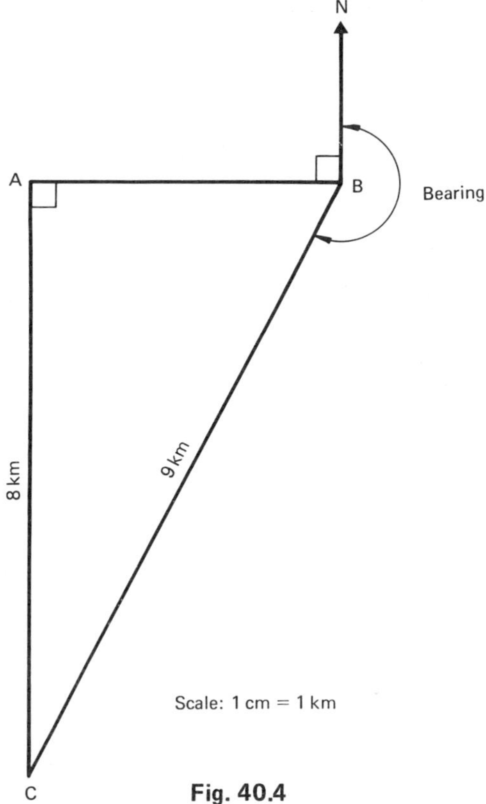

Fig. 40.4

(b) The bearing of a point B from A is 065°. What is the bearing of A from B?

The situation is shown in Fig. 40.5 where it can be seen that the bearing of A from B is 245°.

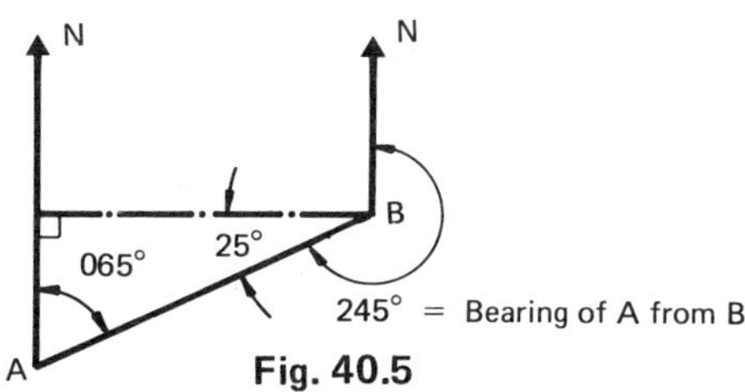

Fig. 40.5

Exercise 40.1 (All of the type found in CSE examination papers)

1. A bearing is shown in Fig. 40.6. What is the bearing of A from B? (AL)

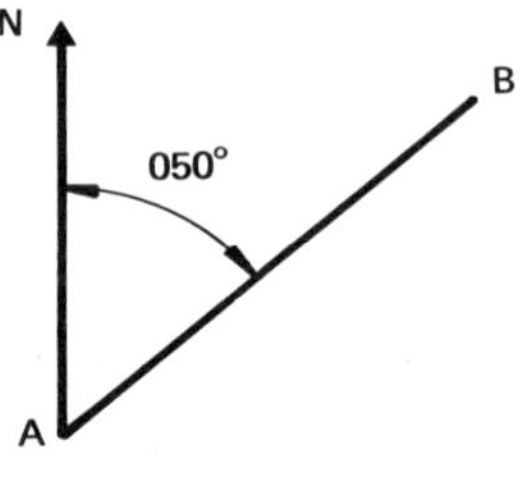

Fig. 40.6

2. What is the next compass direction in this series?

N NE E SE S ... (AL)

3. (a) Find the larger angle, in degrees, between the directions east and north-west.

   (b) If AN is due north in Fig. 40.7, write down the bearing of A from B. (EA)

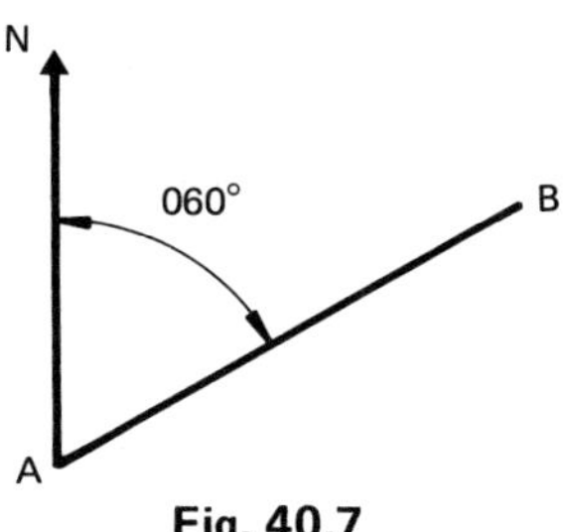

Fig. 40.7

4. From the diagram in Fig. 40.8, calculate the bearing of X from Y.

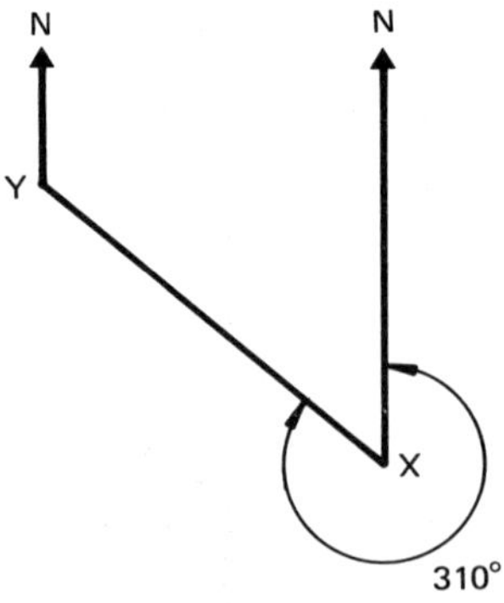

Fig. 40.8

5. A ship is on a bearing of 070° from a lighthouse. What is the bearing of the lighthouse from the ship? (Y)

6. The bearing of B from A is 195°. What is the bearing of A from B? (Y)

7. A dog runs 60 m away from its owner in a direction 060°, then chases a ball for 120 m in a direction of 150°. Make a scale drawing of the dog's movements and find out how far it must run, and in what direction, in order to return the ball to its owner. (EM)

8. B and C in Fig. 40.9 are both 100 km from A. C is on a bearing of 225° from B.
   (a) Calculate the bearing of A from B.
   (b) Calculate the size of angle ABC.
   (c) Calculate the bearing of C from A. (EA)

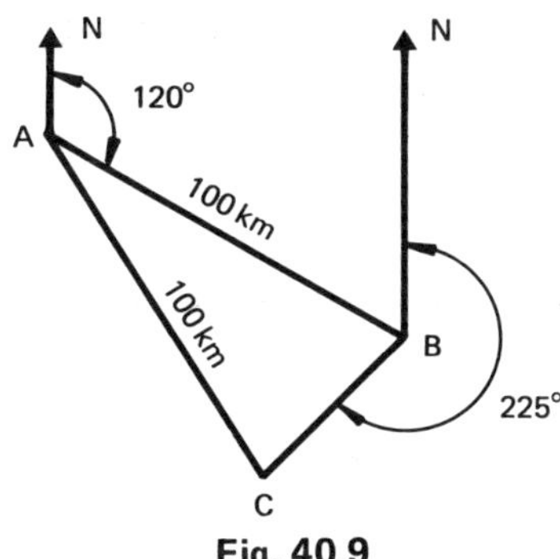

Fig. 40.9

9. A ship sails from P on a course of 035° for 8 km to Q, then alters course to 155° for 13 km to R. Using a scale of 1 cm to represent 1 km, make a scale drawing of the course of the ship from P to Q to R. Find the course to be steered to go directly from R to P. (S)

10. The diagram in Fig. 40.10 shows the positions of three towns A, B and C, which are such that CA points due north.
    (a) Find angle ACB.
    (b) Calculate the bearing of B from A.
    (c) Calculate the bearing of C from B.

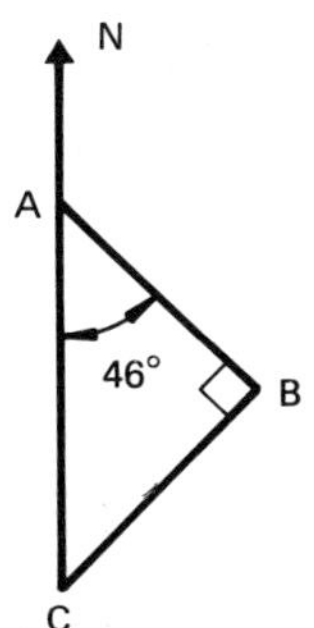

Fig. 40.10

11. A boat leaves a harbour A on a course of S 60°E and sails 80 km in this direction until it reaches point B. Using a scale of 1 cm = 10 km, make a scale drawing of the movement of the boat. Hence find how far B is east of A and what distance B is south of A.

12. A point C is 50 m from A on a bearing of N 40° W. Point B is 60 m from A on a bearing of S 30° W. Make a scale drawing showing the relative positions of A, B and C using a scale of 1 cm = 10 km. Hence find the distance between points B and C and state the bearing of B from C.

Multi-choice questions 40

1. In Fig. 40.11 find the bearing of P from Q.

   A 025° **B** 065° **C** 245° **D** 295°
   E 345° (AL)

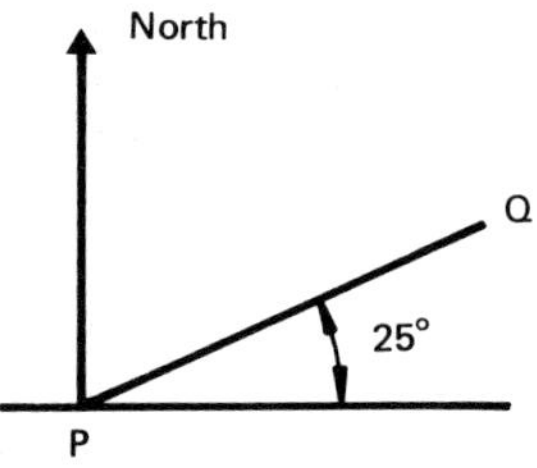

Fig. 40.11

On a map, town T is due south of town V and town V is on a bearing of N 68° E (068°) from town X. ∠VXT is 77° (Fig. 40.12). Use this information to answer questions 2, 3, 4 and 5.

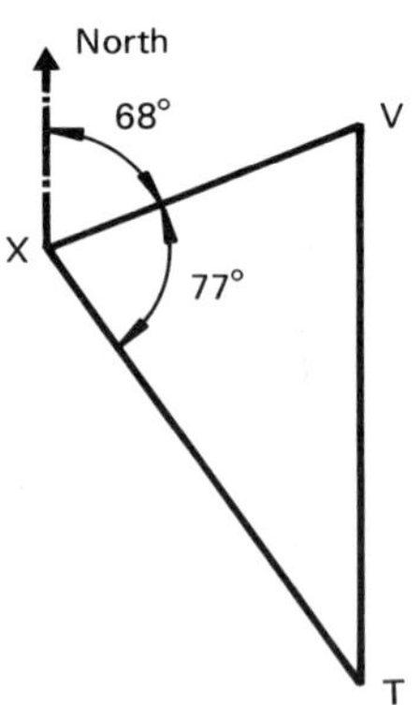

Fig. 40.12

2. Determine the bearing of X from V.

   A N 77° E (077°) **B** S 68° E (112°)
   C S 68° W (248°) **D** N 68° W (292°) (WY)

3. What is the bearing of V from T?

   A Due N (000°) **B** N 77° E (077°)
   C Due S (180°) **D** N 77° W (283°) (WY)

4. The bearing of T from X is

   A N 35° E (035°) **B** S 35° E (145°)
   C N 77° W (283°) **D** N 35° W (325°) (WY)

5. Determine the bearing of X from T.

   A N 35° E (035°) **B** S 35° E (145°)
   C N 77° W (283°) **D** N 35° W (325°) (WY)

Use Fig. 40.13 to answer questions 6 and 7.

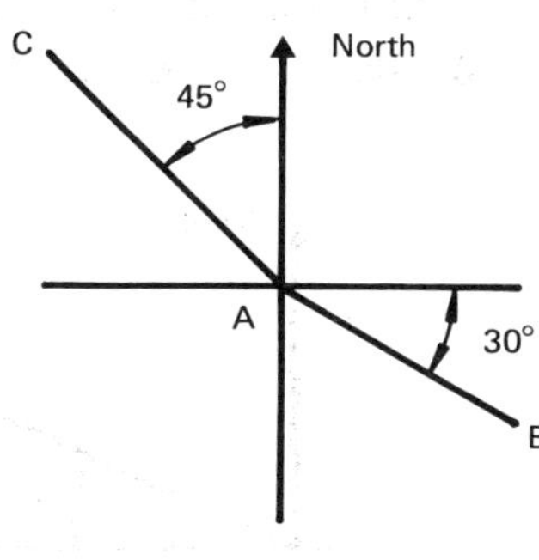

Fig. 40.13

6. The bearing of B from A is

   A 030° **B** 060° **C** 120° **D** 150° (AL)

7. The bearing of C from A is

   A 045° **B** 135° **C** 225° **D** 315° (AL)

Use Fig. 40.14 to answer questions 8 and 9.

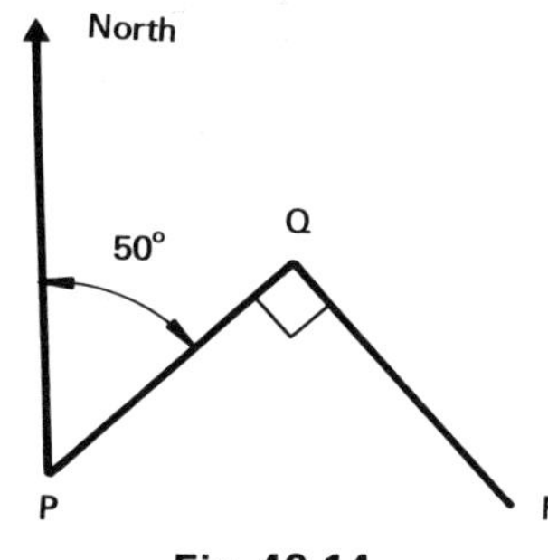

Fig. 40.14

8. The bearing of P from Q is

   A 040° **B** 050° **C** 140° **D** 230°
   E 320° (AL)

9. The bearing of R from Q is

A 040° **B** 050° **C** 140° **D** 230°

E 320° **(AL)**

10. Town A is on a bearing of 040° from town B. What is the bearing of town B from town A?

A 130° **B** 140° **C** 220° **D** 310°

E 400° **(WY)**

11. A ship sails on a bearing of 330° and then turns anticlockwise through an angle of 90°. Its new course is on a bearing of

A 030° **B** 060° **C** 230° **D** 240°

E 420° **(WY)**

12. I am travelling south-west when I turn anticlockwise through an angle of 150°. What is the direction of my new bearing?

A 015° **B** 075° **C** 285° **D** 345°

E 380° **(WY)**

ANSWERS

Exercise 1.1

1. (a) 42 (b) 468 (c) 18 231 (d) 2862
   (e) 115 594
2. (a) 15 (b) 106 (c) 103 (d) 9451
3. (a) 63 (b) 180 (c) 4811 (d) 1785
   (e) 4985
4. (a) 22 (b) 11 587 (c) 96 (d) 4722

Exercise 1.2

1. 18 2. 19 3. 28 4. 38 5. 23
6. 5 7. 18 8. 35

Exercise 1.3

1. 7000 2. 3050 3. 0 4. 0 5. 56
6. 9 7. 16 8. 9 9. 12 10. 3

Exercise 1.4

1. 250, 1250 2. 17, 21 3. 27, 33 4. 22, 11
5. 11, 7 6. 15, 21 7. 21, 34 8. 7, 9
9. 6, 8 10. 21, 55, 89 11. 31, 255 12. 82, 244
13. 34 14. 67 15. 63 16. 81
17. 15, 34, 55, 63 18. 15, 55 19. 36, 45 20. 16, 18, 20

Exercise 1.5

1. (a) 1, 3, 5, 15 (b) 1, 2, 4, 8, 16, 32
   (c) 1, 2, 3, 4, 6, 7, 12, 14, 21, 28, 42, 84
2. 1, 2, 3, 4, 6, 9, 12, 18
3. 20, 25, 30, 35, 40, 45, 50, 55
4. (a) $2^2 \times 3$ (b) $2^2 \times 3^2$ (c) $2 \times 3^2 \times 5$ (d) $3 \times 5^2 \times 7$
   (e) $2^3 \times 3 \times 5 \times 11$
5. 19, 23, 29
6. (a) 30 (b) 48 (c) 30 (d) 60
   (e) 1260 (f) 3780 (g) 26 460
7. (a) 10 (b) 13 (c) 28 (d) 126
8. 12, 16, 20, 24, 28, 32, 36
9. (a) $2^2 \times 3 \times 11$ (b) 0 (c) 42

Exercise 1.6

1. (a) 9, 5, 1 (b) 21, 28, 36
2. 3^4 greater by 17
3. (a) 2, 4, 6, 8 (b) 3, 6, 9 (c) 2, 3, 5, 7
4. (a) 81 and 139 (b) 16, 64 and 81 (c) 16, 32 and 64
5. 5 and 7
6. (a) 1, 3, 9 (b) 10, 15
7. 29, 49
8. 37, 41
9. 900 000
10. 5
11. 26
12. 25
13. 780

14. (a) 55 (b) 15

15. (a) 5, 7, 19, 31 (b) 2, 5, 7, 19, 31 (c) 14 (d) 81

17. 1, 3, 21

18. (a) 4, 8, 12, 16, 20, 24 (b) $2^2 \times 3 \times 7$ (c) 1, 3, 6, 10, 15, 21

19. (a) 49, no (b) 2639, no (c) 170 (d) 1500

20. (a) 24, 35 (b) 49, 36 (c) 1, 2, 3, 5, 7 (d) (i) 56 (ii) 196

21. 5, 80, 160

22. (a) 11, 17, 26 (b) 21, 28, 45 (c) 9, 25, 36 (d) 81, 729, 2187 (e) 40, 5

23. (a) 52 (b) 9

24. (a) 51, 71 (b) 32, 16 (c) 135, 405

25. (a) 22 (b) 5

26. (a) 28 (b) 3 (c) 21 (d) 3, 6, 15

27. (a) 63, 83 (b) 2, $\frac{1}{8}$

Multi-choice questions 1

1. E	**2.** E	**3.** D	**4.** D	**5.** D
6. E	**7.** D	**8.** B	**9.** B	**10.** B
11. A	**12.** D	**13.** C	**14.** B	**15.** B
16. B	**17.** B	**18.** D	**19.** D	**20.** C
21. D	**22.** D	**23.** B	**24.** B	**25.** A
26. B	**27.** C			

Exercise 2.1

1. $3\frac{1}{2}$	**2.** $2\frac{1}{8}$	**3.** $5\frac{1}{5}$	**4.** $2\frac{4}{7}$	**5.** $2\frac{5}{9}$
6. $\frac{16}{5}$	**7.** $\frac{7}{4}$	**8.** $\frac{41}{7}$	**9.** $\frac{19}{8}$	**10.** $\frac{69}{20}$
11. $\frac{1}{2}$	**12.** $\frac{5}{7}$	**13.** $\frac{1}{3}$	**14.** $\frac{3}{4}$	**15.** $\frac{2}{5}$

Exercise 2.2

1. $\frac{3}{5}, \frac{7}{10}$	**2.** $\frac{1}{2}, \frac{2}{3}, \frac{5}{6}$	**3.** $\frac{17}{32}, \frac{9}{16}, \frac{5}{8}, \frac{3}{4}$	**4.** $\frac{3}{8}, \frac{5}{9}, \frac{4}{7}, \frac{3}{5}$
5. $\frac{8}{15}$	**6.** $\frac{7}{8}$	**7.** $\frac{43}{72}$	**8.** $1\frac{11}{12}$
9. $1\frac{107}{120}$	**10.** $3\frac{15}{16}$	**11.** $8\frac{7}{15}$	**12.** $10\frac{17}{56}$
13. $11\frac{2}{3}$	**14.** $9\frac{33}{40}$	**15.** $\frac{1}{6}$	**16.** $\frac{13}{24}$
17. $\frac{1}{15}$	**18.** $1\frac{1}{12}$	**19.** $1\frac{123}{160}$	**20.** $2\frac{1}{16}$
21. $\frac{3}{28}$	**22.** $\frac{9}{14}$	**23.** $\frac{25}{27}$	**24.** $3\frac{17}{20}$
25. $\frac{3}{14}$	**26.** $2\frac{2}{5}$	**27.** 6	**28.** $\frac{1}{3}$
29. 108	**30.** $4\frac{1}{4}$	**31.** 200	**32.** 12
33. $2\frac{2}{3}$	**34.** $\frac{3}{4}$	**35.** $3\frac{1}{2}$	**36.** $\frac{6}{25}$
37. $\frac{1}{6}$	**38.** $5\frac{1}{3}$	**39.** $\frac{3}{4}$	**40.** 6

Exercise 2.3

1. $\frac{1}{10}$	**2.** $1\frac{1}{14}$	**3.** $1\frac{1}{2}$	**4.** $\frac{11}{20}$	**5.** $1\frac{1}{10}$
6. $\frac{2}{3}$	**7.** $8\frac{1}{2}$	**8.** $3\frac{7}{16}$		

Exercise 2.4

1. (a) $\frac{1}{7}$ (b) $\frac{1}{20}$ (c) $\frac{11}{35}$ (d) $\frac{1}{35}$ (e) $1\frac{1}{4}$

2. (a) $\frac{2}{7}$ (b) $\frac{5}{22}$ (c) $\frac{5}{24}$ (d) $4\frac{1}{12}$

3. $\frac{1}{4}$

4. (a) $\frac{7}{12}$ (b) $\frac{2}{3}$ (c) $5\frac{1}{2}$

5. $\frac{5}{23}, \frac{5}{22}, \frac{5}{21}$

6. (a) 3 (b) 15 (c) $10\frac{1}{2}$

7. $2\frac{1}{2}, 2\frac{3}{4}, 3$

8. $\frac{1}{3}$

9. $\frac{2}{3}, \frac{5}{8}, \frac{7}{12}$

10. 20
11. $\frac{3}{4}, \frac{11}{16}, \frac{5}{8}$
12. (a) $5\frac{3}{8}$ (b) 12
13. $\frac{1}{36}, \frac{1}{49}$
14. $\frac{11}{32}$
15. $1\frac{11}{16}$

Multi-choice questions 2

1. B 2. D 3. B 4. B 5. B
6. A 7. C 8. B 9. C 10. D

Exercise 3.1

1. 0.9 2. 0.38 3. 0.672 4. 0.06
5. 0.209 6. 0.003 7. 498.2 8. 54.079
9. $\frac{3}{10}$ 10. $3 + \frac{8}{10}$ 11. $7 + \frac{9}{10} + \frac{8}{100}$
12. $567 + \frac{2}{10} + \frac{3}{100} + \frac{4}{1000}$ 13. $\frac{2}{1000}$ 14. $\frac{5}{100} + \frac{9}{1000}$
15. $\frac{6}{10} + \frac{3}{1000}$ 16. $\frac{1}{10} + \frac{7}{100} + \frac{3}{10\,000}$

Exercise 3.2

1. 2.714 2. 17.36 3. 895.619 4. 0.615
5. 1.12 6. 12.146 7. 0.027 8. 26.033

Exercise 3.3

1. 3.5, 35, 350 2. 59.83, 598.3, 5983
3. 0.38, 3.8, 38 4. 982.345, 9823.45, 98 234.5
5. 81.624, 816.24, 8162.4 6. 18.9, 1.89, 0.189
7. 1.813, 0.1813, 0.018 13 8. 52.731, 5.2731, 0.527 31
9. 0.003, 0.0003, 0.000 03 10. 0.0325, 0.003 25, 0.000 325

Exercise 3.4

1. (a) 19.37 (b) 19.4
2. (a) 0.007 52 (b) 0.008 (c) 0.01
3. (a) 4.970 (b) 4.97
4. (a) 153.262 (b) 153.26 (c) 153.3
5. (a) 24.9358 (b) 24.94 (c) 25
6. (a) 0.007 326 (b) 0.007 33 (c) 0.0073
7. (a) 35.60 (b) 35.6
8. (a) 35 680 (b) 35 700 (c) 36 000
9. (a) 17 359 000 (b) 17 000 000
10. (a) 0.0078 (b) 0.008

Exercise 3.5

1. 12 2. 10 3. 0.14 4. 3 5. 3
6. 24 7. 0.3 8. 30

Exercise 3.6

1. 0.75 2. 0.2 3. 0.875 4. 0.8125
5. 0.4286 6. 0.8333 7. 1.594 8. 3.234
9. 0.5556 10. 0.8889 11. 0.1778 12. 0.4556
13. 0.3535 14. 0.2121 15. 0.5626 16. 0.7317
17. $\frac{3}{10}$ 18. $\frac{13}{20}$ 19. $\frac{7}{16}$ 20. $2\frac{31}{50}$
21. $1\frac{3}{4}$ 22. $9\frac{37}{200}$ 23. 0.0175 24. 1.748
25. 0.095

Exercise 3.7

1. (a) 28.8 (b) 0.8 (c) 2.88 (d) 8
2. (a) 341.1 (b) 3.74 (c) 826.4

3. 0.875
4. (a) $\frac{3}{4}$ (b) 0.75
5. (a) 2873 (b) 0.2873
6. 0.86
7. (a) 73 800 (b) 0.809
8. 23 by 17.4
9. (a) 0.0018 (b) 0.1743
10. 5.25
11. (a) $\frac{8}{1000}$ (b) 49.95
12. 7.4, 2900
13. 48 400
14. 0.05, 0.22, 0.5
15. (a) 171.306 (b) (i) 167.1 (ii) 170
16. (a) 56.798 (b) (i) 738.06 (ii) 740
17. (a) 0.625 (b) 0.6875
18. (a) $\frac{7}{10}$ (b) $\frac{3}{25}$

Multi-choice questions 3

1. C 2. B 3. D 4. C 5. D
6. B 7. D 8. C 9. C 10. B

Exercise 4.1

1. £153.52 2. £722.43 3. £96.42 4. £2.00
5. £2.19 6. £1.14 7. £46.51 8. £4.46
9. £6.95 10. £0.94 11. £156 12. £24.03
13. £5.85 14. £168.82 15. 37 p 16. 23 p

Exercise 4.2

1. 418.8 2. 16 490 3. 252 000 4. £226.97
5. £73.09 6. £93.96

Exercise 4.3

1. £15.08 2. £1.47 3. £57.18 4. £150
5. £15.08 6. £5 7. £6.24 8. £20.25
9. £8.00 10. 81 450 11. £3.80 12. £54
13. £42 14. (a) 264 (b) £2.50 15. 8 p
16. £1.35

Multi-choice questions 4

1. A 2. C 3. B 4. A 5. C
6. C 7. C 8. B 9. C 10. B
11. D 12. D

Exercise 5.1

1. (a) 6780 m (b) 790 m (c) 6.93 m
 (d) 0.053 m (e) 7.395 m (f) 0.007 m
2. (a) 9.375 km (b) 0.368 km (c) 0.002 75 km
 (d) 0.039 41 km (e) 7.356 82 km
3. (a) 1715 cm (b) 39.5 cm (c) 178 000 cm
 (d) 86.4 cm (e) 0.52 cm
4. (a) 58 000 mm (b) 235 mm (c) 160 000 mm
 (d) 392 mm (e) 5.9 mm
5. (a) 0.680 kg (b) 37.8 kg (c) 0.002 987 kg
 (d) 0.459 kg
6. (a) 78 g (b) 0.045 g (c) 19 100 g (d) 590 g
7. 27 t
8. 7320 kg
9. (a) 6.315 m (b) 93 467.36 m (c) 0.2668 m
10. 0.687 m
11. (a) 2.072 kg (b) 26.243 kg

12. 98.05 kg
13. 5.75 m
14. 120.65 m
15. 17.85 m
16. 47 lengths; 9 cm remains
17. 83
18. 32 000
19. 468
20. 41 kg

Exercise 5.2

1. 20 miles	2. 40 km	3. 2.53	4. 585.2
5. 4.4	6. 0.55 lb	7. 2.27	8. 11 000

Exercise 5.3

1. (a) 15 320 g (b) 2.935 m (c) 2.895 km
2. (a) 8.539 kg (b) 94 600 m (c) 84 600 m
3. (a) 1000 (b) 3400
4. 570 cm
5. (a) 13 (b) (i) 0.70 m (ii) 700 mm
6. 750 g
7. 1.10 lb
8. 18 p
9. 700 kg
10. 1.36 m
11. 14
12. £86.80
13. 200
14. 870 g
15. (a) £2.70 (b) 54 p
16. 30 cm
17. 55 lb
18. 45 minutes
19. (a) 5200 (b) 50 cm
20. 5; 50 cm

Multi-choice questions 5

1. C	2. D	3. D	4. D	5. B
6. A	7. A	8. D	9. C	10. C

Exercise 6.1

1. $\frac{9}{7}$	2. $\frac{1}{2}$	3. $\frac{2}{1}$	4. $\frac{4}{5}$	5. $\frac{2}{3}$
6. $\frac{1}{8}$	7. $\frac{1}{160}$	8. $\frac{1}{25}$	9. $\frac{20}{1}$	10. $\frac{10}{1}$
11. $\frac{1}{6}$	12. $\frac{3}{28}$	13. $\frac{72}{5}$		

Exercise 6.2

1. £1000; £600
2. 56 kg; 24 kg
3. 24 m; 36 m; 60 m
4. 168 mm; 588 mm; 924 mm
5. (a) 25 cm (b) 60 cm
6. £32; £16

Exercise 6.3

1. £1.20	2. £120	3. £10.50	4. 31.7 litres
5. £28	6. 8 days	7. 20 days	8. 40

Exercise 6.4

1. £2.50; £7.50	2. 60 p	3. 31.5 km	4. 4.5 km
5. £2.24	6. £455	7. 4 : 5	8. (a) £18 (b) £9
9. £150; £200	10. £3.75	11. (a) 15 (b) £28.50	

12. $\frac{3}{5000}$

Multi-choice questions 6

1. C	**2.** C	**3.** D	**4.** A	**5.** B
6. A	**7.** B	**8.** B		

Exercise 7.1

1. 90%	**2.** 65%	**3.** 56%	**4.** 58%	**5.** 25%
6. 80%	**7.** 94%	**8.** 5%	**9.** 56.2%	**10.** 75.2%
11. 0.44	**12.** 0.15	**13.** 0.09	**14.** 0.083	**15.** 0.952

Exercise 7.2

1. (a) 21 (b) 12 (c) 6 (d) 3.24
2. (a) 6.25% (b) 20% (c) 20% (d) $33\frac{1}{3}$%
3. 75%; 24
4. 50 cm
5. (a) £9.60 (b) £18 (c) £406
6. 2160 kg
7. 400
8. 4700

Exercise 7.3

1. 25%	**2.** 20%	**3.** 50%	**4.** 75%	**5.** 25%

Exercise 7.4

1. £180	**2.** £52.80	**3.** £77	**4.** £700

Exercise 7.5

1. 7.2
2. 50%
3. 46.74%
4. £284
5. (a) £2.08 (b) 25%
6. (a) 16 (b) £5250
7. (a) £42 (b) 25%
8. 3.5%
9. (a) $\frac{3}{4}$ (b) 75% (c) 0.6 (d) 60%
(e) $\frac{1}{8}$ (f) 0.125
10. (a) $\frac{12}{25}$ (b) $\frac{1}{16}$
11. (a) £15 (b) £16
12. (a) 60% (b) 12
13. £61.92
14. £247.50
15. (a) £500 (b) £200

Multi-choice questions 7

1. C	**2.** C	**3.** D	**4.** D	**5.** B
6. B	**7.** C	**8.** C	**9.** B	**10.** D

Exercise 8.1

1. 13	**2.** −9	**3.** −32	**4.** −16	**5.** −30
6. −4	**7.** −8	**8.** 4	**9.** 2	**10.** −23
11. 6	**12.** 13	**13.** 4	**14.** −5	**15.** 15
16. −30	**17.** −6	**18.** −20	**19.** 8	**20.** −60
21. 84	**22.** 25	**23.** −4	**24.** −3	**25.** 4
26. −5	**27.** 3	**28.** −2	**29.** −2	**30.** 4

Exercise 8.2

1. (c), (d), (g) **2.** (c), (d) **3.** (a), (b) **4.** (a), (b), (d)
5. (a), (b), (d)

Exercise 8.3

1. (a) 40 (b) -13 (c) 13 (d) 40
2. 49
3. 3
4. (a) 15 (b) -1
5. 22
6. $-4, -2, 0, 6$
7. $-2, -5, -9, -42$
8. (a) 44 (b) -26
9. (a) $-$ (b) $-$ (c) $-$ (d) $+$
10. 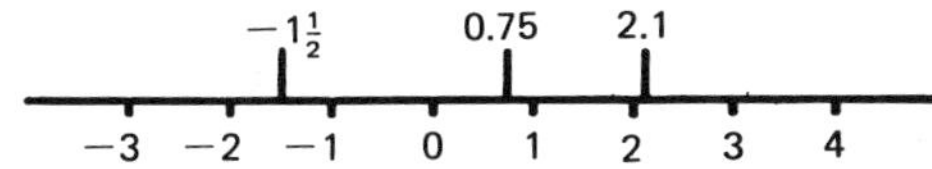

Multi-choice questions 8

1. C	2. D	3. D	4. D	5. C
6. C	7. A	8. C		

Exercise 9.1

1. $6 \times x$	2. $4 \times a - 3$	3. $5 \times y + z$	4. $x \times y \times z$
5. $5 \times m \times n$			

Exercise 9.2

1. 9	2. 0	3. 5	4. 15
5. 21	6. 42	7. 126	8. 90
9. 4	10. 2	11. -7	12. -35

Exercise 9.3

1. 8	2. 81	3. 25	4. 162	5. 48
6. 48	7. 200	8. 750	9. 360	10. 13

Exercise 9.4

1. $8x$	2. $4y$	3. $10p$	4. $-3q$
5. $-6x$	6. $-2m$	7. $-6x$	8. $5p + 9q$
9. $13x - 9y + 10z$			

Exercise 9.5

1. $6yp$	2. $-6mp$	3. $20ab$	4. $18abc$
5. $12xyz$	6. $-14mnp$	7. $-30abcd$	8. $-120yzq$

Exercise 9.6

1. $3a + 6b$	2. $10x - 15y$	3. $20p - 15q$	4. $-p + q$
5. $-6x + 8y$	6. $10m$	7. $13a - 21b$	8. $-4p - 7q$
9. $25n$	10. $7x - 9y$		

Exercise 9.7

1. (a) $3x$ (b) $18x^2$ (c) 2
2. (a) $10x - y$ (b) $-2x - y$ (c) x
3. 25
4. (a) -11 (b) 0
5. $x + 4y$
6. (a) $5x - 6y$ (b) $30a^2b$
7. (a) 0 (b) 0 (c) 3
8. (a) 18 (b) -1.2
9. (a) 49 (b) 101
10. (a) 2 (b) $8\frac{2}{3}$

Multi-choice questions 9

1. C 2. A 3. A 4. D 5. B
6. A 7. B 8. B 9. D 10. D

Exercise 10.1

1. $2(x+y)$ 2. $3(p-q)$ 3. $5(x+3y)$
4. $b(r-s)$ 5. $2(2x-3y)$ 6. $x(ax+b)$
7. $(a-b)(x+y)$ 8. $(x+y)(p-q)$

Exercise 10.2

1. x^2+5x+6 2. x^2-6x+5 3. x^2+x-6
4. $2x^2-5x-3$ 5. $6x^2+5x-6$ 6. $5x^2-17x+6$
7. x^2+2x+1 8. x^2-4x+4 9. x^2-4
10. $4x^2-9$

Exercise 10.3

1. $(x+1)(x+2)$ 2. $(x-1)(x+2)$ 3. $(x+3)(x+5)$
4. $(x+5)(x-3)$ 5. $(x-5)(x+3)$ 6. $(x-4)(x+2)$
7. $(x-4)(x-2)$ 8. $(x-4)(x-8)$ 9. $(3x+2)(x+1)$
10. $(2x+3)(x+2)$ 11. $(5x-1)(x-2)$ 12. $(3x+2)(x-3)$

Exercise 10.4

1. x^2+4x+4 2. x^2-6x+9 3. $4x^2+4x+1$
4. $9x^2-24x+16$ 5. $4x^2+28x+49$ 6. x^2-1
7. $4x^2-25$ 8. $9x^2-16$ 9. $(x+2)^2$
10. $(x-2)^2$ 11. $(3x-2)^2$ 12. $(2x+3)^2$
13. $(x-2)(x+2)$ 14. $(3x-4)(3x+4)$ 15. $(5x-7)(5x+7)$

Exercise 10.5

1. $3pq(1-4q)$
2. 3200
3. $p=-8$
4. (a) $9x^2-12x+4$ (b) $(y-4)(y+4)$
5. $(x+1)(x+11)$
6. x^2+5x+6
7. (a) 44 (b) x^2-4y^2 (c) $4x^2-20x+25$
8. (a) q^2-p^2 (b) 200 000
9. $5x(x-2)$
10. $6x^2+13x-5$
11. (a) $(x+5)(x-5)$ (b) $ab(5b^2-2a)$
12. (a) $5yz(x-3)$ (b) $(3-p)(a-b)$ (c) $(x-2)(x+1)$
(d) $(2x+3y)(2x-3y)$
13. (a) $2x+8y$ (b) $a^2+6ab+9b^2$ (c) a^2-9b^2
14. (a) $2x(x-2)$ (b) $(3-x)(3+x)$
15. $p=-3$; $q=4$

Multi-choice questions 10

1. C 2. D 3. D 4. D 5. B
6. A 7. B 8. C 9. D 10. D

Exercise 11.1

1. q 2. $\dfrac{y}{x}$ 3. $\dfrac{2c^2}{b}$ 4. $\dfrac{3xy}{2z}$
5. $\dfrac{6an^2}{bm}$ 6. $\dfrac{4qs}{r}$ 7. $\dfrac{6a^2d^3}{bc}$ 8. $\dfrac{b^2c}{a}$
9. $\dfrac{yb}{xa}$ 10. $\dfrac{np}{mq^2}$

Exercise 11.2

1. $\frac{5a+4b}{20}$ 2. $\frac{3p-2q}{6}$ 3. $\frac{11a}{20}$ 4. $\frac{13x}{12}$
5. $\frac{13}{6x}$ 6. $\frac{7m}{6x}$ 7. $\frac{14x}{15y}$ 8. $\frac{x+5}{3}$
9. $\frac{10x-3}{12}$ 10. $\frac{9m+2}{2}$ 11. $-\frac{x+9}{12}$ 12. $\frac{9}{4x}$
13. $\frac{x+6}{6}$ 14. $\frac{a+12b}{6}$ 15. $\frac{25-7x}{10}$

Exercise 11.3

1. $\frac{5x}{8}$ 2. $\frac{x}{3y^2}$ 3. $\frac{x+12}{20}$ 4. $\frac{3y+2a}{ay}$
5. $x+5$ 6. $\frac{2x^3y^2}{3}$ 7. $\frac{31x}{12}$ 8. $\frac{11}{3p}$

Exercise 12.1

1. 11 2. 7 3. 6 4. 4 5. 12

Exercise 12.2

1. 7 2. 1 3. $\frac{1}{2}$ 4. 25 5. 25
6. 6

Exercise 13.1

1. a^7
2. p^{10}
3. y^{23}
4. 3^9, i.e. 19 683
5. 2^8, i.e. 256
6. $24a^6$
7. p^2
8. 4
9. q^4
10. m^3
11. t^5
12. m
13. $27b^6$
14. $8x^6y^3$
15. $625a^4b^8c^{12}$
16. $\frac{81p^8}{16q^{12}}$
17. (a) $\frac{1}{10}$ (b) $\frac{1}{9}$ (c) $\frac{1}{16}$ (d) $\frac{1}{125}$
18. (a) 25 (b) 2 (c) 2 (d) 3
19. (a) 2^6 (b) 2^{20} (c) 2^{18}
20. (a) $a^{1/2}$ (b) $a^{2/3}$ (c) $a^{4/5}$

Exercise 13.2

1. 8.27×10^2 2. 1.73×10^3 3. 1.7362×10^4
4. 8.036×10^6 5. 3×10^{-2} 6. 5.6×10^{-3}
7. 6×10^{-4} 8. 7×10^{-6} 9. 320
10. 5000 11. 1 870 000 12. 0.002
13. 0.567 14. 0.032

Exercise 13.3

1. 3.5×10^6
2. 2.15×10^2
3. 2.79
4. 0.005 01

5. 2×10^{20}
6. 9.43×10^{-4}
7. 8.192
8. 4
9. (a) 6×10^5 (b) 3.2×10^3
10. (a) 9 (b) 1 (c) $\frac{1}{10}$
11. 8.5×10^6
12. 3.2×10^4 by 24 410
13. 10^3
14. 8
15. (a) 8.7×10^4 (b) 7.3×10^{-2}
16. 5
17. 4
18. 0.009
19. 8.2×10^4
20. (a) 3 (b) 2 (c) 4

Multi-choice questions 13

1. D	2. C	3. B	4. B	5. C
6. A	7. D	8. B	9. A	10. B

Exercise 14.1

1. 0.9031; 1.9031; 2.9031; 3.9031; 4.9031
2. 0.8129; 1.8129; 2.8129; 3.8129; 4.8129
3. 0.7185; 1.7185; 2.7185; 3.7185; 4.7185
4. 0.4950; 1.4950; 2.4950; 3.4950; 4.4950
5. $\bar{1}.4962$; $\bar{3}.4962$; $\bar{4}.4962$
6. $\bar{3}.4295$; $\bar{5}.4295$; $\bar{6}.4295$
7. 3.802; 38.02; 380.2; 3802
8. 185 300; 1853; 18.53; 1.853
9. 0.2422; 0.024 22; 0.002 422
10. 0.000 011 99; 0.001 199; 0.011 99; 0.1199

Exercise 14.2

1. 8.999	2. 452.0	3. 16 790	4. 11.19
5. 0.007 450	6. 4.666	7. 0.2449	8. 21.44
9. 10.15	10. 3.552	11. 2.632	12. 1.345

Exercise 14.3

1. 1.967
2. 2.932
3. (a) 0.9524 (b) 2.892 (c) $\bar{1}.7709$
4. 3.345
5. 13 040
6. 7.574
7. 1.321
8. 232.8
9. (a) 442.3 (b) 6128 (c) 3.797 (d) 222.3
10. (a) 143.7 (b) 41 500 (c) 73.35 (d) 68.98

Multi-choice questions 14

1. C	2. B	3. D	4. D	5. D
6. B	7. A	8. A		

Exercise 15.1

1. (a) 17.64 (b) 2.0494 (c) 0.2381
2. (a) 54.46 (b) 2.7166 (c) 0.1355
3. (a) 40.05 (b) 2.5158 (c) 0.1580
4. (a) 79.90 (b) 2.9898 (c) 0.1119
5. (a) 1521 (b) 6.245 (c) 0.025 64
6. (a) 1866 (b) 6.573 (c) 0.023 15

7. (a) 3327 (b) 7.594 (c) 0.017 34
8. (a) 8087 (b) 9.484 (c) 0.011 12
9. (a) 160 000 (b) 20 (c) 0.0025
10. (a) 541 700 (b) 27.129 (c) 0.001 359
11. (a) 861 800 (b) 30.468 (c) 0.001 078
12. (a) 513 100 (b) 26.764 (c) 0.001 396
13. (a) 0.067 60 (b) 0.5099 (c) 3.846
14. (a) 0.1239 (b) 0.5933 (c) 2.841
15. (a) 0.003 844 (b) 0.2490 (c) 16.13
16. (a) 0.000 010 099 (b) 0.056 37 (c) 314.7
17. (a) 0.100 62 (b) 0.5632 (c) 3.153
18. (a) 0.000 038 82 (b) 0.078 94 (c) 160.5
19. (a) 0.000 000 078 96 (b) 0.016 763 (c) 3559
20. (a) 0.000 000 000 036 (b) 0.002 449 5 (c) 166 700
21. 15
22. 11
23. 28
24. $\frac{1}{4}$
25. $\frac{3}{5}$
26. $\frac{7}{6}$

Exercise 15.2

1. 211.2; 14
2. 0.1142
3. 28
4. (a) 4 (b) 15
5. (a) 2.25 (b) 15
6. (a) 0.1 (b) 5
7. (a) 34 (b) 67 (c) 49 (d) 63
   (e) 15 (f) 55
8. (a) 8.98 (b) 0.288
9. 277.8
10. (a) 39.46 (b) 7.255
11. 4.2
12. 13.78

Multi-choice questions 15

1. D 2. B 3. C 4. D 5. D
6. D 7. A 8. A

Exercise 16.1

1. $x = 10$ 2. $x = 21$ 3. $x = 4$ 4. $x = 3$
5. $x = 8$ 6. $x = 4$ 7. $x = 2$ 8. $x = 5$
9. $x = 6$ 10. $p = \frac{7}{3}$ 11. $x = \frac{5}{3}$ 12. $x = 13$
13. $m = 2$ 14. $x = \frac{6}{5}$ 15. $m = \frac{15}{4}$ 16. $x = 6$
17. $x = \frac{87}{56}$ 18. $x = \frac{24}{65}$

Exercise 16.2

1. $x = -6$
2. (a) $x = 2$ (b) $x = 14$
3. $m = 5$
4. (a) $x = -\frac{2}{3}$ (b) $x = 18$
5. (a) $x = 3$ (b) $x = \frac{45}{8}$ (c) $x = 6$
6. (a) $n = -5$ (b) $x = -4$
7. (a) $x = 5$ (b) $x = 2$
8. $x = 7$
9. $p = 3$
10. (a) $a = 6$ (b) $b = \frac{2}{3}$ (c) $c = 1$ (d) $d = 1.5$
    (e) $d = 1.5$ (e) $e = 4.5$
11. 14
12. (a) $x = 12$ (b) $x = 20$

Multi-choice questions 16

1. B	2. A	3. B	4. D	5. D
6. D	7. C	8. C	9. B	10. D

Exercise 17.1

1. $x = 7, y = 2$
2. $x = 4\frac{2}{7}, y = 2\frac{4}{7}$
3. $x = 1, y = 5$
4. $x = 3, y = 2$
5. $x = 6, y = 2\frac{2}{3}$

Exercise 17.2

1. $x = 3, y = 1$
2. $x = 4, y = 2$
3. $x = 2, y = 3$
4. $x = 5, y = 3$
5. $x = 7, y = 4$
6. $p = 12$
7. (a) $x + y = 9$ (b) $x - y = 1$ (c) $x = 5, y = 4$
8. $x = 4, y = 3$

Multi-choice questions 17

1. B	2. C	3. B	4. B	5. B
6. B	7. C	8. C		

Exercise 18.1

1. $x = \pm 3$
2. $x = \pm 5$
3. $x = 0, x = 5$
4. $x = 0, x = -3$
5. $x = 0, x = 3$
6. $x = 3, x = -4$
7. $x = 1\frac{1}{2}, x = -1\frac{1}{3}$
8. $x = -2, x = -3$
9. $x = 2, x = 3$
10. $x = 4, x = -5$
11. $x = -\frac{1}{2}, x = -3$
12. $x = -\frac{2}{3}, x = 5$

Exercise 18.2

1. $x = 5$ (twice)
2. (a) $(x-4)(x-3)$ (b) $x = 4, x = 3$
3. $x = \pm 5$
4. (a) $(x-1)(x+4)$ (b) $x = 1, x = -4$
5. $x = 1, x = 2\frac{1}{2}$
6. $x = -4, x = 3$
7. $x = \pm 6$
8. $x = 4, x = -3$
9. (a) $x = 2, -5$ (b) $x = 7, -5$
10. $x = \pm 4$

Multi-choice questions 18

1. C	2. B	3. A	4. B	5. A
6. A	7. B			

Exercise 19.1

1. 32
2. 80
3. 360
4. 144
5. 704

Exercise 19.2

1. $d = \frac{C}{\pi}$
2. $P = \frac{c}{V}$
3. $T = \frac{I}{PR}$
4. $E = IR$
5. $s = \frac{ST}{t}$
6. $q = \frac{p}{x}$
7. $V = \frac{RT}{P}$
8. $P = p - 15$
9. $u = v - at$
10. $r = \frac{n-p}{c}$
11. $x = \frac{7}{a} - 3$
12. $r = R - \frac{2R}{V}$
13. $P = \frac{dh}{x} + Q$
14. $n = \frac{p(x+3)-5}{2}$

Exercise 19.3

1. (a) 1080 (b) 6 (c) $n = \frac{S}{180} + 2$
2. (a) 11.8 (b) $10\frac{1}{4}$ (c) $b = \frac{P-2a}{2}$
3. 536.9
4. (a) $s = \frac{3p}{q+r}$ (b) $q = \frac{3p}{s} - r$
5. (a) $x = \frac{5}{4}$ (b) $x = 3 - \frac{7}{k}$
6. $x = \frac{y+2}{5}$
7. 250
8. $R = \frac{100I}{PT}$
9. $x = \frac{y-c}{m}$
10. (a) 22 (b) 64

Multi-choice questions 19

1. A 2. C 3. D 4. B 5. A
6. A 7. C

Exercise 20.1

1. £462 2. £72 3. 4 years 4. 7% 5. £160

Exercise 20.2

1. £336
2. (a) £337.50 (b) £500
3. £340
4. 6 months
5. £90

Exercise 21.1

1. (a) 28 cm (b) 48 cm²
2. 696 m²
3. 6 m
4. 7 m
5. 9600
6. 40 cm²
7. 20 cm²
8. 5 m
9. 72 cm²
10. (a) 616 cm² (b) 88 cm
11. (a) 56.52 mm (b) 254.34 mm²
12. (a) 808.3 cm² (b) 76.98 cm
13. 236.4 mm²
14. (a) 19.17 cm (b) 86.25 cm²

Exercise 21.2

1. 7.065 cm²
2. (a) 64 cm (b) 208 cm²
3. (a) 10 cm (b) 100 cm²
4. 36 cm²
5. (a) 7.75 cm (b) 51.6 cm²
6. 7 cm
7. 3 cm²
8. (a) 33.6 cm (b) 51.84 cm²

9. 3 m^2
10. (a) 10 000 m^2 (b) 330 mm
11. 351 cm^2
12. 14.4 cm
13. 36 cm
14. (a) 38.5 cm^2 (b) 14 cm

Multi-choice questions 21

1. C	2. C	3. D	4. D	5. A
6. D	7. B	8. A	9. D	10. A
11. C	12. B			

Exercise 22.1

1. (a) 24 000 cm^3 (b) 6200 cm^2
2. 179.2 cm^3
3. (a) 12 320 cm^3 (b) 2992 cm^2
4. 38 500 cm^3
5. (a) 340.6 cm^2 (b) 195.2 cm^3
6. 5544 cm^3
7. 144 cm^3
8. (a) 131 cm^3 (b) 124.7 cm^2
9. 6912 litres
10. 763 litres

Exercise 22.2

1. (a) 1250 mm^2 (b) 1800 mm^2 (c) 45 000 mm^3
   (d) 8100 mm^2
2. (a) 154 cm^2 (b) 6 cm
3. (a) 600 cm^2 (b) 1000 cm^3
4. 40 litres
5. (a) 216 cm^3 (b) 72 cm
6. (a) 154 cm^2 (b) 1540 cm^3
7. 15 cm^3
8. (a) 36 000 cm^3 (b) 125 (c) 25 000
9. (a) 20.10 cm^2 (b) 4.59 cm^3
10. 76.19 m

Multi-choice questions 22

1. D	2. D	3. E	4. A	5. D
6. C	7. B	8. A	9. B	10. B
11. C	12. C			

Exercise 23.1

1. 800 mm
2. 10 m
3. 4 m × 6 m
4. 5.4 km
5. (a) 12.5 m (b) 22.5 m^2 (c) 14 cm
6. (a) 80 m (b) 5 m^2 (c) 8 m^3
7. 10 km^2
8. 1.536 m^3

Exercise 23.2

1. 500 m
2. 270 cm^2
3. (a) 4 m (b) 35 m^3 (c) 7 m × 8 m
4. (a) 10 m (b) 14 cm (c) 20 m^2
5. (a) 46 m (b) 60 m^2 (c) 0.5 m^3
6. (a) 0.8 m (b) 37.5 m (c) 120 cm^2

Exercise 24.1

1. (a) $x = 4, y = 11;\ x = 7, y = 17;\ x = 11, y = 25$
   (b) $y = 7, x = 2;\ y = 11, x = 4;\ y = 23, x = 10$
2. (a) 10 m, 18 m, 40 m (b) 1 s, 4 s, 6 s
3. (a) $P = 5, Q = 27$ (b) $Q = 37, P = 7$
4. (a) £18 (b) 500
5. (a) $t = 7$ s, $v = 9.7$ m/s (b) $v = 9.5$ m/s, $t = 6.75$ s

Exercise 24.2

1. (a)

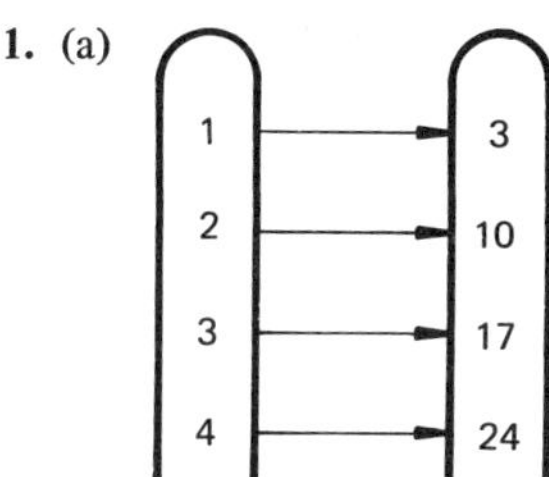

(b)

1 → 3
2 → 6
3 → 9
4 → 12

2. 125
3. (a)

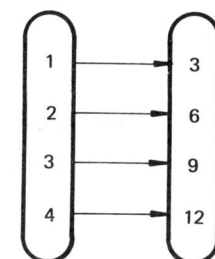

1 → 5
3 → 1
4 → −1
7 → −7

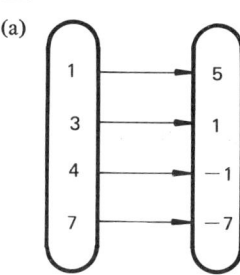

(b)

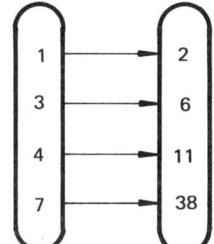

1 → 2
3 → 6
4 → 11
7 → 38

4. (a) −19 (b) −4 (c) 16
5. (a) 27 (b) 5 (c) 0 (d) 2
   (e) 35

6.

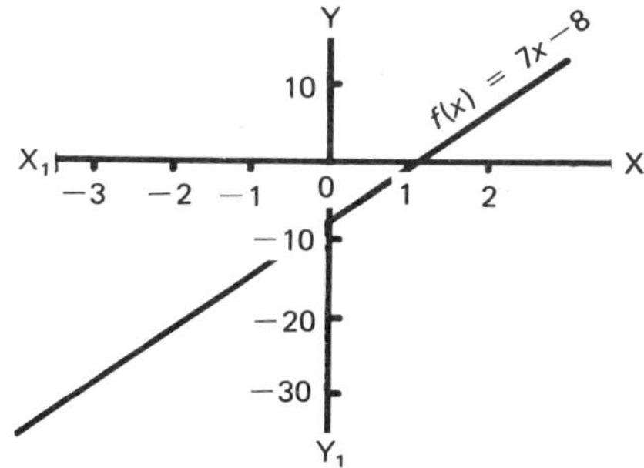

7.

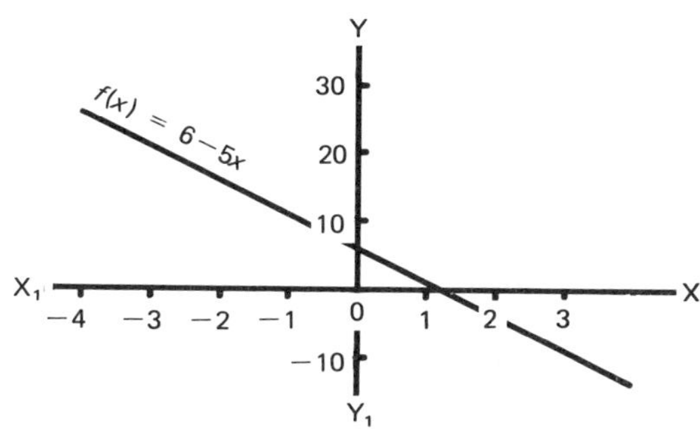

8. $y = 3x + 1$
9. $y = 2x - 3$
10. $c = -2$
11. (a)

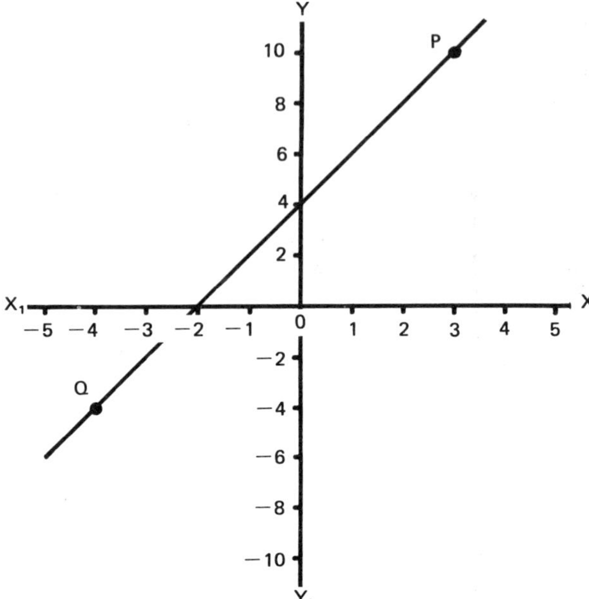

(b) $y = 2x + 4$

12. $m = 1, c = 2$

Exercise 24.3

1. (a) £19.25 (b) 210 (c) £8.50
2. (a)

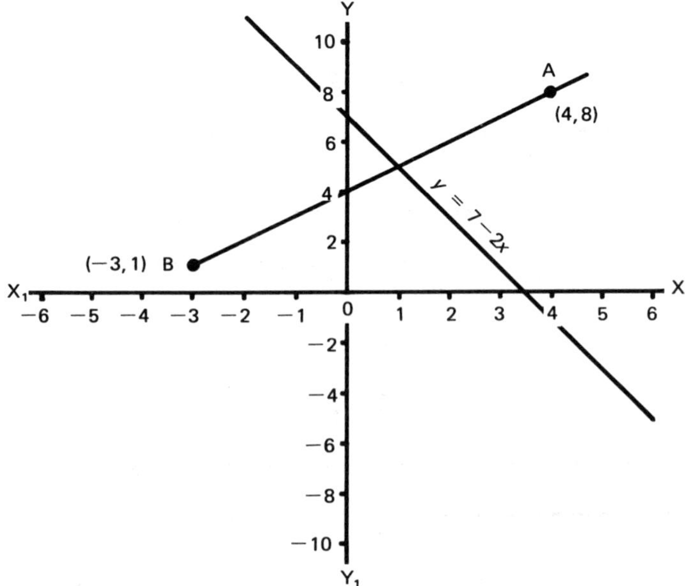

(b) (i) (0, 4) (ii) $m = 1$ (c) (ii) (1, 5)

3. (a) (i) 3 (ii) $\frac{1}{2}$ (b) 9 square units
4. (a) 14 units (b) $-\frac{2}{3}$ square units

5. 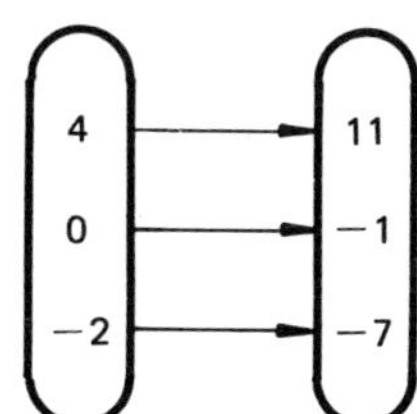

6. $c = 2$

7. (a) A$(-2, 8)$ (b) $m = -2$ (c) $y = -2x + 4$

(d)

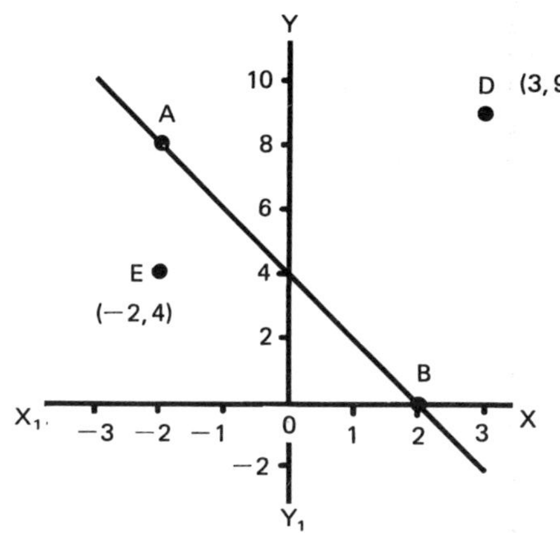

8. (a) A$(0, 5)$ (b) B$(-20, 0)$ (c) 50 square units

Multi-choice questions 24

1. D 2. A 3. C 4. C 5. C
6. A 7. C 8. D

Exercise 25.1

1. (a) $<$ (b) $>$ (c) $=$ (d) $>$ (e) $>$

2. (a)

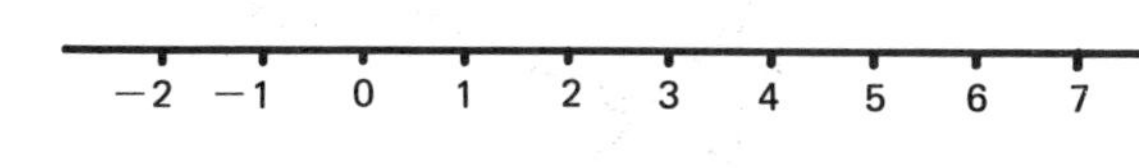

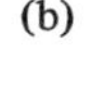
(b)

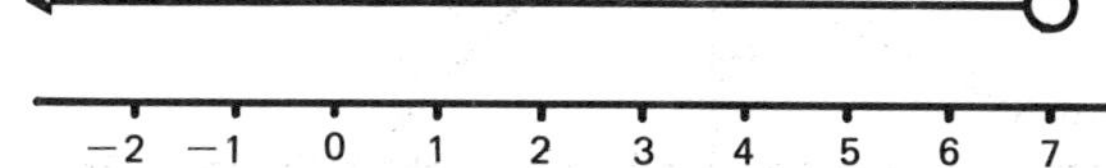

(c)

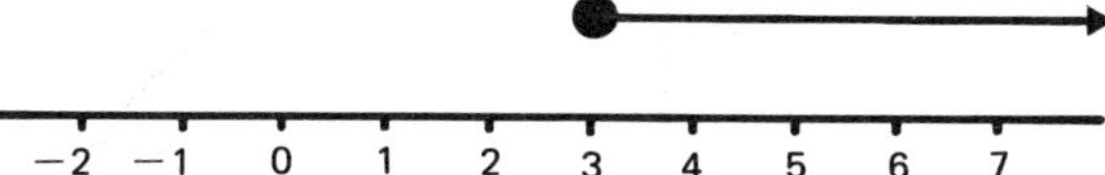

(d)

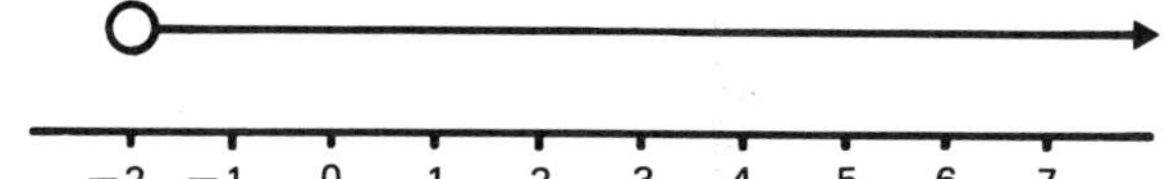

3. (a) $\{5, 6, 7, 8\}$ (b) $\{-2, -1, 0, 1, 2\}$ (c) $\{5, 6, 7, 8\}$
   (d) $\{-3, -2, -1, 0, 1, 2, 3, 4\}$

4. (a) $x \geqslant 4$ (b) $x \leqslant 5$ (c) $x > 12$ (d) $x > 6$
   (e) $x < 5$ (f) $x \geqslant 2$ (g) $x < -2$

5. (a) $y \leqslant 15$ (b) $y > -3$

6. (a)

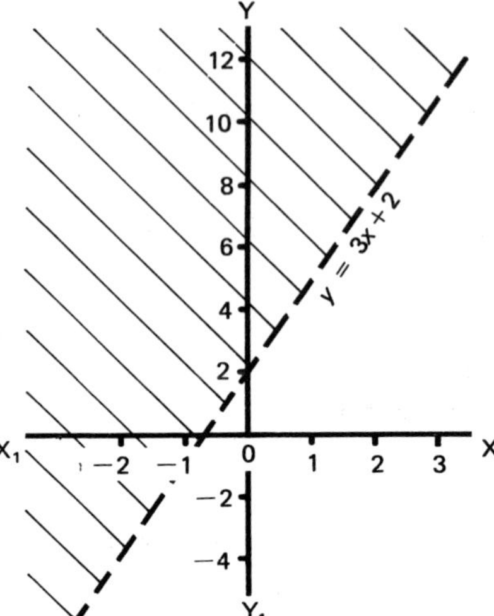

(b)

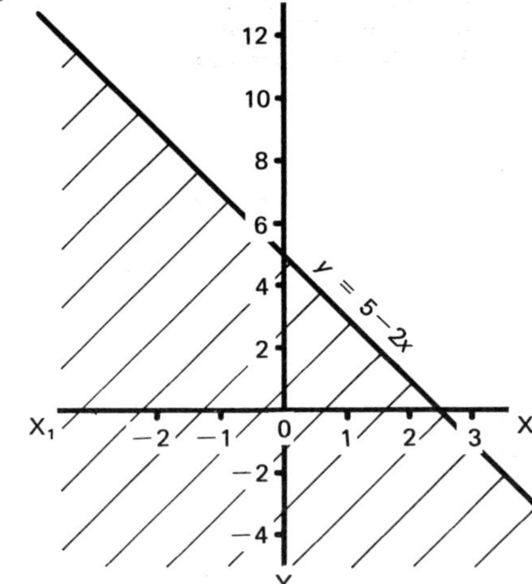

(c)

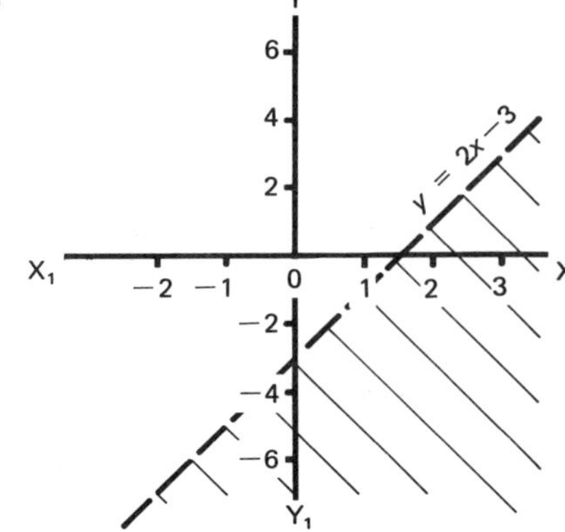

Exercise 25.2

1. (a) $y \leqslant -1$ (b) $y \geqslant x-1$
2. (a) $=$ (b) $<$
3. 11
4. (a) $=$ (b) $<$ (c) $<$ (d) $>$ (e) $>$
5. 6
6. $x > 3$
7. $x < -1$
8. $\{2, 3, 4\}$
9. (a) $\{4, 5, 6, \ldots\}$ (b) $\{3, 2, 1\}$
10. $\{1, 2\}$
11. $\{1, 2, 3\}$
12. $\{2, 3, 4\}$
13. $x = 9$
14. (a) $x > 3$ (b) $x < 4$ (c) $x = 3$
15. $x = 5$

Multi-choice questions 25

1. D 2. C 3. C 4. B 5. D
6. C 7. B 8. C

Exercise 26.1

1. 6 hours 9 minutes
2. 11 hours
3. 16 hours 50 minutes
4. 19 hours 8 minutes
5. 4 hours 46 minutes
6. 14 hours 31 minutes
7. 10 hours 21 minutes
8. 14 hours 41 minutes

Exercise 26.2

1. 80 km/h 2. 50 km/h 3. 5 h
4. 4 h 5. 240 km 6. 300 km
7. 9 km/h 8. 38.4 km/h

Exercise 26.3

1. (a) 20 15 (b) 1 hour 30 minutes (c) 40 km/h
2. 19 44 hours
3. (a) 3 hours 42 minutes (b) 12 53 hours (c) 1 hour 10 minutes
4. (a) 6 hours 30 minutes (b) 65 km/h
5. (a) 10 km (b) 4.4 km/h
6. 1 hour 35 minutes
7. (a) 48 miles (b) 13 15 hours
8. (a) 3 hours 35 minutes (b) 21 minutes (c) 1 hour 40 minutes
9. (a) 30 km/h (b) 155 km (c) 44.4 km/h
   (d) 3 hours 30 minutes (e) 19 00 hours
10. 160

Multi-choice questions 26

1. D 2. C 3. B 4. D 5. C
6. A

Exercise 27.1

1. $45°44'$ 2. $83°36'$ 3. $71°58'25''$ 4. $80°25'36''$
5. $36°29'$ 6. $8°14'$ 7. $22°22'29''$ 8. $28°31'28''$

Exercise 27.2

1. $103°$
2. $x = 104°, y = 45°$
3. $m = 40°, n = 40°, p = 50°, q = 310°$
4. $127°$
5. $p = 70°, q = 90°, r = 70°, s = 110°$
6. $x = 84°, y = 123°$
7. $85°$
8. $u = 62°, v = 72°, w = 53°, x = 55°, y = 118°, z = 108°$

Multi-choice questions 27

1. C 2. E 3. A 4. B 5. C
6. C 7. C 8. D 9. D 10. D

Exercise 28.1

1. $x = 59°, y = 121°$ 2. $x = 85°, y = 43°$ 3. $x = 11°, y = 47°$
4. $x = 109°, y = 34°$ 5. $x = 112°, y = 29°$

Exercise 28.2

1. 13
2. 11.1
3. 10.2
4. 16.6
5. 9.14
6. 11.9
7. 12.3
8. 13.12
9. (a) 58° (b) 84° (c) 66°
10. (a) $x = 44°, y = 31°$ (b) $x = 72°, y = 36°$

Exercise 28.3

1. 9 cm
2. (a) 5 (b) 30 square units
3. (a) $x = 42°$ (b) $y = 42°$ (c) 138°
   (d) obtuse-angled isosceles triangle
4. 4 cm
5. (a) 59° (b) 57°
6. 9 cm
7. (a) 4 cm, 4 cm, 4 cm (b) 2 cm, 5 cm, 5 cm
   (c) 3 cm, 4 cm, 5 cm
8. (a) 103° (b) 26° (c) 51°
9. $m = 40°, n = 40°, p = 50°, q = 310°$
10. $p = 70°, q = 20°, r = 70°, s = 110°$

Multi-choice questions 28

1. B 2. A 3. B 4. A 5. C
6. B 7. D 8. B

Exercise 29.1

1. $x = 109°, y = 127°$
2. $\angle a = 116°$, $\angle b = 64°$
3. ∠DAC = 65°, ∠BCD = 130°, ∠ABC = 50°, ∠DOC = 90°
4. (a) 90° (b) 6 cm (c) 120 cm^2
5. 11.3 cm
6. 10.2 cm
7. 154°
8. ∠ADB = 40°, ∠BAD = 100°, ∠BDC = 71°

Exercise 29.2

1. 10.47 m
2. (b) rectangle (d) square
3. (a) kite (b) parallelogram (c) rhombus
4. 105°
5. (a) 64° (b) rhombus (c) QRST
6. (a) (i) 90° (ii) 3 cm (b) 30 cm^2 (c) trapezium
7. (a) trapezium (b) 61° (c) 119°
8. (a) 42° (b) 42° (c) 96° (d) obtuse-angled
9. (a) AO = 12 cm, BO = 5 cm (b) 13 cm (c) 120 cm^2
10. 116°

Multi-choice questions 29

1. E 2. C 3. D 4. B 5. D
6. C

Exercise 30.1

1. (a) 720° (b) 1260° (c) 1620°
2. 135°
3. (a) 30° (b) 150°
4. 108°
5. 6

Exercise 30.2

1. 125°
2. (a) 40° (b) 9
3. (a) 72° (b) 108°
4. $x = 120°$, $y = 60°$
5. 15
6. 60°
7. 15
8. (b)

Exercise 31.1

1. (a) 48° (b) 42°
2. 108°
3. 110°
4. (a) 78° (b) 102°
5. (a) 64° (b) 128°
6. (a) 5 cm (b) 6 cm^2
7. (a) 54° (b) 54° (c) 27°
8. ∠ABO = 65°, ∠OAC = 25°
9. (a) 90° (b) 21° (c) 48°
10. (a) 84° (b) 42°

Multi-choice questions 31

1. B 2. D 3. C 4. A 5. C
6. A 7. B 8. D 9. C 10. C

Exercise 32.1

1. 2
2. (a) 2 (b) 1
3. 5
4. (a)

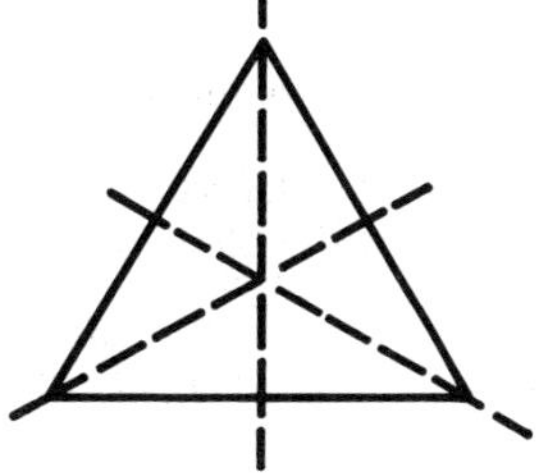

(b) e.g.:

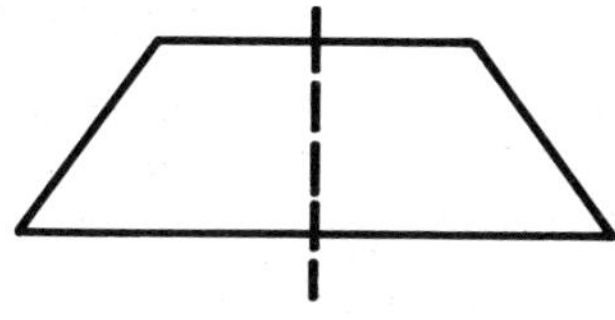

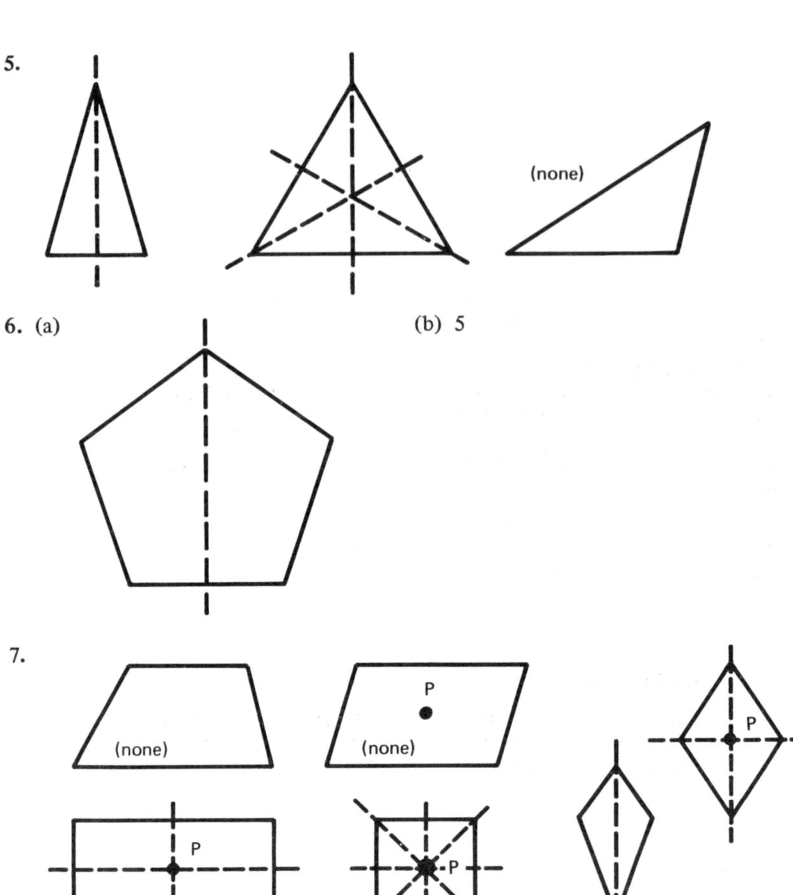

8. (a) 8 (b) 6

Multi-choice questions 32

1. C 2. D 3. D 4. D 5. E
6. B

Exercise 33.1

8. (b) 8.1 cm
9. (b) 3.2 cm
10. (b) 4.4 cm
11. (b) 13.4 cm

Exercise 34.1

1. (a) 0.2419 (b) 0.3305 (c) 0.9739
2. (a) 0.9511 (b) 0.7965 (c) 0.0729
3. (a) 0.4663 (b) 1.1605 (c) 5.145
4. (a) 49°41′ (b) 55°2′ (c) 68°10′
5. (a) 10 cm (b) 11.38 cm (c) 10.96 cm
6. (a) 36°52′ (b) 32°14′ (c) 44°26′
7. 23.09 cm
8. 14.0 cm
9. (a) 36°52′ (b) 48°11′ (c) 51°19′
10. (a) 2.736 cm (b) 4.636 cm (c) 8.302 cm
11. 70°32′; 70°32′; 38°56′

12. (a) 13.15 cm (b) 12.36 cm
13. (a) 8.402 cm (b) 9.004 cm (c) 2.924 cm
14. (a) 33°41′ (b) 30°58′ (c) 29°45′
15. 20.92 cm

Exercise 34.2

1. (a) 11°32′ (b) 4.9 m
2. (a) 80° (b) 10.28 cm (c) 6.128 cm
3. (a) 30.50 cm (b) 39.62 cm (c) 0.3050
   (d) 16°58′
4. 81.0 cm; 58.64 cm; 55°42′
5. (a) 7.75 cm (b) 51.6 cm² (c) 31°48′
6. (a) 27°52′ (b) 66°28′
7. (a) 30°12′ (b) 0.1357
8. (a) 0.4431 (b) 0.4431 (c) 0.4942
9. (a) 0.6 (b) 36°52′ (c) 20 cm
10. (a) $\frac{40}{41}$ (b) 12°41′
11. (a) $\frac{9}{40}$ (b) $\frac{9}{41}$
12. (a) 59° (b) 6.858 cm

Multi-choice questions 34

1. C	2. A	3. C	4. D	5. D
6. B	7. A	8. A	9. C	10. A
11. C	12. B	13. C	14. C	

Exercise 35.1

1. (a) $\{2, 3, 5, 7, 11, 13, 17, 19\}$ (b) $\{5, 10, 15, 20, 25, 30, 35, 40\}$
   (c) $\{9, 11, 13, 15\}$ (d) $\{3, 4, 5, 6, 7\}$
   (e) $\{1, 2, 3, 4, 5, 6, 7, 8, 9\}$
2. finite B, C; infinite A, F; null D, E
3. (a), (b), (c), (f)
4. (a) $\{3, 5, 7, 9, 13, 15, 19\}$ (b) $\{8, 10, 12, 18, 22\}$
   (c) $\{3, 5, 7, 13, 19\}$ (d) $\{3, 5, 10, 15\}$
5. 64
6. none
7. {square, rhombus, kite}
8. $\{1, 4, 9, 16\}$
9. (a) $\{2, 3, 4, 5, 6, 7, 8, 9\}$ (b) $\{3, 5, 7, 9\}$ (c) $\{4, 6, 7, 9\}$
   (d) $\{2, 4, 6, 8\}$ (e) $\{7, 9\}$ (f) $\{3, 4, 5, 6, 7, 9\}$
   (g) $\{2, 8\}$

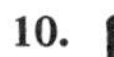

10.

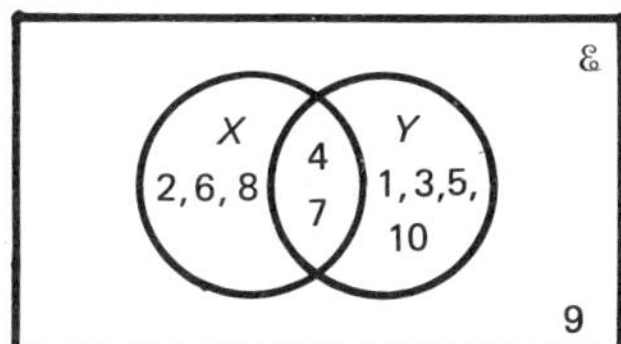

11. (a) Y (b) $A \cap B$ (c) $(A \cup B)'$ (d) $A \cap B'$
12. (a) 16 (b) 21 (c) 30 (d) 7

Exercise 35.2

1. (a) $T = \{2,3,5\}$; $S = \{2,5,7\}$

(b)

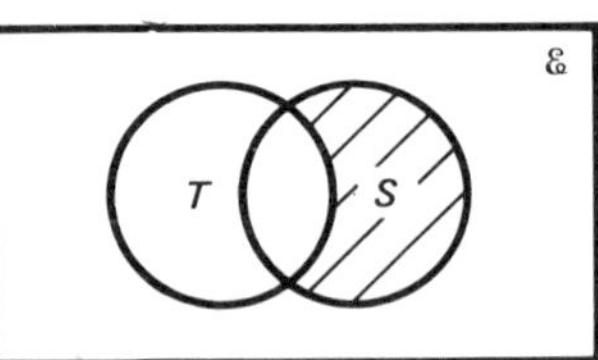

2.

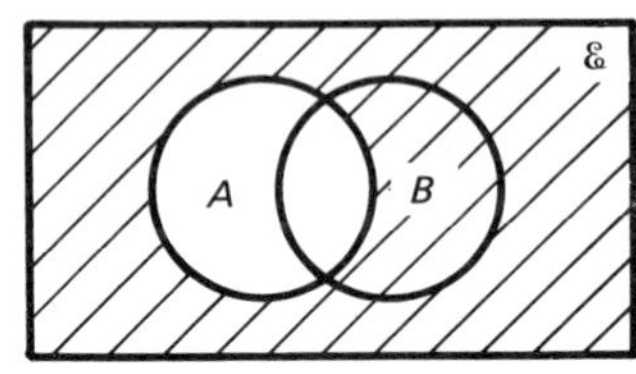

3. (a) (i) (b) (iii) (c) (iii) (d) (ii)
4. (a) $\{2\}$ (b) $\{1,3,5,7,9\}$ (c) 9

5.

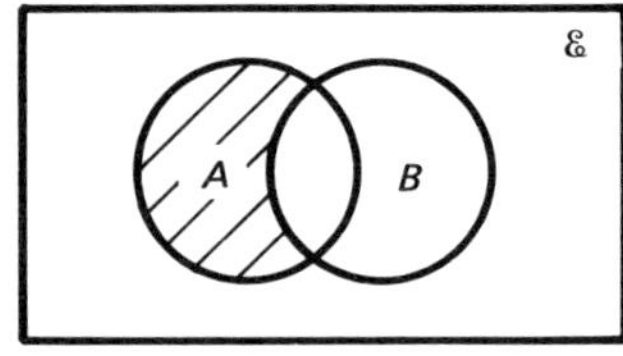

6. $\{G, E, O, F, B, Y, C, T\}$
7. $\{b, d\}$
8. (a) $\{1,2,3,4,6,9,12,18,36\}$ (b) $\{3,6,9,12,18,36\}$
9. (a) $\{1,2,3,4,6,12\}$ (b) $\{5,10,15,20,25\}$ (c) $\{\quad\} = \emptyset$
10. (a) 12 (b) 9 (c) 3 (d) 4
11. (a) $\{2,3,5\}$ (b) $\{7,8\}$
12. (a) 4 (b) $\{2,10\}$ (c) $\{1,3,5,7,9\}$ (d) $\{3,7,9\}$

Multi-choice questions 35

1. A 2. D 3. C 4. B 5. A
6. D 7. B 8. C 9. D

Exercise 36.1

1. 7 2. 252 3. 23 4. 69
5. 1003 6. 120_5 7. 1043_6 8. 11011_2
9. 2104_7 10. 22413_8

Exercise 36.2

1. 10101_2 2. 1001_2 3. 1233_5 4. 1143_6
5. 634_8 6. 21_3 7. 12_4 8. 1210_3
9. 213_4 10. 4_5 11. 110_2 12. 10101_2
13. 10220_3 14. 221001_3 15. 200010_4

Exercise 36.3

1. 26_{10} 2. 1111_2 3. 357_8 4. 38_{10}
5. 1143_5 6. 5 7. 121_8 8. 44_{10}
9. 11010_2 10. 8 11. 1002_4
12. (a) 210_3 (b) 30_8 13. (a) 1120_3 (b) 1022_3
14. 111_2 15. 3

Multi-choice questions 36

1. D **2.** E **3.** E **4.** B **5.** C
6. C **7.** E

Exercise 37.1

1. $\begin{pmatrix} 3 & 7 \\ 5 & 9 \end{pmatrix}$ **2.** $\begin{pmatrix} 1 & -5 \\ 1 & 1 \end{pmatrix}$ **3.** $\begin{pmatrix} -1 & 5 \\ -1 & -1 \end{pmatrix}$

4. $\begin{pmatrix} 4 & 16 \\ 13 & 38 \end{pmatrix}$ **5.** $\begin{pmatrix} 20 & 31 \\ 16 & 22 \end{pmatrix}$ **6.** $\begin{pmatrix} \frac{5}{7} & -\frac{1}{7} \\ -\frac{3}{7} & \frac{2}{7} \end{pmatrix}$

7. $\begin{pmatrix} -\frac{1}{2} & \frac{3}{4} \\ \frac{1}{4} & -\frac{1}{8} \end{pmatrix}$ **8.** $\frac{1}{2}$ **9.** $x = -6, y = 3$

10. P.Q = Q; Q.P = Q

Exercise 37.2

1. $\begin{pmatrix} 5 \\ -1 \end{pmatrix}$

2. $\begin{pmatrix} 3 & 0 \\ 6 & 9 \end{pmatrix}$

3. (a) $\begin{pmatrix} 2 & -4 \\ 1 & -1 \end{pmatrix}$ (b) $\begin{pmatrix} 7 & 6 \\ 3 & 6 \end{pmatrix}$

4. (a) 2×3 (b) $\begin{pmatrix} 10 & 4 \\ -2 & 12 \end{pmatrix}$

5. $\begin{pmatrix} 0 & -2 \\ 0 & -4 \end{pmatrix}$

6. (a) $\begin{pmatrix} 1 & 4 \\ 3 & 9 \end{pmatrix}$ (b) $\begin{pmatrix} 7 & 21 \\ 13 & 38 \end{pmatrix}$ (c) $\begin{pmatrix} 2 & -5 \\ -1 & 3 \end{pmatrix}$

7. $\begin{pmatrix} 13 \\ -3 \end{pmatrix}$

8. $\begin{pmatrix} 9 \\ -1 \end{pmatrix}$

9. $\begin{pmatrix} 7 \\ 0 \end{pmatrix}$

10. (a) $\begin{pmatrix} 1 & 2 \\ -4 & 2 \end{pmatrix}$ (b) $\begin{pmatrix} 9 & -3 \\ 4 & 1 \end{pmatrix}$ (c) $\begin{pmatrix} 7 & 0 \\ 9 & 3 \end{pmatrix}$

(d) $\begin{pmatrix} \frac{2}{7} & -\frac{1}{7} \\ \frac{1}{7} & \frac{3}{7} \end{pmatrix}$

Exercise 38.1

1.

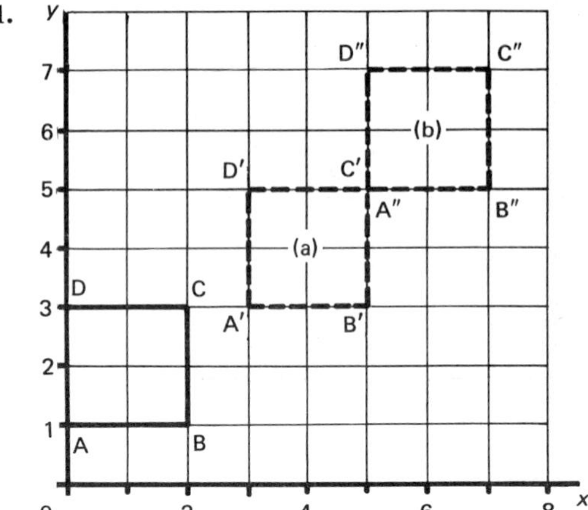

2.

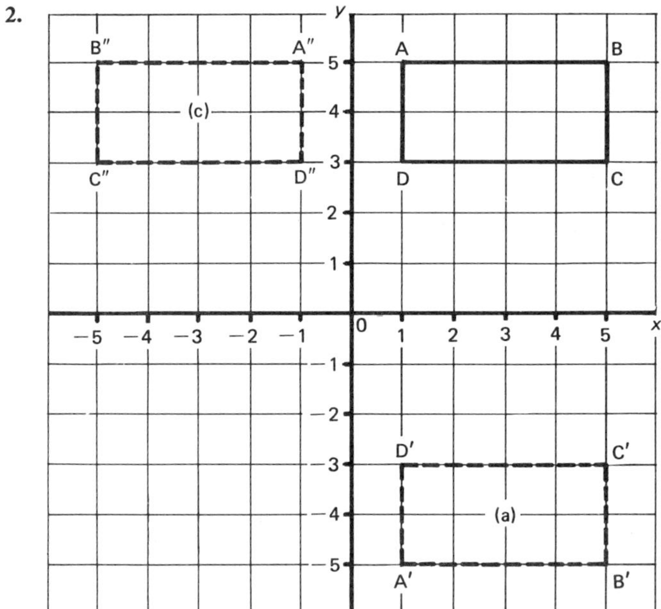

3. (a) S(2, 6)

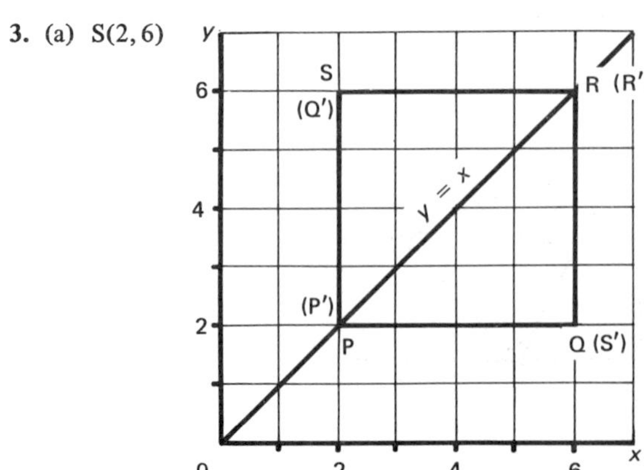

(e) P′(2, 2)
Q′(2, 6)
R′(6, 6)
S′(6, 2)

4.

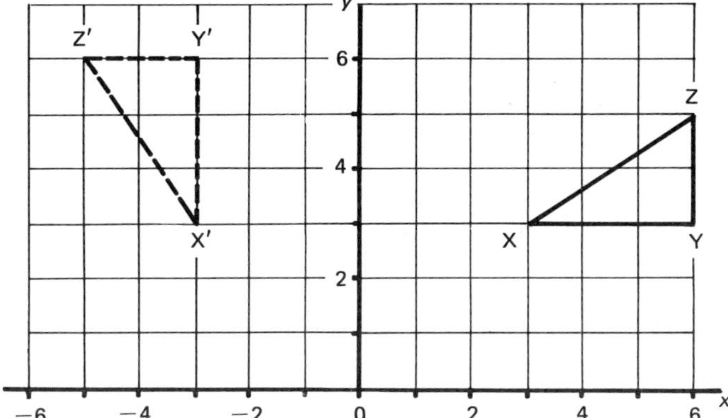

5.

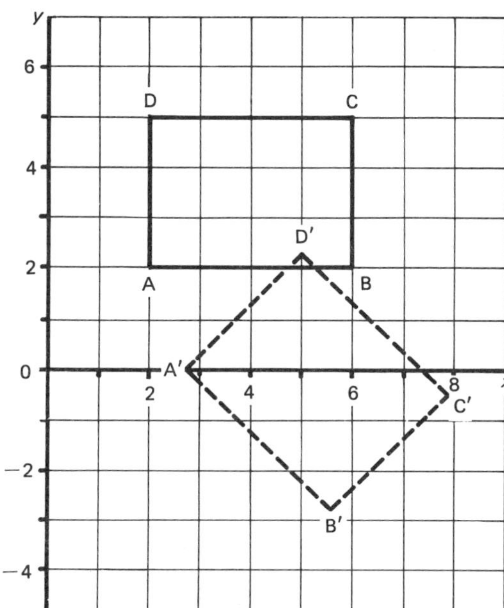

A′(2.8, 0)
B′(5.6 — 2.8)
C′(7.7 — 0.7)
D′(4.9 — 2.2)

Exercise 38.2

1. (2, −4)

2. $y = 2$
3. (a) (5, −2) (b) (−5, 2)

4.

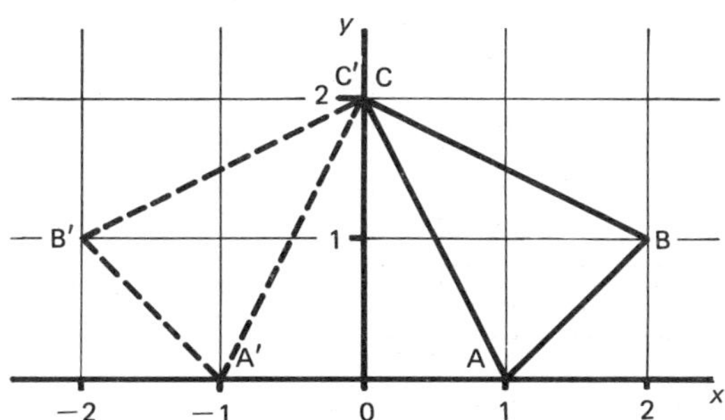

5. (6, 3)
6. (3, 5)

7.

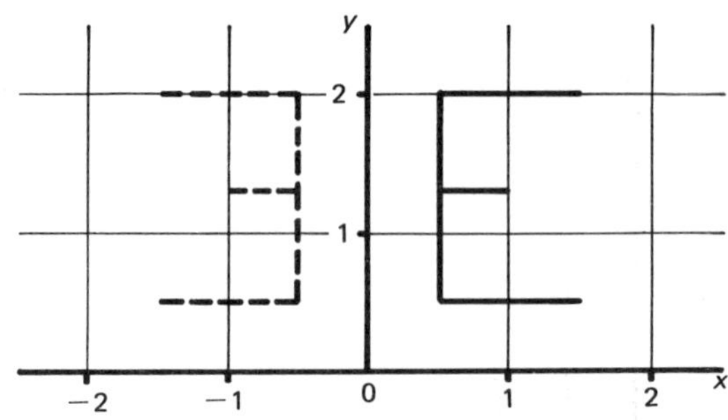

8. (5.0, −0.6)
9. (−1, 3)
10. (−4, −3)

Exercise 39.1

1.

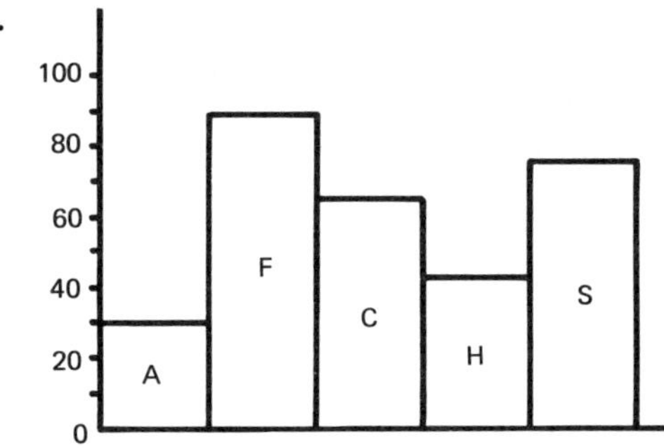

2.

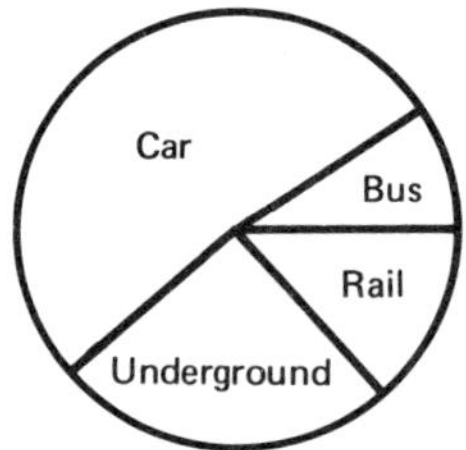

3. Food and drink £38.06; Housing £16.94;
Transport £11.94; Clothing £13.06;
Other £20.00

6.

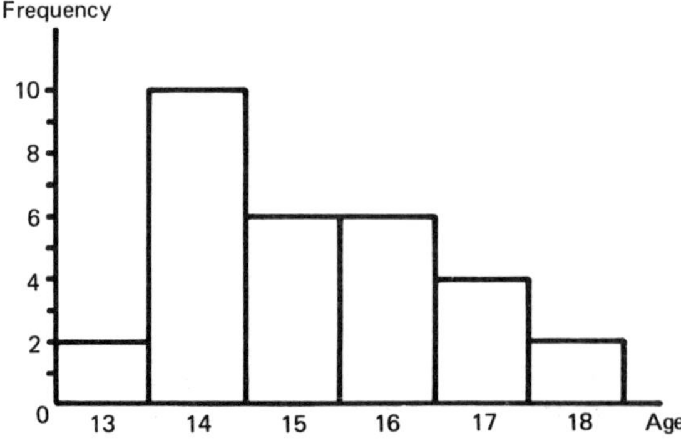

7.

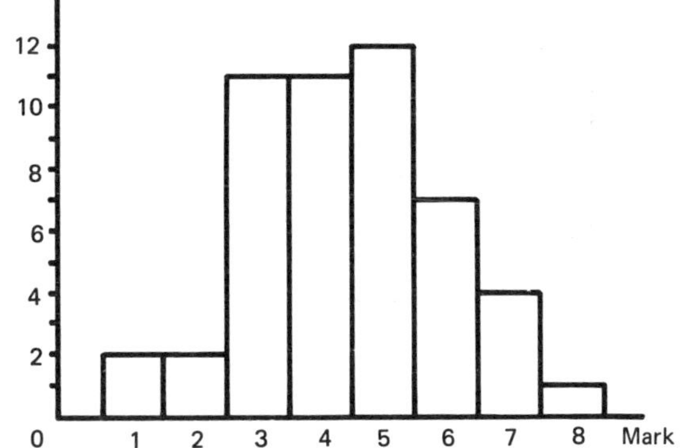

8. 176.7 cm
9. 8
10. 14
11. 5
12. 4.5
13. 5
14. (a) $3\frac{1}{2}$ kg (b) $2\frac{1}{2}$ kg (c) $2\frac{1}{2}$ kg
15. (a) $\frac{1}{6}$ (b) $\frac{1}{2}$ (c) $\frac{1}{2}$
16. (a) $\frac{1}{8}$ (b) $\frac{1}{4}$ (c) $\frac{3}{8}$ (d) $\frac{5}{8}$
17. (a) $\frac{1}{5}$ (b) $\frac{1}{2}$ (c) $\frac{7}{10}$

Exercise 39.2

1. $\frac{1}{3}$
2. 5
3. 120°
4.

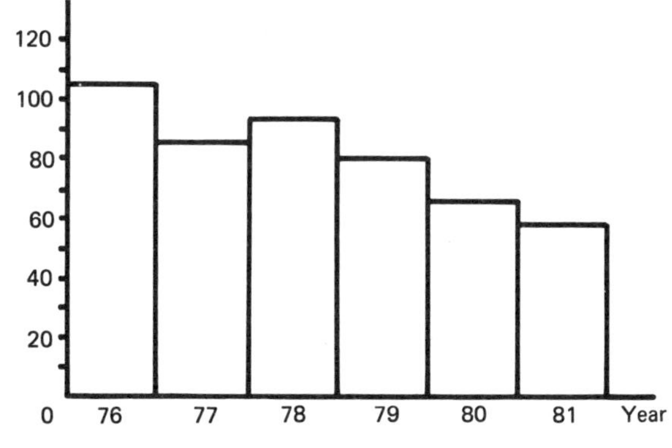

5. (a) 25 (b) 5 (c) 7 (d) 10
6. 9
7. 4
8. 88
9. $\frac{3}{7}$
10. (a) 200 (b) 40%
11. 5.9 hours
12. (a) Food (b) £15

Multi-choice questions 39

1. B 2. D 3. D 4. E 5. B
6. B 7. D 8. D 9. D 10. B

11. A 12. D 13. C 14. B 15. D
16. B 17. C

Exercise 40.1

1. 230°
2. SW
3. (a) 225° (b) 240°
4. 130°
5. 250°
6. 015°
7. 134 m in a direction 303°
8. (a) 300° (b) 75° (c) 150°
9. 297°
10. (a) 44° (b) 134° (c) 224°
11. 69.3 km east; 40 km south
12. 90.5 km; 179°

Multi-choice questions 40

1. C 2. C 3. A 4. B 5. D
6. C 7. D 8. D 9. C 10. C
11. D 12. B

IMPORTANT DATA AND FORMULAE

Mathematical symbols

- $<$ is less than
- $\leqslant$ is less than or equal to
- $>$ is greater than
- $\geqslant$ is greater than or equal to

Arithmetic

- The ratio $a:b$ may be written in the form $\frac{a}{b}$.
- If a quantity Q is to be divided in the ratio $a:b:c$ then amount of each part $= \frac{Q}{a+b+c}$.
- Profit % $= \frac{\text{selling price} - \text{cost price}}{\text{cost price}} \times 100$.

  Loss % $= \frac{\text{cost price} - \text{selling price}}{\text{cost price}} \times 100$.

Algebra

- $a^2 + 2ab + b^2 = (a+b)^2$
- $a^2 - 2ab + b^2 = (a-b)^2$
- $a^2 - b^2 = (a+b)(a-b)$
- $a^m \times a^n = a^{m+n}$
- $a^m \div a^n = a^{m-n}$
- $(a^m)^n = a^{mn}$
- $a^{m/n} = \sqrt[n]{a^m}$
- $a^0 = 1$

- $a^{-m} = \dfrac{1}{a^m}$
- $\log(ab) = \log a + \log b$
- $\log\left(\dfrac{a}{b}\right) = \log a - \log b$
- $\log a^m = m \log a$
- $\log \sqrt[m]{a} = \dfrac{1}{m} \log a$

SI units

- *Length*: 1 metre (m) = 100 centimetres (cm)
  = 1000 millimetres (mm)
  1 kilometre (km) = 1000 metres
- *Mass*: 1000 grams (g) = 1 kilogram (kg)
  1000 milligrams (mg) = 1 gram
  1000 kilograms = 1 tonne (t)

Mensuration

- *AREAS AND PERIMETERS*

Table A

Figure	*Diagram*	*Formulae*
Rectangle	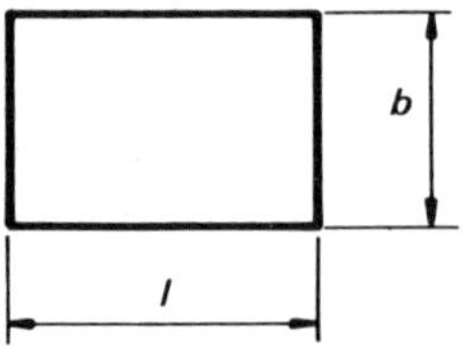 *l*, *b*	Area = $l \times b$ Perimeter = $2l + 2b$
Parallelogram	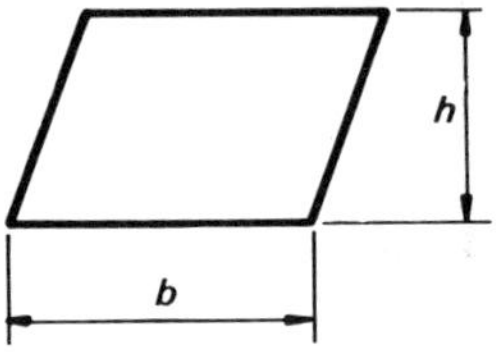 *b*, *h*	Area = $b \times h$

Table A continued

Figure	*Diagram*	*Formulae*
Triangle	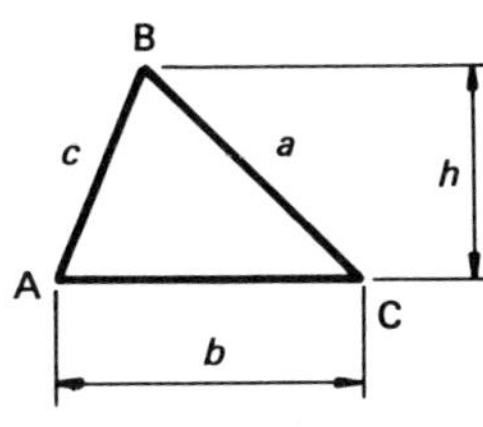	Area $= \frac{1}{2} \times b \times h$ Area $= \sqrt{s(s-a)(s-b)(s-c)}$ where $s = \dfrac{a+b+c}{2}$
Trapezium	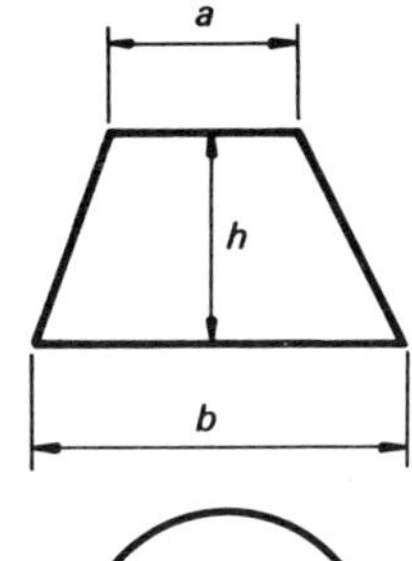	Area $= \frac{1}{2} \times h \times (a+b)$
Circle	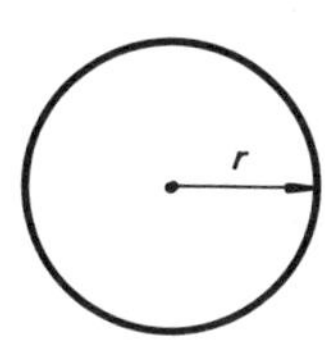	Area $= \pi r^2$ Circumference $= 2\pi r = \pi d$ $\left(\pi = 3.142 \quad \text{or} \quad \dfrac{22}{7}\right)$
Sector of a circle	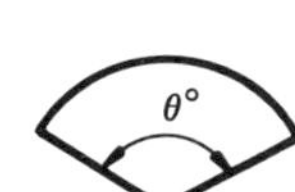	Area $= \pi r^2 \times \dfrac{\theta}{360}$ Length of arc $= 2\pi r \times \dfrac{\theta}{360}$

- *VOLUMES AND SURFACE AREAS*

Table B

Solid	*Volume*	*Surface area*
Any solid having a uniform cross-section	Cross-sectional area × length of solid	Lateral surface + ends, i.e. (perimeter of cross-section × length of solid) + (total area of ends)
Cylinder	$\pi r^2 h$	$2\pi r(h+r)$

Table B continued

Solid	*Volume*	*Surface area*
Cone	$\frac{1}{3}\pi r^2 h$ (h is the vertical height)	$\pi r l$ (l is the slant height)
Frustum of a cone	$\frac{1}{3}\pi h(R^2+Rr+r^2)$ (h is the vertical height)	Curved surface area $=\pi l(R+r)$ Total surface area $=\pi l(R+r)+\pi R^2+\pi r^2$ (l is the slant height)
Sphere	$\frac{4}{3}\pi r^3$	$4\pi r^2$
Pyramid	$\frac{1}{3}Ah$	Sum of the areas of the triangles forming the sides plus the area of the base (A = area of base)

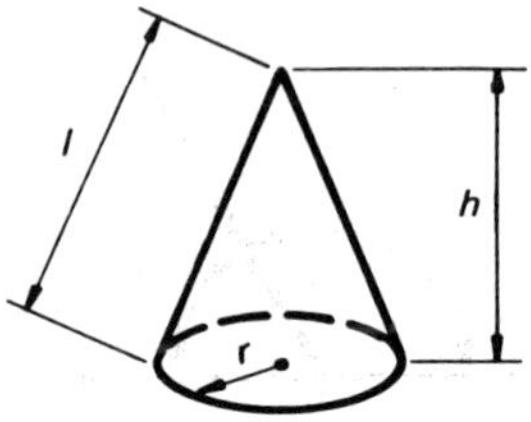

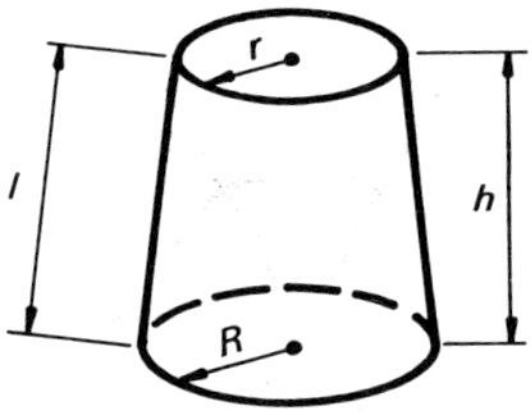

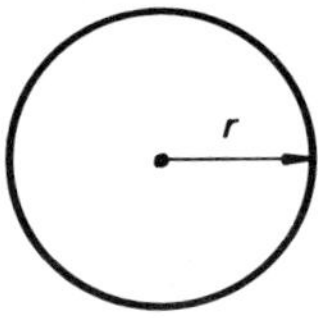

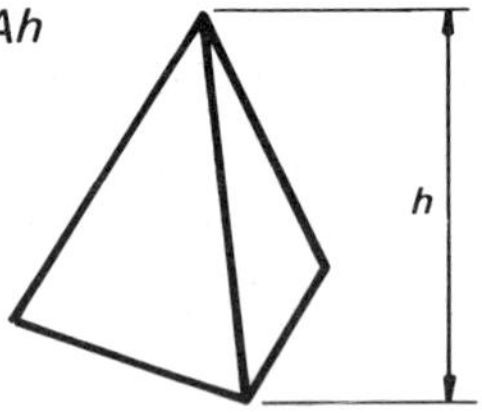

Equation of a straight line

- $y = mx + c$ where m is the gradient and c is the intercept on the y-axis.

Time, distance and speed

- Average speed $= \dfrac{\text{total distance travelled}}{\text{total time taken}}$.

Geometry

- Complementary angles are angles whose sum is 90°.
- Supplementary angles are angles whose sum is 180°.
- The total angle on a straight line is 180° (Fig. 1).

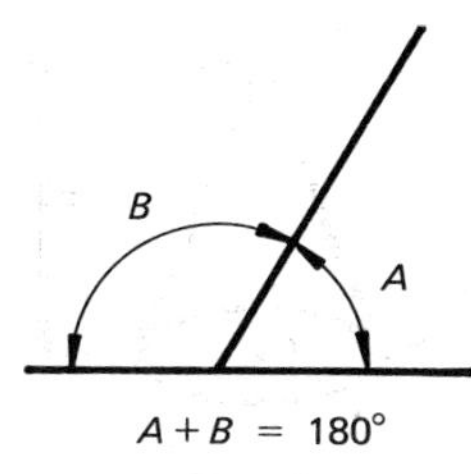

Fig. 1

- When two straight lines intersect the vertically opposite angles are equal (Fig. 2).

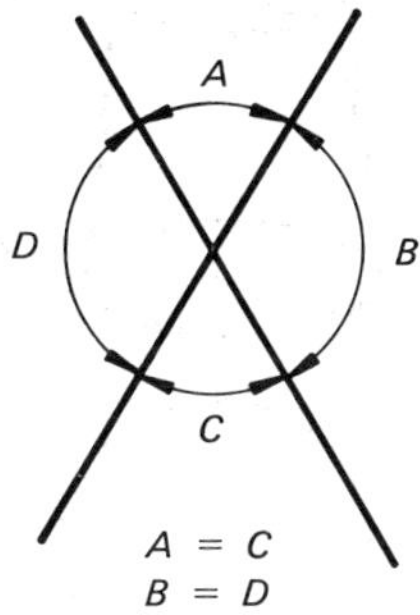

Fig. 2

- When two parallel lines are cut by a transversal (Fig. 3).
  - (i) The corresponding angles are equal: $a = l$, $b = m$, $c = p$ and $d = q$.
  - (ii) The alternate angles are equal: $d = m$ and $c = l$.
  - (iii) The interior angles are supplementary: $d + l = 180°$ and $c + m = 180°$.

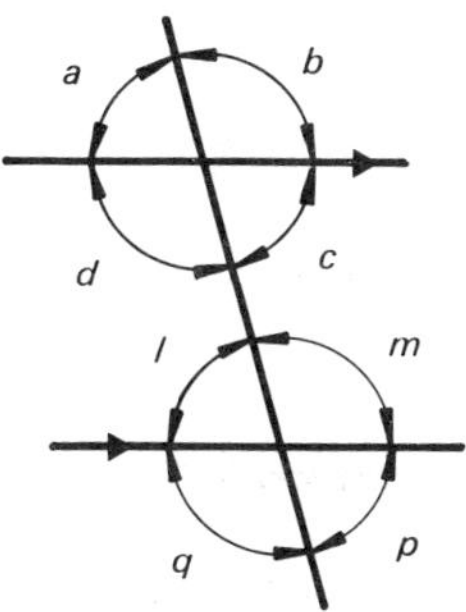

Fig. 3

- The sum of the angles of a triangle is 180° (Fig. 4).

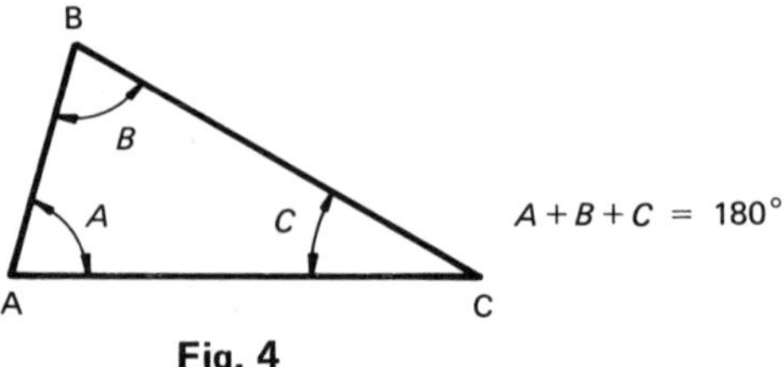

Fig. 4

- When the side of a triangle is produced, the exterior angle so formed is equal to the sum of the opposite interior angles (Fig. 5).

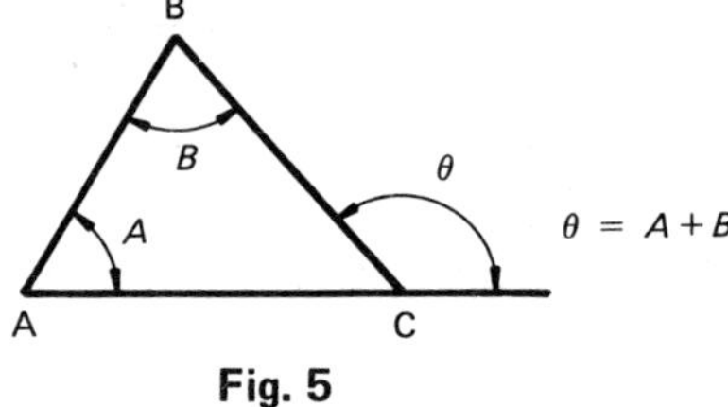

Fig. 5

- Pythagoras' theorem states that the square on the hypotenuse is equal to the sum of the squares of the other two sides. Thus in the diagram (Fig. 6)

$$a^2 = b^2+c^2$$

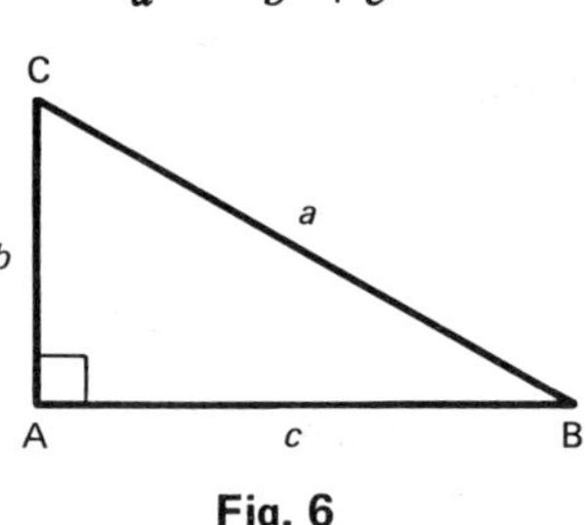

Fig. 6

- In an isosceles triangle the perpendicular dropped from the apex to the unequal side (Fig. 7)
  (i) bisects the unequal side
  (ii) bisects the apex angle.

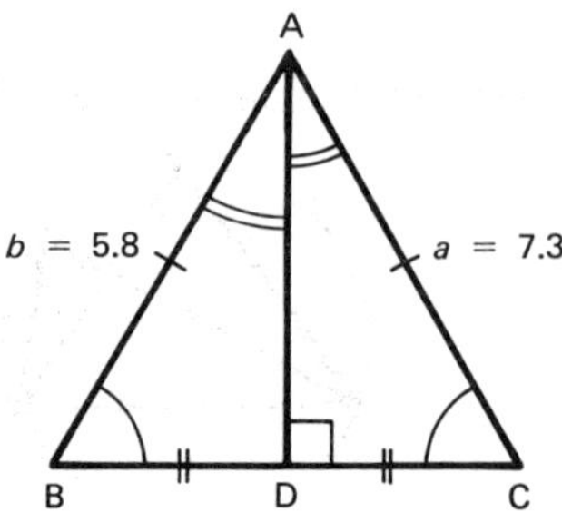

Fig. 7

- The sum of the angles of a quadrilateral is 360° (Fig. 8).

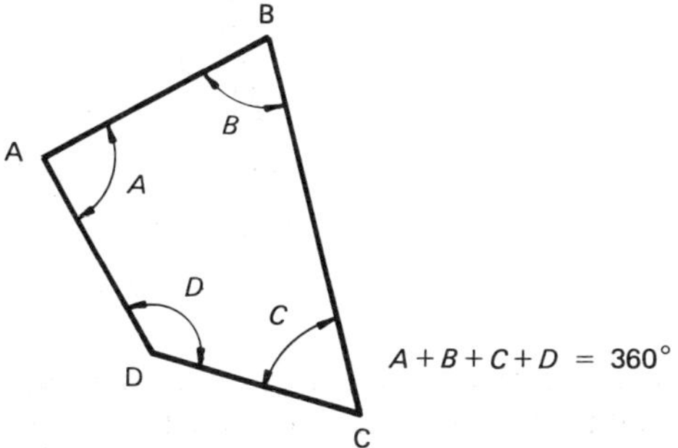

Fig. 8

- In a convex polygon the sum of the interior angles is $(2n-4)$ right angles, where n is the number of sides.
- The sum of the exterior angles of a polygon is 360°, no matter how many sides the polygon has.
- The angle which an arc of a circle subtends at the centre is twice the angle which the arc subtends at the circumference (Fig. 9).

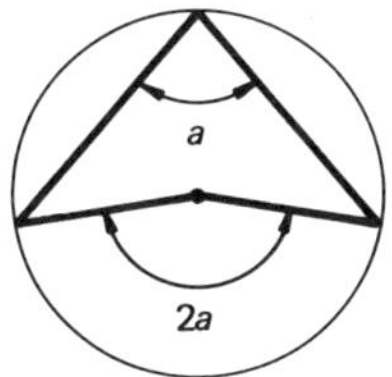

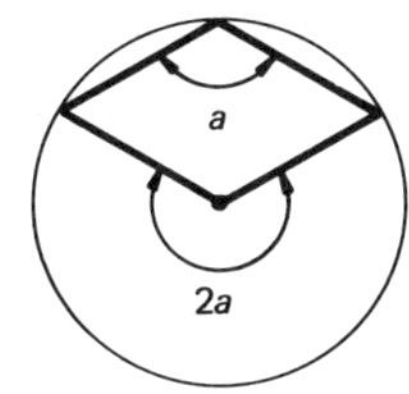

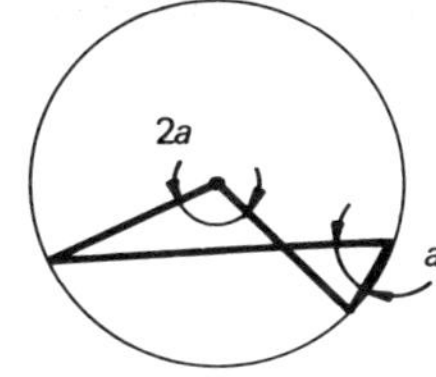

Fig. 9

- If a triangle is drawn in a semi-circle, the angle opposite to the diameter is a right angle (Fig. 10).

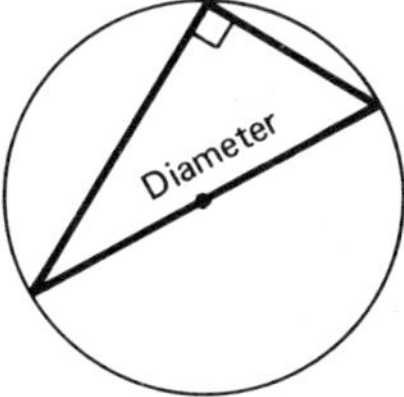

Fig. 10

- Angles in the same segment of a circle are equal (Fig. 11).

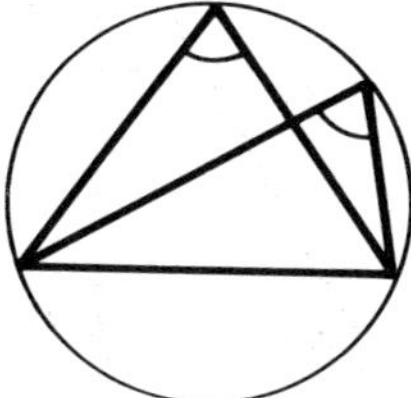

Fig. 11

- The opposite angles of a cyclic quadrilateral are equal to 180° (Fig. 12).

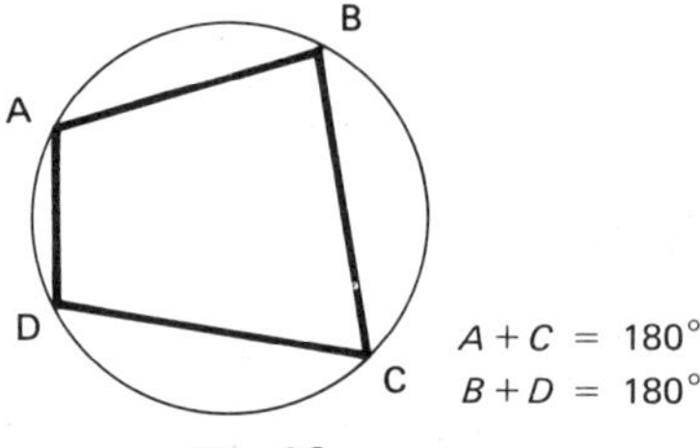

Fig. 12

Trigonometry

- $\sin A = \dfrac{\text{side opposite to } A}{\text{hypotenuse}} = \dfrac{BC}{AB}$

  $\cos A = \dfrac{\text{side adjacent to } A}{\text{hypotenuse}} = \dfrac{AC}{AB}$

  $\tan A = \dfrac{\text{side opposite to } A}{\text{side adjacent to } A} = \dfrac{BC}{AC}$

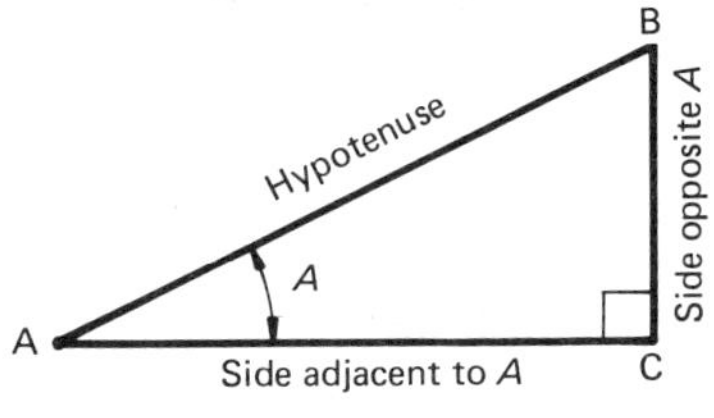

Fig. 13

Sets

- & stands for the universal set.
- $A = \{1, 3, 5\}$ means that A is the set of the first three odd integers.
- $5 \in A$ means that 5 is an element of A.
- $A \subset B$ means that B is a subset of A.
- $B \supset A$ means that B includes A (i.e. A is a subset of B).
- A' is the complement of A (i.e. the elements of & less the elements of A).
- $X \cap Y$ is the intersection of the sets X and Y.

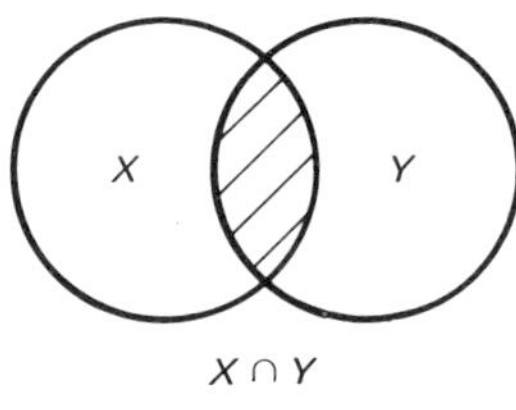

$X \cap Y$

Fig. 14

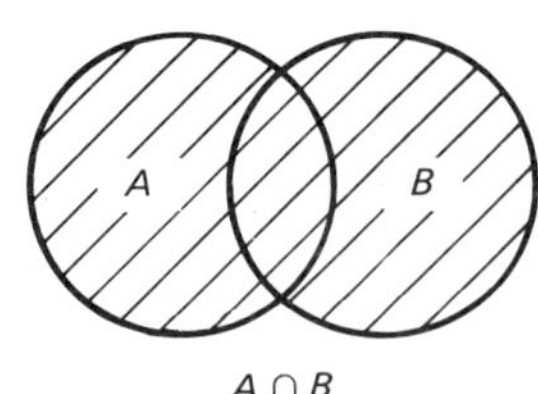

$A \cap B$

Fig. 15

- $A \cup B$ is the union of the sets A and B.
- $n(X)$ means the number of elements in the set X.

Number Bases

- The number 4325 in the base x means

$$5 \times x^0 + 2 \times x^1 + 3 \times x^2 + 4 \times x^3$$

Matrices

- $\begin{pmatrix} a & b \\ c & d \end{pmatrix} + \begin{pmatrix} u & v \\ w & x \end{pmatrix} = \begin{pmatrix} a+u & b+v \\ c+w & d+x \end{pmatrix}$
- $\begin{pmatrix} a & b \\ c & d \end{pmatrix} \begin{pmatrix} u & v \\ w & x \end{pmatrix} = \begin{pmatrix} au+bw & av+bx \\ cu+dw & cv+dx \end{pmatrix}$
- If $A = \begin{pmatrix} a & b \\ c & d \end{pmatrix}$ $A^{-1} = \dfrac{1}{ad-bc} \begin{pmatrix} d & -b \\ -c & a \end{pmatrix}$

Statistics

- Arithmetic mean $= \dfrac{\text{sum of all the values}}{\text{the number of values}}$.
- When a set of values is arranged in ascending (or descending) order the *median* is the value which lies half way along the series. When there are an even number of values in the set the median is found by taking the mean of the two middle values.
- The *mode* of a set of values is the value which occurs most frequently.
- Probability = (number of equi-probable events which produce a favourable outcome) ÷ (total number of equi-probable events).

  Total probability, covering all possible events is 1. That is, probability of favourable outcomes + probability of unfavourable outcomes = 1.

INDEX

ACKNOWLEDGEMENTS

The author and the publishers wish to thank the following examination boards for granting permission to use questions taken from their past examination papers. The letters in brackets after the name of the board show the abbreviations used in the text.

Associated Lancashire Schools Examining Board (AL)

East Anglian Examinations Board (EA)

East Midland Regional Examinations Board (EM)

North West Regional Examinations Board (NW)

Southern Regional Examinations Board (S)

South Western Examinations Board (SW)

West Midlands Examinations Board (WM)

West Yorkshire and Lindsey Regional Examinations Board (WY)

Yorkshire Regional Examinations Board (Y)

Welsh Joint Education Committee (W)